Photochemical Energy Conversion

Photochemical Energy Conversion

Proceedings of the Seventh International Conference on Photochemical Conversion and Storage of Solar Energy, held July 31–August 5, 1988 in Evanston, Illinois, U.S.A.

Editors

James R. Norris, Jr.

and

Dan Meisel

Chemistry Division
Argonne National Laboratory
Argonne, Illinois

Elsevier
New York • Amsterdam • London

Elsevier Science Publishing Co., Inc.
655 Avenue of the Americas, New York, New York 10010

Sole distributors outside the United States and Canada:

Elsevier Science Publishers B.V.
P.O. Box 211, 1000 AE Amsterdam, The Netherlands

ISBN 0-444-01477-2

Current printing (last digit):
10 9 8 7 6 5 4 3 2 1

Manufactured in the United States of America

CONTENTS

Part II: Organized Assemblies and Heterogeneous Systems

Part III: Photosynthesis

Part IV: Photoelectrochemistry and Photocatalysis

Foreword

These Proceedings include the plenary talks presented at the Seventh International Conference on Photochemical Conversion and Storage of Solar Energy (IPS-7). This series of meetings, which was initiated by a few dozen scientists 15 years ago, currently encompasses nearly 300 scientists representing a variety of physical and chemical disciplines all focused on the problem of utilizing solar energy to produce fuels, electricity, or useful chemicals. Substantial advances can be recognized in this field since the energy crisis of the early 1970's. The prospects of conversion and storage in photoelectrochemical cells are now discussed in terms of economical viability. Water cleavage to hydrogen and oxygen has been demonstrated, albeit still at low efficiencies. Conversion of carbon dioxide to methane, other alternative fuels, or organic chemicals has been reported and is being actively pursued. Introduction of various microenvironments and heterogeneous systems has been shown to lead to efficient charge separation, persistent to long periods of time. The field of photocatalysis, an outgrowth of this endeavor, has established itself in a variety of applications from mineralization of pollutants to recovery of trace amounts of precious metals. This broad range of subjects contributes to the main strength of this conference, namely, its ability to unite a variety of disciplines towards a common goal.

Some of the most outstanding advances during recent years, in our own biased opinion, are in the understanding of several basic physical principles that have stirred the scientific community. Among these, two advances are rather evident. The principles that control the rate of electron transfer in the ground and excited states are now much better understood than a decade ago. This leads to a more rational design and synthesis of a variety of complex molecular systems aimed at increasing the efficiency of charge transfer and charge separation. The other major development was culminated in the recent crystallization and direct structure determination of the in vivo bacterial photosynthetic reaction centers. This structure of the basic photosynthetic apparatus confirmed much of the work encountering this problem almost 20 years ago. As a result of these developments, intensive studies, both experimentally and theoretically, of the sequence of electron transfer at the molecular level in natural bacterial and green plant photosynthetic systems are presently undertaken in many laboratories. Undoubtedly, both topics will provide guidelines to the design of man-made artificial systems, which mimic the natural systems. Our deliberate emphasis of these two subjects is reflected in the content of this book. Currently, a misleading feeling of abundant inexpensive fossil energy sources prevails. We believe that this sentiment is based on a false and temporary reality. Clearly, the appropriate strategy now is to spend the talent and resources available on the basic scientific problems before the pressures of energy shortages return.

This volume is divided into four parts, proceeding from the very basic to the more applied subjects of research. In the first part, the principles of electron and energy transfer are discussed in light of the most updated theories, along with their experimental verification. The study of the organization of molecules in microassemblies or in small heterogeneous particles are discussed in the second part. The application to photosynthesis of the basic principles, which were revealed in the first part, is presented in the third part. The fourth and last part focuses on recent advances in the fields of photoelectrochemistry and the closely related field of photocatalysis. The efforts of the many scientists who contributed to these advances, in particular those presented in this book, are greatly appreciated.

This conference could not have taken place without the support of several agencies. The consistent support of the U.S. Department of Energy, Office of Basic Energy Sciences, Division of Chemical Sciences, of this area of research, as well as the special grant for this conference are much appreciated. At our home institute, Argonne National Laboratory, the Office of Energy, Environmental, and Biological Research, the Chemistry Division and the Division of Educational Programs were the most forthcoming at various stages of the organization of the meeting. A generous grant from the Gas Research Institute, as well as its support of the field over the years, is gratefully acknowledged by us and the scientific community we represent. Thanks are due also to Amoco Technology Company for its special grant. Last, but not least, it is our pleasure to acknowledge the exceptional dedication and efficiency of the conference secretary, Ms. Kay Foreman, in all stages of the conference organization and the preparation of this volume.

J. R. Norris
D. Meisel
Argonne National Laboratory
August, 1988

SOLVATION DYNAMICS AND SOLVENT-CONTROLLED ELECTRON TRANSFER

Ilya Rips, Joseph Klafter and Joshua Jortner

School of Chemistry

Tel Aviv University, 69978 Tel-Aviv, Israel

I. INTRODUCTION

Electron transfer in polar solvents [1] is accompanied by the solvation of the donor-acceptor pair, which proceeds via the reorientation of the solvent molecules. Traditional descriptions of outer sphere electron transfer (ET) in polar solvents rest [2.3] on the implicit assumption that either (a) the microscopic electronic process, rather than the dielectric relaxation, constitutes the rate determining step, or (b) the solvation occurs fast enough, so that the orientational part of solvent polarization is in equilibrium with the charge distribution at any time instant. For case (a) the reaction is non-adiabatic [3] and its rate is given by the Fermi Golden Rule, while for case (b) the transition state theory is applicable [2b,c]. The non-adiabatic theory dominated the field for almost two decades [1,3b]. The ET rate, k_{ET} , in the classical limit is given by the Arrhenius type expression

$$k_{ET} = A \exp(-E_a/k_BT) \tag{I.1}$$

where the preexponential frequency factor, A, is independent of the dissipative properties of the medium, while the activation energy is

$$E_a = (E_r - \Delta E)^2/4E_r \tag{I.2}$$

Published 1989 by Elsevier Science Publishing Co., Inc.
Photochemical Energy Conversion
James R. Norris, Jr. and Dan Meisel, Editors

with ΔE being the (free) energy gap and E_r is the medium reorganization energy. Solvent influence on the rates of non-adiabatic ET is reduced to the dependence of the energetic parameters ΔE and E_r on the dielectric properties of the solvent. Recent progress in the field of ET kinetics [4-12] rests on the notion that the solvation process can constitute the rate determining step. We shall refer to this regime as solvent-controlled ET. The criterion for the realization of solvent-controlled ET is determined by the magnitude of the adiabaticity parameter [4,9]

$$\kappa_A = 4\pi V^2 \tau_L / \hbar E_r \tag{I.3}$$

where V is the electronic coupling, and τ_L is the longitudinal dielectric relaxation time of the solvent. For a solvent characterized by a Debye type dielectric susceptibility the latter is given by

$$\tau_L = (\epsilon_\infty / \epsilon_S) \tau_D \tag{I.4}$$

where ϵ_S and ϵ_∞ are the static and the high frequency dielectric constants, respectively and τ_D is the Debye relaxation time. When $\kappa_A >> 1$ the solvent controlled ET situation is realized. Previous studies of the solvent-controlled ET [4-12] were based on the following assumptions:

(1) The polar solvent is described as a dielectric continuum. Molecular structure of the solvent and, particularly, of the first solvation shells is disregarded.

(2) Applicability of the linear response theory. The dielectric saturation, associated with the large electrical field in the vicinity of the donor-acceptor pair, is neglected.

Within this framework the ET rate is given by Eq.(I.1) with the frequency factor

$$A = A_{NA} / [1 + \kappa_A h(\Delta E, E_r)] \tag{I.5}$$

where

$$A_{NA} = 2\pi V^2/\hbar(4\pi E_r k_B T)^{1/2} \tag{I.6}$$

is the non-adiabatic frequency factor [3], while $h(\Delta E,E_r)$ is a form factor. The latter depends upon the actual character of the crossing of the diabatic potential surfaces and varies in the

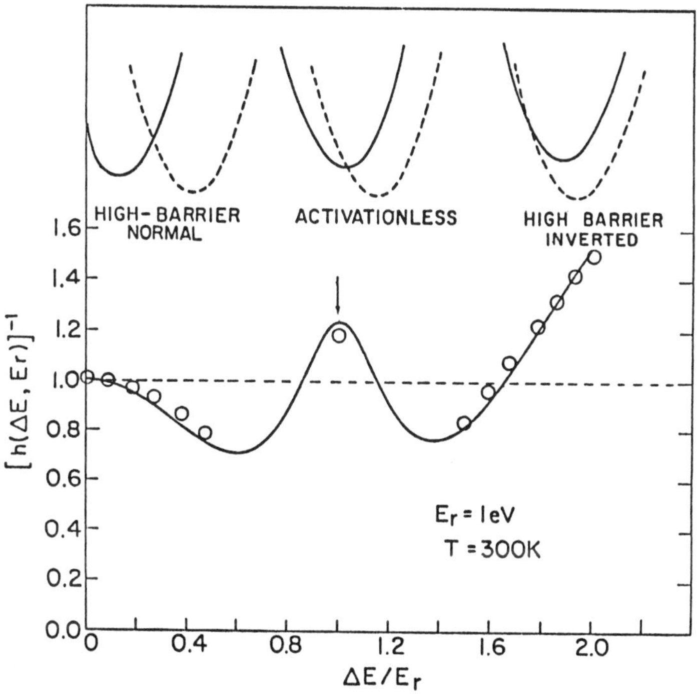

Fig. 1. The dependence of the inverse form factor, $[h(\Delta E,E_r)]^{-1}$, on character of the crossing of the diabatic potential surfaces. The physical parameters are E_r = 1eV and T = 300K. The solid line and open points represent the general result of the stochastic Liouville equation and its limiting solutions, respectively (Ref.[9b]). The dashed line corresponds to interpolation $h(\Delta E,E_r) = 1$.

range $h(\Delta E,E_r) = 0.7$-1.5 over the physically relevant region $0 \leq \Delta E/E_r \leq 2.0$ (Figure 1). When $\kappa_A >> 1$ the solvent-controlled limit is realized, whereupon,

$$A \simeq \delta/\tau_L \tag{I.7}$$

with

$$\delta = (E_r / 16\pi k_B T)^{1/2} \left[h(\Delta E, E_r)\right]^{-1} \tag{I.8}$$

As for ET in a number of polar solvents $E_r \simeq 1eV$ one expects that $(E_r/16\pi k_B T)^{1/2} \simeq 1$ at 300K, which together with the data of Figure 1 imply that δ is close to unity (within 50%), so that $A \approx \tau_L^{-1}$ in the solvent controlled limit. This simple relation has been shown to provide a reasonable description of activationless intramolecular ET (where $k_{ET} = A$) in mono-alcohols [13,14], where $k_{ET} \simeq \tau_{1L}^{-1}$. τ_{1L} is the longitudinal relaxation time due to relaxation of clustered alcohol molecules [15]. In more recent studies [16,17] deviations from this result have been observed. Specific interactions complicate the solvation kinetics and the solvent-controlled ET kinetics in associated solvents. It is quite plausible that the adherence to the $k = \tau_{1L}^{-1}$ relation for intramolecular activationless ET in long aliphatic alcohols originates from cancellation between the effects of structure breaking in the hydrogen bonded solvent on the one hand, and the molecular structure of the solvent and dielectric saturation effect on the other hand. The former tend to shorten the effective relaxation time, while the latter lengthen it. Therefore, a critical test of the theory of solvent-controlled ET requires experimental data in nonassociated polar solvents. A recent experimental study of intramolecular ET in bianthryl molecule in a series of Debye-type nonassociated polar solvents [18], which correspond to solvent controlled activationless ET, reveal that the ET rate is lower by a numerical factor of 2-6 from the prediction of Eq. (I.7) for an activationless process. This result indicates that the effective relaxation time, which determines the solvent controlled ET rate according to Eq. (I.6), may be longer than the continuum dielectric model prediction, τ_L. The theory of solvent controlled ET should be extended to account for the effects of the molecular structure of the solvent on the ET dynamics.

There has been recent considerable progress in the understanding of microscopic solvation dynamics induced by the (instantaneous) formation of an electronically excited "giant dipole" in a polar solvent [19-25]. Such solvent relaxation proceeds on a single nuclear potential surface. The continuum dielectric model predicts an exponential solvation kinetics with a timescale τ_L. Recent subpicosecond time-resolved spectroscopy data [26-28] revealed that the solvation kinetics is nonexponential with an average solvation time τ_s exceeding τ_L. This deviation has been attributed [27,28] to the effects of the solvent molecular structure on the solvation kinetics. The current theoretical description of microscopic solvation dynamics rests on the Onsager "inverted snowball" picture [19], which asserts that the solvent relaxation time in the immediate vicinity of the ion or dipole is τ_D, while the relaxation time in the bulk is τ_L (with $\tau_L << \tau_D$). Solvent relaxation thus proceeds from the exterior solvent towards the solute, involving multiple time scales from τ_L to τ_D. A quantification of the "inverted snowball" picture was provided within the framework of the mean spherical approximation (MSA) [22-24], being applied both for ion [22,23] and dipole [24] solvation.

In this paper we shall utilize the microscopic theory of solvation to describe solvent-controlled ET. Eq. (I.6) will be modified to surpass the continuum approximation, by replacing τ_L by a solvent relaxation time, which accounts for molecular size effects, i.e., the solvent and solute sizes and for dielectric screening effects. We shall invoke the first passage time approximation for reaching the crossing point of the potential surfaces in the ET process. Accordingly, expression for the frequency factor, Eq.(I.7), in the solvent controlled limit will be replaced by

$$A \simeq 1/\tau_{ET} \qquad (I.9)$$

where τ_{ET} is the average solvent relaxation time for ET process, which characterizes the solvent dynamics during the solvent controlled ET. τ_{ET} will depend on

(i) the energetic parameters ΔE and E_r, which determine the crossing of the diabatic potential surfaces for ET.

(ii) size effects of the ions and solvent.

(iii) dielectric screening effects.

We shall proceed now to the evaluation of τ_{ET}, which will allow us to estimate the solvent controlled ET rate.

II. ION AND DIPOLE SOLVATION DYNAMICS

We shall briefly summarize the results for ion [23] and dipole [24] solvation dynamics within the framework of the MSA. The model [22] consists of a hard sphere solute of radius R embedded in a polar solvent of hard polarizable spheres of radius r_S. A charge Z, or a dipole moment μ is created on the solute at time t = 0, which induces orientational relaxation of the solvent molecules. Solvation kinetics can be studied via the solvation time correlation function (TCF)

$$S(t) = [E_S(t)-E_S(\infty)] \; / \; [E_S(0)-E_S(\infty)] \tag{II.1}$$

where $E_S(t)$ is the orientational part of the ion (dipole) solvation energy at time t. Invoking the linear response theory the Laplace transform of the solvation TCF

$$\hat{S}(p) = \int_0^\infty dt \; e^{-pt} \; S(t) \tag{II.2}$$

can be expressed in terms of the admittance $\hat{\chi}(p)$ of the system [23]

$$\hat{S}(p) = [\hat{\chi}(p) - \hat{\chi}(0)]/p[\hat{\chi}(\infty) - \hat{\chi}(0)] \tag{II.3}$$

Of physical interest is also the average solvation time

$$\tau_s = \int_0^\infty dt\, S(t) = \hat{S}(0) = \hat{\chi}(0)/[\hat{\chi}(\infty) - \hat{\chi}(0)] \tag{II.4}$$

For the case of ion the admittance of the system is within the MSA [22]

$$\hat{\chi}_{ion}(p) = (1/2R_i)\{1-1/\hat{\epsilon}(p)\}/[1+\hat{\Delta}(p)] \tag{II.5}$$

$\hat{\Delta}(p)$ is the dynamical correction to the ionic radius, which can be well approximated by [23]

$$\hat{\Delta}(p) \simeq (3r_s/R_i)\left\{4.7622\left[\hat{\epsilon}(p)\right]^{1/6} - 2\right\}^{-1} \tag{II.6}$$

Dipole solvation kinetics has been studied only in the particular case of equally sized solute and solvent molecules, $r_s = R$. The admittance is given by [24]

$$\hat{\chi}_{dip}(p) \propto \frac{1}{2}\left\{1 - \frac{9}{\hat{h}(p)+4}\right\} \tag{II.7}$$

Explicit expression for $\hat{h}(p)$ in terms of the macroscopic dielectric susceptibility function of the solvent has been derived [23]. For practical purposes $\hat{h}(p)$ can be well approximated by

$$\hat{h}(p) \simeq 4.7622\left[\hat{\epsilon}(p)\right]^{1/6} \tag{II.8}$$

In Figure 2 we compare the solvation time correlation function for dipole, $S_d(t)$, with that for an ion, $S_i(t)$, in a solvent characterized by a Debye macroscopic dielectric susceptibility [29]

$$\hat{\epsilon}(p) = \epsilon_\infty + \frac{\epsilon_s - \epsilon_\infty}{1 + p\tau_D} \tag{II.9}$$

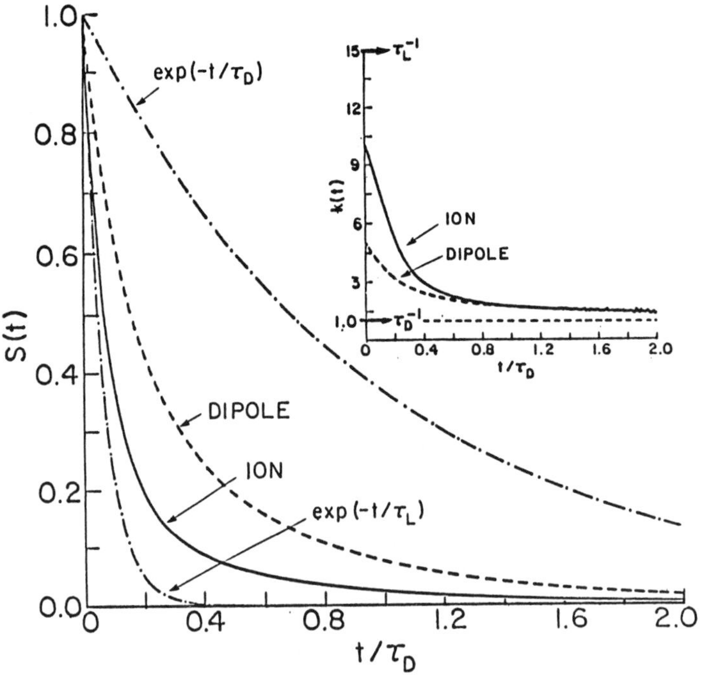

Fig. 2. Comparison of the ion and the dipole solvation dynamics in a Debye solvent within the MSA. The curves were evaluated for the solvent/solute radius ratio $r_S/R = 1.0$. The static and optical dielectric constants of the solvent were taken equal to $\epsilon_S = 30$ and $\epsilon_\infty = 2$, respectively. The solid curve corresponds to the ion solvation while the dashed curve to the dipole solvation. Also exhibited are the continuum dielectric model results for the constant charge (timescale τ_L) and constant field (timescale τ_D) relaxation. Upper insert shows the "time-dependent solvation rate" $k(t) = -\dot{S}(t)/S(t)$ of the ion (solid curve) and the dipole (dashed curve).

Inspection of Fig.2 shows that

(1) Solvation kinetics of the ion and of the dipole is non-exponential, exhibiting hierarchy of relaxation times.

(2) The dipole solvation proceeds slower than that of the ion with the same radius.

(3) Deviation of the short-time "rate" from the continuum model prediction, τ_L^{-1}, is more

pronounced for dipole solvation, resulting in a narrower range of solvation "rates" in that case.

(4) On the long time scale the dipole and ion solvation rates converge to single molecule relaxation rate, namely τ_D^{-1}.

Observations (2) and (3) are associated with shorter spatial range of solute-solvent interaction in the case of the dipole solvation relative to that of the ion. Accordingly, dipole solvation is characterized by a larger contribution to the total solvation energy from the slowly relaxing first solvation shells.

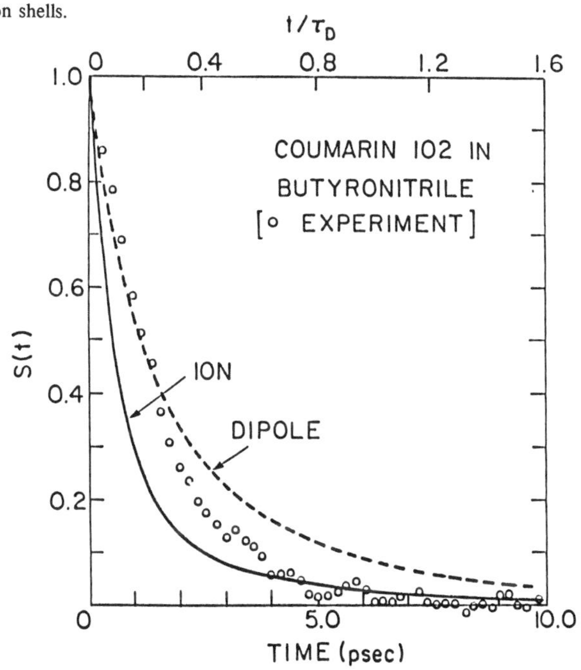

Fig. 3. Calculated ion (solid curve) and dipole (dashed curve) solvation functions together with the experimental data (open points) on the coumarin 102 solvation kinetics in butyronitrile (Ref.[28]). Butyronitrile is a Debye solvent with $\epsilon_\infty = 1.9$, $\epsilon_S = 22.23$ and $\tau_L = 0.53$ psec [30].

A sensitive test of the theory rests on the confrontation of the experimental solvation kinetics with the predictions of the MSA theory. In Figures 3 and 4 we exhibit the experimental solvation TCF for the time-dependent fluorescence shift of the (rigid) coumarin 102 in butyronitrile [28] (Fig. 3) and of coumarin 153 in propylene carbonate [27](Fig. 4).

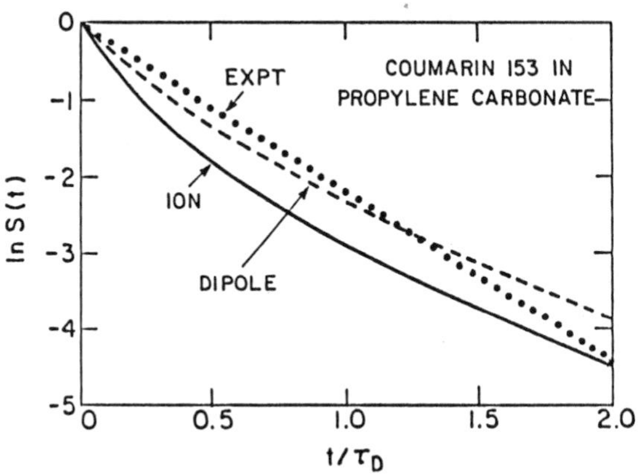

Fig. 4. Logarithmic plot of the calculated ion (solid curve) and dipole (dashed curve) solvation functions and experimental data (dotted curve) for coumarin 153 solvation kinetics in propylene carbonate at 252°K (Ref. [27]). Theoretical curves were evaluated using $r_S/R = 1$ and $\epsilon_\infty = 10$, $\epsilon_S = 77.3$.

These experimental data for solvation dynamics in Debye solvents are confronted with the calculated $S_d(t)$ and $S_i(t)$, both for $r_S/R = 1$. At short times the dipole solvation TCF describes very well the experimental results. However, at intermediate times dipole solvation kinetics is appreciably slower than the experimental data, which are better described by the ion solvation function in this time domain. This crossover from short-time dipole solvation to the intermediate-time ion solvation can be rationalized in terms of Onsager's picture [19]. On the short time-scale the exterior solvent molecules provide the dominant contribution to the solvation of the probe molecule. For these distant molecules the charge distribution within the solute molecule is well represented by a point dipole. On the other hand, the intermediate and long-time behaviour of the solvation function is determined by the reorientation of the solvent molecules belonging to the first shells. At these short distances the solvent molecules "see" the actual charge distribution within the solute molecule. Relaxation of these molecules occurs therefore in the field of charge(s) rather than of a point dipole.

III. AVERAGE RELAXATION TIME FOR SOLVENT CONTROLLED ELECTRON TRANSFER

The time-dependent solvation "rate"

$$k_S(t) = -d\ell n\, S(t)/dt \tag{III.1}$$

can be utilized to derive an explicit expression for the average relaxation time τ_{ET} for the solvent-controlled ET, which was introduced in Eq. (I.8)

$$\tau_{ET}^{-1} = (1/t^*) \int_0^{t^*} dt\, k_S(t) \tag{III.2}$$

t^* is the characteristic time necessary to reach the crossing point of the diabatic potential surfaces. This time can be expressed in terms of the ratio of the coordinate of the crossing point, Q_0, and the displacement between the minima of the potential surfaces, ΔQ. For the two parabolic potential surfaces

$$U_1(Q) = Q^2 /4E_r$$
$$U_2(Q) = (Q - 2E_r)^2 /4E_r - \Delta E \tag{III.3}$$

these are given by $Q_0 = E_r - \Delta E$ and $\Delta Q = 2E_r$. It is assumed that the system is initially on the first surface. From the definition of the solvation time correlation function $1\text{-}S(t^*) = Q_0/\Delta Q$, so that

$$S(t^*) = (E_r + \Delta E)/2E_r \tag{III.4}$$

for the normal crossing of the diabatic potential surfaces. Eq. (I.9) together with Eqs. (III.2) and (III.4) allow to evaluate the frequency factor for the solvent-controlled ET in general case. In the presence of activation barrier determination of the frequency factor requires solution of Eq.(III.4). Since, however, the expression for the solvation TCF, S(t), is not known this case can be handled only numerically. As an alternative, one can assume analytic form of the S(t) function. As a particular example one can take the solvation TCF in the form of stretched exponent

$$S(t) = \exp\{-(t/\tau)^{\alpha}\} \qquad \text{(III.5)}$$

where α is a parameter ($\alpha \leq 1$). A certain justification for this particular form can be provided on the basis of hierarchical relaxation. The solvation rate, Eq. (III.1), should be limited from both above and below by ~ τ_L^{-1} and by τ_D^{-1}, respectively. Since the stretched exponent does not satisfy these conditions, it cannot provide a good description of the overall solvation kinetics. Still it reproduces the actual decrease of the solvation "rate", with increasing time, $k_s(t) \propto t^{\alpha-1}$.

The average solvation time in the case of stretched exponent is given by

$$\tau_s = \tau\,\Gamma(1/\alpha + 1) \qquad \text{(III.6)}$$

where $\Gamma(x)$ is the gamma function. This can be compared to the ET time, for which a closed analytic expression can be derived

$$\tau_{ET} = \tau\left\{\ln\left[\frac{2E_r}{E_r + \Delta E}\right]\right\}^{1/\alpha - 1} \exp\{E_a/kT\} \qquad \text{(III.7)}$$

For the normal ET with $E_r > \Delta E$ one expects that $\tau_{ET}/\tau_s = \gamma \exp(E_a/k_BT)$ with $\gamma < 1$. As already mentioned the stretched exponential relaxation function implies diverging short-time solvation "rate". The latter is of considerable interest for the description of the solvent-controlled activationless ET. In this case the ET rate is given by initial solvation rate

$$k_{ET}^{0} \simeq k_s(0) \qquad \text{(III.8)}$$

The initial solvation rate depends in general on the nature of the charge distribution. The electron exchange reaction between the ions of the same sign should be described in terms of the ion solvation dynamics, while for ET between ions of opposite signs dipole solvation dynamics applies. The dipole solvation dynamics should be used also for the description of intramolecular ET within a large organic molecule. Since we are interested only in the short time behaviour the dipole approximation for the charge distribution is valid. The problem is now reduced to the calculation of the initial solvation rate.

IV. INITIAL SOLVATION RATE

Even at very short times there is a certain contribution from the slowly relaxing first solvation shells. This contribution, of course, depends upon the spatial range of the solute-solvent interaction (ion-dipole or dipole-dipole), and it explains the deviation of the short-time solvation rate predicted by the MSA from the continuum model value τ_L^{-1}. The latter can be observed from Fig.2. For the given interaction the initial solvation rate depends on the solvent/solute radius ratio and dielectric constants of the solvent. We now show how the short-time solvation rate can be extracted from the asymptotic behaviour of the Laplace transform of solvation TCF, $\hat{S}(p)$. To calculate the initial solvation rate

$$k_S(0) = -\dot{S}(0) \equiv - \left. d\,S(t)/dt \right|_{t=0} \tag{IV.1}$$

we use the short time expansion of the solvation TCF

$$S(t) = 1 + \dot{S}(0)t + O(t^2) \tag{IV.2}$$

The Laplace transform of Eq.(IV.2)

$$\lim_{p\to\infty} \hat{S}(p) = 1/p + \dot{S}(0)/p^2 + O(1/p^3) \qquad \text{(IV.3)}$$

leads to expression for the initial solvation rate

$$k_S(0) = \lim_{p\to\infty} p^2\{\hat{S}(p) - 1/p\} \qquad \text{(IV.4)}$$

This result can be rewritten in terms of the admittance of the system

$$k_S(0) = -\lim_{p\to\infty} p\,[\hat{\chi}(p) - \hat{\chi}(\infty)] / [\hat{\chi}(0)-\hat{\chi}(\infty)] \qquad \text{(IV.5)}$$

According to the Onsager picture of solvation [19] dominant contribution at short times comes from distant solvent molecules, for which the point dipole approximation for the solute molecule is justified. Therefore, in evaluation of the frequency factor one can use the initial dipole solvation rate. Utilizing Eq.(IV.5) together with (II.7) and (II.9) we derive an explicit expression for this case

$$k_S^d(0) \simeq \frac{1}{\tau_L} \frac{[4.76\epsilon_S^{1/6} + 4]}{6[4.76\epsilon_\infty^{1/6}+4][(\epsilon_S/\epsilon_\infty)^{1/6} - 1]} \qquad \text{(IV.6)}$$

According to the previous reasoning the inverse initial solvation rate gives the time of the solvent-controlled activationless ET, τ^0_{ET} .It is larger than the longitudinal dielectric relaxation time, τ_L , and shorter than the average dipole solvation time, τ_S . In Figure 5 we plot the ratio of τ^0_{ET}/τ_L as a function of $\epsilon_S/\epsilon_\infty$ inferred from Eq.(IV.6). The result exhibits a weak dependence on $\epsilon_S/\epsilon_\infty$ and changes from about unity in weakly polar solvents to $\simeq 3$ in highly polar solvents. The ratio is also insensitive to the actual value of the optical dielectric constant, ϵ_∞ , within the realistic range of the values of this parameter. From Eqs. (II.4) and (IV.6) we can evaluate the ratio τ^0_{ET}/τ^d_S . The latter is exhibited in Figure 6. In the

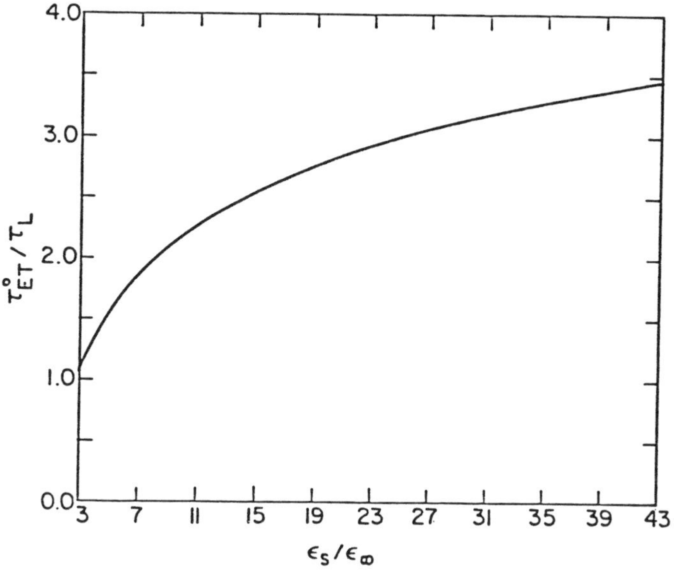

Fig. 5. The MSA correction to the activationless solvent-controlled ET time , τ^0_{ET}/τ_L , as a function of solvent polarity ($\epsilon_\infty/\epsilon_s$). Evaluated for r_s = R=1.

acceptable range of the parameter $\epsilon_s/\epsilon_\infty$ the activationless ET time is shorter than the dipole solvation time. The difference between the two is only minor in accordance with the fact that the distribution of the relaxation times for the dipole solvation is narrow.

V. CONFRONTATION WITH EXPERIMENT

The rate of the solvent-controlled activationless ET process is given by Eq. (III.8) in conjunction with Eq. (IV.6) for the initial dipole solvation rate. We compare now theoretical results with the recent experimental data of Barbara et al [18] for the intramolecular ET rates in the electronically excited state of bianthryl in polar non-associated solvents. This ET process is induced by a strong (V > 100cm^{-1}) intramolecular coupling, so that the adiabaticity parameter, Eq. (I.3), is apparently sufficiently large to make the solvent-controlled ET limit applicable. Finally, this ET process is nearly activationless. The experimental data of Barbara et al [18] show that for the intramolecular ET in series of nitriles and acetates $\tau_{ET}/\tau_L > 1$, i.e., the experimental ET timescale, exceeds τ_L by a numerical factor of 2-6.

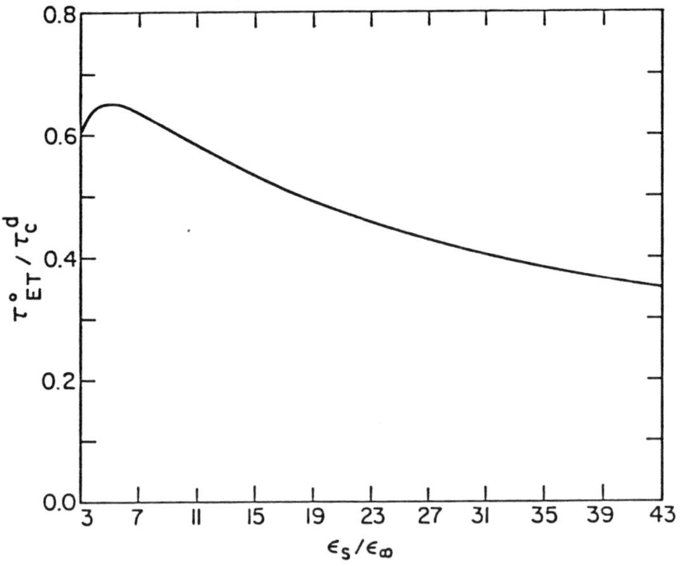

Fig. 6. The dependence of the ratio of the initial solvation time and the average solvation time for a dipole on $\epsilon_S/\epsilon_\infty$.

This deviation of the timescale for the activationless ET from the prediction of the continuum dielectric model can be attributed to molecular size effects. In Table 1 we compare these experimental values of τ_{ET}/τ_L with theoretical predictions. The calculations were performed for the dipole solvation with $r_S/R = 1$. From these results we infer that

(1) The experimental ET time is shorter than the averaged solvation time, τ_S, as expected from the truncation of the dipole solvation process at short times (Fig. 6).

(2) The increase of the τ_{ET}/τ_L ratio above the continuum model prediction is due to the molecular size and dielectric screening effects on the short time solvation process.

(3) The experimental data for τ_{ET}/τ_L are in reasonable, although not perfect, agreement with theory. The predicted enhancement of τ_{ET}/τ_L beyond the continuum model value of unity is borne out by experiment.

(4) Most important, the ET time is shorter (except in pentanitrile) than the averaged solvation times in contradiction with the prediction of the continuum dielectric model.

TABLE 1

ACTIVATIONLESS ET vs SOLVATION

(Theory and Experiment)

Solvent	ϵ_∞	ϵ_S	τ_L (psec)	τ_S/τ_L	τ^0_{ET}/τ_L (calc) (b)	τ_{ET}/τ_L (exp) (a)	τ^0_{ET}/τ_S (calc) (b)	τ_{ET}/τ_S (exp) (a)
acetonitrile	1.81	36.23	0.19	5.89	2.8	3.5	0.47	0.78
propionitrile	1.87	28.64	0.31	4.92	2.6	4.0	0.53	0.8
butyronitrile	1.90	22.23	0.53	4.11	2.36	4.0	0.57	0.95
pentanitrile	1.93	19.56	0.74	3.75	2.23	6.0	0.6	1.2
PC c)	10 d)	77.3	5.56	3.09	1.94	2.0	0.15	0.7
	2.63e)	65	1.74	6.76	3.0		0.44	

a) evaluated using experimental ET times in bianthryl (Ref. [18]);

b) calculated from Eqs. (IV.6) and (II.4).

c) PC = propylene carbonate

d) from extrapolation of the Cole-Cole plot (Ref. [31]).

e) the optical dielectric constant determined from $\epsilon_\infty = 1.3\ n^2$.

Limitations of the theory are reflected in its inability to reproduce the experimental trend of increasing τ_{ET}/τ_L with the increase of the length of the alkyl chain. This is not surprising as the theoretical estimates in Table 1 were performed assuming that the solvent/solvent radius ratio equal to unity. Since the initial solvation rate of a dipole (ion) increases with decreasing r_S/R we can expect that for the proper values of this ratio the calculated values of τ_{ET}/τ_L will show an increase with increasing size of the solvent molecule. It should be also noted that the comparison between theory and experiment rests on the assumption that the ET process is strictly activationless. Existence of a low energy barrier ($E_a < k_B T$) for the solvent-controlled ET will enhance the calculated value of τ_{ET}/τ_L by the reciprocal Arrhenius factor, washing up the discrepancy between theory and experiment. The possibility of a plausible contribution of a small energy barrier renders the interpretation of experimental results somewhat ambivalent. In this context the solvation kinetics considered in section III presents a better test for the theory.

VI. CONCLUSIONS

The successful application of the MSA theory for the solvation kinetics inspires confidence in the validity of the present theory of solvent-controlled ET. While the continuum model provides a reasonable first-order description of the frequency factor, molecular size effects have to be incorporated within the framework of a quantitative description of solvent-controlled ET. The most important conclusion which emerges from an analysis is that the continuum limit frequency factor for the solvent controlled ET has to be replaced by

$$A \simeq 1/f\tau_L \qquad \text{(VI.1)}$$

where the correction factor $f = f(r_s, R, \epsilon_s, \epsilon_\infty, \Delta E, E_r)$ depends on the size effects and dielectric screening and on the energetics of the ET process. Calculation of this factor was performed within the framework of MSA. A slight generalization of this result to incorporate the "transition" from the non-adiabatic to the solvent-controlled ET will result in replacing Eq. (I.5) of the continuum limit by

$$A = A_{NA} / [1 + f\kappa_A \, h(\Delta E, E_r)] \qquad \text{(VI.2)}$$

Obviously, as $h(\Delta E, E_r) \simeq 1$ (Fig. 1) the refined condition for solvent-controlled ET is

$f\kappa_A >> 1$. Thus the molecular size effects result in the enhancement of the effective adiabaticity parameter.

Up to this point we have assumed that the contribution of the high-frequency quantum modes of the donor and acceptor centres is negligible. It is a straightforward matter to extend the formalism to incorporate these effects. For normal and activationless ET in the limit $h\omega_c > k_B T$ the role of the quantum modes reduces essentially to the polaron "dressing" of the electronic coupling. Thus V^2 should be replaced by $V^2 \exp(-\Lambda)$ where Λ is the electron-phonon coupling constant for the high-frequency modes. Accordingly, the frequency factor is now

$$A = A_{NA} / [1 + f\kappa_A \, h(\Delta E, E_r) \exp(-\Lambda)] \qquad \text{(VI.3)}$$

resulting in partial cancellation between the size factor and dressing factor.

The present theory is applicable to ET in Debye solvents. Deviations from the Debye dielectric susceptibility are quite common in polar solvents [29]. These are characterized by a distribution, $g(\tau)$, of dielectric relaxation times, which may originate from the static of dynamic disorder. In the case of static disorder the spatial inhomogeneity leads to non-exponential ET kinetics. The short-time ET rate is then expressed in terms of the statistical average $\bar{k} = \int d\tau \, k(\tau) \, g(\tau)$. We have studied ET in solvents characterized by the Cole-Davidson relaxation [33]:

$$\epsilon(p) = \epsilon_\infty + \frac{\epsilon_s - \epsilon_\infty}{(1 + p\tau_0)^\beta} \qquad \text{(VI.4)}$$

τ_0 is the upper limit of the distribution of the relaxation times and β is the parameter ($\beta \leq 1$). The frequency factor for the initial ET rate obtained previously can be modified to include the solvent size effects

$$A = A_{NA} \Big/ \left(1 + f \cdot \kappa_A\right)^{\beta} \tag{VI.5}$$

This extension will modify the magnitude of the effective adiabaticity parameter, κ_A, but will not modify the $k \propto \tau_0^{-\beta}$ relation which provides a signature of ET in a microscopically inhomogeneous medium.

Eqs. (VI.2), (VI.3) and (VI.5) constitute "intelligent guesses" regarding the extension of the theory of solvent-controlled ET dynamics. The rigorous extension of the theory will rest on the treatment of the correlation function of the reaction coordinate, which should be identified with the solvation time correlation function.

REFERENCES

[1] J.Ulstrup Charge Transfer Processes in Condensed Media. Springer Verlag, Berlin, 1979.

[2] (a) R.A.Marcus J.Chem.Phys. 24, 966 (1956); Ibid. 24, 979 (1956);
(b) Annu.Rev.Phys.Chem. 15, 155 (1964).

[3] (a) V.G.Levich, R.R.Dogonadze Dokl.Akad.Nauk SSSR 124, 123 (1959) /Proc.Acad.Sci.USSR Phys.Chem.Sect. 124, 9 (1959)/;
(b) V.G.Levich Adv.Electrochem.Eng. 4, 249 (1965).

[4] L.D.Zusman Chem.Phys. 49, 295 (1980).

[5] I.V.Alexandrov Chem.Phys. 51, 499 (1980).

[6] G.van der Zwan, J.T.Hynes J.Chem.Phys. 76, 2993 (1982); J.T.Hynes J.Phys.Chem. 90, 3701 (1986).

[7] D.F.Calef, P.G.Wolynes J.Phys.Chem. 87, 3387 (1983); J.Chem.Phys. 78, 470 (1983).

[8] H.Sumi, R.A.Marcus J.Chem.Phys. 84, 4894 (1986).

[9] (a) I.Rips, J.Jortner J.Chem.Phys. 87, 2090 (1987);
(b) Ibid. 87, 6513 (1987);

(c) Ibid. 88, 818 (1988).

[10] I.Rips, J.Jortner Chem.Phys.Lett. 133, 411 (1987); Int.J.Quant.Chem. Symp. 21, 313 (1987).

[11] M.Sparpaglione, S.Mukamel J.Chem.Phys. 88, 3263 (1988).

[12] M.D.Newton, H.L.Friedman J.Chem.Phys. 88, 4460 (1988).

[13] E.M.Kosower, D.Huppert Annu.Rev.Phys.Chem. 37, 127 (1986) and references therein.

[14] E.M.Kosower, D.Huppert Chem.Phys.Lett. 96, 433 (1983); F.Heisel, J.A.Miehe Chem.Phys.Lett. 128, 323 (1986); S.G.Su, J.D.Simon J.Phys.Chem. 90, 6475 (1986).

[15] S.K.Garg, C.P.Smyth J.Phys.Chem. 69, 1294 (1965).

[16] D.Huppert, V.Ittah, E.M.Kosower Chem.Phys.Lett. 144, 15 (1988); S.G.Su, J.D.Simon J.Chem.Phys. 89, 908 (1988).

[17] G.E.McManis, M.N.Golovin, M.J.Weaver J.Phys.Chem. 90, 6563 (1986); R.M.Nielson, G.E.McManis, M.N.Golovin, M.J.Weaver J.Phys.Chem. 92, 3441 (1988).

[18] T.J.Kang, M.A.Kahlow, D.Giser, S.Swallen, V.Nagarajan, W.Jarzeba, P.F.Barbara J.Phys.Chem. (1988) in press.

[19] L.Onsager Can.J.Chem. 55, 1819 (1977).

[20] D.F.Calef,P.G.Wolynes J.Chem.Phys. 78, 4145 (1983).

[21] V.Friedrich, D.Kivelson J.Chem.Phys. 86, 6425 (1987); R.F.Loring, S.Mukamel J.Chem.Phys. 87, 1272 (1987);

[22] P.G.Wolynes J.Chem.Phys. 86, 5133 (1987).

[23] I.Rips, J.Klafter, J.Jortner J.Chem.Phys. 88, 3246 (1988).

[24] I.Rips, J.Klafter, J.Jortner J.Chem.Phys. 89 (1988) in press.

[25] M.Maroncelli, G.R.Fleming J.Chem.Phys. 89, 875 (1988).

[26] E.W.Castner Jr., M.Maroncelli, G.R.Fleming J.Chem.Phys. 86, 1090 (1987).

[27] M.Maroncelli, G.R.Fleming J.Chem.Phys. 86, 6221 (1987).

[28] (a) M.A.Kahlow, T.J.Kang, P.F.Barbara J.Phys.Chem. 91, 6452 (1987);
(b) J.Chem.Phys. 88, 2372 (1988).

[29] C.J.F.Bottcher, P.Bordewijk Theory of Electric Polarization. vol.2. Elsevier, Amsterdam,1978. Ch.IX.

[30] Krishnaji, A.Mansingh J.Chem.Phys. 41, 827 (1964).

[31] R.P.Payne, I.E.Theodorou J.Phys. Chem. 76, 2892 (1972).

[32] E.A.S.Cavell J.Chem.Soc. Farad.Trans. II. 74, 78 (1974).

[33] D.W.Davidson, R.H.Cole J.Chem.Phys. 19, 1484 (1952).

A Comparison of Long Distance Electron and Energy Transfer

Gerhard L. Closs,[†,#] Piotr Piotrowiak,[†] John R. Miller[#]

Department of Chemistry, The University of Chicago
Chicago, Illinois 60637
Chemistry Division, Argonne National Laboratory
Argonne, Illinois 60439

Abstract: Intramolecular long distance electron transfer (ET) and triplet energy transfer (TT) are being compared using two similar series of model compounds in which the donor and acceptor distance is varied in an identical way by employing common rigid spacer groups with identical stereochemical attachments in both series. It is found that transfer rates in the two series are correlated in a quantitative way. The results support a simple model which views TT as a simultaneous two-electron two-site exchange. The model is discussed in terms of the corresponding HOMO and LUMO overlap integrals.

In recent years there has been a flurry of activity in the study of long distance electron transfer (ET) [1]. To a large extend, this activity can be linked to the spectacular progress made in photosynthesis research culminating in x-ray structure analyses of reaction centers [2]. This led to the challenge of the mechanistic and theoretical chemists to explain in a quantitative way the kinetics of the ET processes of the early stages of photosynthetic charge separation. A quantitative analysis of ET rates requires a number of parameters which are best determined from experiments, although attempts are being made to derive them ab initio [3].

Much of the work carried out in our and other laboratories has been directed to test ET theories and determine parameters from model system studies [4]. Most recently, we have attempted to correlate electron transfer with energy transfer, specifically triplet energy transfer [5]. The aim is to test, via model systems, the dependence of the rates of these transfer processes on a number of variables, such as energetics, solvent, distance and stereochemistry. At this time, the study is fairly incomplete, but the initial results allow us to draw up a number of hypotheses and draw conclusion on some.

The systems used to study intramolecular transfer of electrons or energy are described schematically by eq.'s 1 and 2, where D and A are the donor and acceptor groups and Sp is a rigid

† The University of Chicago
# Argonne National Laboratory

Published 1989 by Elsevier Science Publishing Co., Inc.
Photochemical Energy Conversion
James R. Norris, Jr. and Dan Meisel, Editors

spacer.

$$^{(-)}\text{D—Sp—A} \longrightarrow \text{D—Sp—A}^{(-)} \qquad (1a)$$

$$\text{D—Sp—A}^{(+)} \longrightarrow {}^{(+)}\text{D—Sp—A} \qquad (1b)$$

$$^{3}\text{D'—Sp—A} \longrightarrow \text{D'—Sp—A}^{3} \qquad (2)$$

Two series of compounds were synthesized, one for electron transfer (ET), eq. 1a, and hole transfer (HT), eq. 1b, and another one for triplet energy transfer (TT). The donor in ET and HT is a 4-biphenylyl group and in TT a 4-benzophenoneyl. The acceptors in both series are 2-naphthyl. Cyclohexane and decalin ring systems serve as rigid spacers to which the donors and acceptors are attached with various stereochemistry. The compounds used for the comparison are listed in Table I. At this time, the HT experiments are at a very preliminary stage and the results are not included in this discussion.

The ET experiments were carried out by converting the bifunctional compounds, dissolved in a suitable solvent, into the ions by radiolysis with a 30 picosecond, 20 MEv electron pulse according to scheme (3) [4]. The initial distribution of the two ions is almost statistical and the conversion to equilibrium was monitored by fast absorption spectroscopy. Extrapolation of the rates to zero concentration yields the intramolecular ET rate constants as listed in Table I.

$$\begin{array}{ccc} & \text{D—Sp—A} & \\ & \swarrow \quad \searrow & \\ ^{(-)}\text{D—SP—A} & \rightleftharpoons & \text{D—Sp—A}^{(-)} \end{array} \qquad (3)$$

The experiments on triplet transfer (TT) were carried out in a similar fashion, except that formation of the donor triplet, the benzophenoneyl group, was accomplished by laser pulse photoexcitation [5]. Here, it is possible to excite only the donor and monitor the triplet transfer to the naphthalene acceptor by absorption spectroscopy. Extrapolation to zero concentration is again necessary to separate intermolecular from intramolecular transfer. The rate constants are listed in Table 1.

Before discussing the results, it may be advantageous to briefly present the theoretical model upon which the comparison will be based. Both types of reactions involve the interaction between donor and acceptor being held apart by the rigid spacer. Since the spacer component of the molecules will neither bind the electron nor support the triplet excitation within the available energy, both transfers will proceed directly from the π-system of the donor to that of the acceptor. The electronic interaction between these groups will be very weak and the processes must be treated as a non-radiate transition from one diabatic surface to another. A common starting point for possible theories describing the kinetics of reactions of this kind is the Golden Rule (eq.4)

$$k = 2\pi/\hbar^{-1}\,|V|^2\ \text{FCWDS} \qquad (4)$$

separating the electronic coupling matrix element, V, from nuclear coordinates as expressed in the Franck Condon weighted density of states (FCWDS).

The separation of electronic and nuclear coordinates, expressed by eq .4, allows one to separate the experimental variables in a corresponding way. For example, the effect of the free energy of the reaction (ΔG^o) on its rate is absorbed in the Franck Condon factors, as are the internal reorganization of bond length and bond angles (λ_v) and the reorganization of the solvent molecules (λ_s) occurring during the transfer step. The dependence of the rate in ET on these three parameters form the basis of Marcus theory and its various extensions. A set of experiments on ET of the type described by scheme 3 on a series of compounds in which the free energy of the reaction was varied by 2.5 eV and rates were measured for several solvents, have generally confirmed the theoretical predictions including the so-called inverted region and drastic solvent effects depending in a highly non-linear fashion on the driving force of the reaction.

The electronic coupling matrix element V, which in ET and HT can be looked upon as the two center resonance integral between reactant and product diabatic states, can be expected to fall off exponentially with the distance separating donor and acceptor. However, it has become clear in recent years that the through space interaction of the wavefunctions falls off too fast to account for the experimentally determined distance dependence. Instead, it seems more reasonable to assume that the coupling is mediated through the σ-bond frame of the spacer. This mechanism, some times called superexchange, involves mixing of the π-orbitals of the donor with the bonding and antibonding σ-orbitals of the spacer. Molecular orbital calculations have shown that this will increase the coupling and moderate the fall-off with distance which will, however, still be exponential. This behavior yields equ. 5a and 5b for the rate constants for through space and through bond couplings, respectively.

$$k = k_o \exp\text{-}[\beta(R - R_o)] \qquad (5a)$$

$$k = k_o' \exp\text{-}[\beta'(N - 1)] \qquad (5b)$$

Eq. 5a describes the distance (R) dependence with R_o being the van der Waals distance, where the rate constant is k_o. In equ. 5b, more appropriate for through-bond coupling, N is the number of intervening σ-bonds and k_o' describes the rate constant for a donor acceptor pair separated by one σ-bond. The susceptibility to distance or number of separating bonds is described by β and β'. It should be noted that the stereochemistry of the attachment to the spacer is reflected in k_o and k_o' while the particular structure of the spacer in the through-bond model will affect β' but should have no influence in the through space model except defining the distance. Experimental tests for the through-bond model include transfer experiments on stereoisomers in which the number of intervening σ-bonds is the same but drastic changes in the direct distance occur.

So far, this discussion is equally valid for ET and energy transfer occurring by the exchange (Dexter) mechanism [6]. Another mechanism of energy transfer, by coupling of the transition dipole moments (Förster) will show a different distance dependence and is excluded from this discussion.

While transfer of singlet excitation may occur by both mechanisms, with the Förster mechanism predominating in most cases, triplet energy transfer in molecules with light atoms, such as organic molecules, is limited to the Dexter mechanism because of spin restrictions.

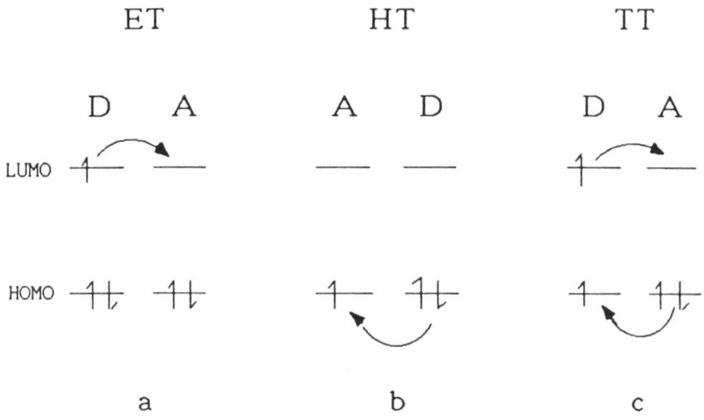

Figure 1. HOMO, LUMO diagram of electron shifts in a) electron transfer, b) hole transfer and c) triplet energy transfer.

The similarity of the Dexter mechanism of energy transfer with ET and HT is shown with the aid of Figure 1. In ET of the kind discussed here, one electron undergoes a two-site exchange process from donor LUMO to acceptor LUMO (Figure 1a). An equivalent process is in operation in HT except that the exchange is between HOMO's (Figure 1b). Using the same simple model, energy transfer by exchange is a two electron, two site exchange involving both the LUMO and the HOMO (Figure 1c). To relate ET and HT to TT by the Dexter mechanism, one has to compare the matrix elements V, going into eq. 4 for the three reactions. In ET and HT these are simply the resonance integrals (6a) and (6b) between the diabatic states, as pointed out above. For the Dexter mechanism V is given by the exchange integral (6c) in which the spatial functions have to be multiplied by the appropriate spin functions to distinguish between singlet and triplet energy transfer.

$$\langle \psi^D{}_{LU}(1)|H|\psi^A{}_{LU}(1)\rangle = V_{ET} \qquad (6a)$$

$$\langle \psi^D{}_{HO}(1)|H|\psi^A{}_{HO}(1)\rangle = V_{HT} \qquad (6b)$$

$$\langle \psi^D{}_{LU}(1)\psi^A{}_{HO}(2)|H|\psi^D{}_{HO}(2)\psi^A{}_{LU}(1)\rangle = V_{TT} \qquad (6c)$$

The magnitude of the matrix elements may be quite different. What is more relevant for this discussion is their distance dependence, or number of bond dependence in the through-bond model. Also, in a through-space model the Hamiltonian of 6c takes on the simple form of e^2/R. This is no longer the case for superexchange.

In the case of weak coupling, the distance dependence of the matrix elements (eq. 6) approximately parallel the distance dependence of the overlap integrals which in both through-space and through-bond mechanism fall off exponentially. This admittedly over-simplified picture leads to the relations expressed in eq. 7.

$$\langle \psi^{D}_{LU} | \psi^{A}_{LU} \rangle \propto V_{ET} \quad (7a)$$

$$\langle \psi^{D}_{HO} | \psi^{A}_{HO} \rangle \propto V_{HT} \quad (7b)$$

$$\langle \psi^{D}_{LU} | \psi^{A}_{LU} \rangle \langle \psi^{D}_{HO} | \psi^{A}_{HO} \rangle \propto V_{TT} \quad (7c)$$

If eq.'s 7 are valid, it should be possible to relate the distance dependence of triplet energy transfer rates to the electron (k_{ET}) and hole transfer (k_{HT}) rate constants in comparable systems.

TABLE 1

Compound		$k_{ET}^{cor\ a}$ [s^{-1}]	k_{TT}^{b} [s^{-1}]
	D-2,6 ae	9.3×10^{6}	1.3×10^{5}
	D-2,6 ea	2.9×10^{7}	7.0×10^{5}
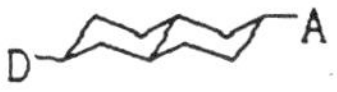	D-2,6 ee	5.0×10^{7}	3.1×10^{6}
	D-2,7 ae	8.0×10^{7}	1.1×10^{7}
	D-2,7 ee	1.6×10^{8}	9.1×10^{7}
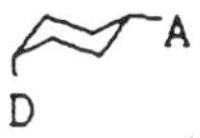	C-1,4 ae	1.4×10^{8}	4.0×10^{7}
	C-1,4 ee	9.1×10^{8}	1.3×10^{9}
	C-1,3 ee	1.6×10^{9}	7.7×10^{9}

a) Measured at room temperature in methyltetrahydrofuran and corrected for changes of solvent reorganization with distance [5].

b) Measured at room temperature in benzene.

$$k_{ET} = k_0^- \exp -[\beta^-(R - R_0)] \tag{8a}$$
$$k_{HT} = k_0^+ \exp -[\beta^+(R - R_0)] \tag{8b}$$

Making use of eq. 7 we obtain for the triplet energy rate constants (k_{TT})

$$k_{TT} = k_{0TT} \exp -[(\beta^- + \beta^+)(R - R_0)] \tag{8c}$$

The corresponding equations for the through bond mechanism are obtained from eq.'s 8 simply by substituting ($R - R_0$) with ($N - 1$).

If this simple model is correct, eq.'s 8a-c predict that the exponent describing the fall-off of the triplet transfer rate constants is simply the sum of the exponents in electron and hole transfer. This prediction is subject to experimental test. The experiments described above and the results listed in Table I. constitute a major part of such a test.

Using the same set of spacers, a series of electron transfer rate constants and another for triplet transfer rate constants have been measured. What is missing at this time is the corresponding series of hole transfer rate constants, although a few preliminary, as yet unpublished data points are available.

As has been noticed previously, neither the ET nor the TT data can be correlated successfully with distance or number of bonds. This failure is shown quite clearly in Figure 2a and 2b. The reason for the scatter is due to the fact that compounds of different stereochemistry are compared with one another. Eq. 5 shows that a linear correlation with distance or number of bonds will only be successful if all compounds have the same k_0 value. Different stereochemical attachments, however, will correspond to different k_0's because they correspond to different geometries of overlapping wavefunctions. Reasonable linear correlations can be obtained when only compounds of comparable stereo attachments are compared as for example the four isomers with equatorial-equatorial (ee) substitution patterns. Values for the exponents for both ET and TT distance dependencies have been obtained from those limited sets ($\beta_E{}^T = 1.0$ Å^{-1}; $\alpha_{TT} = 2.2$ Å^{-1}).

However, the comparison of ET, HT, and TT making use of eq. 6 and 7 do include the relationship of k_0's as well. From eq 8. it is easy to derive a relationship between ET and TT, or HT and TT rate processes. For the experimentally more complete set of ET and TT rate constants we obtain

$$k_{ET}(R)^{\alpha/\beta-} / k_{TT}(R) = k_{0ET}{}^{\alpha/\beta-} / k_{0TT} \tag{9}$$

with $$\beta^- + \beta^+ = \alpha$$

Of course, equivalent equations can be written for a through bond mechanism, replacing $R - R_0$ in eq 8 by $N - 1$. If the model is correct, eq. 9 suggests that a logarithmic plot of <u>all</u> the ET rate constants versus the TT rate constants should give a straight line, regardless of stereochemistry.

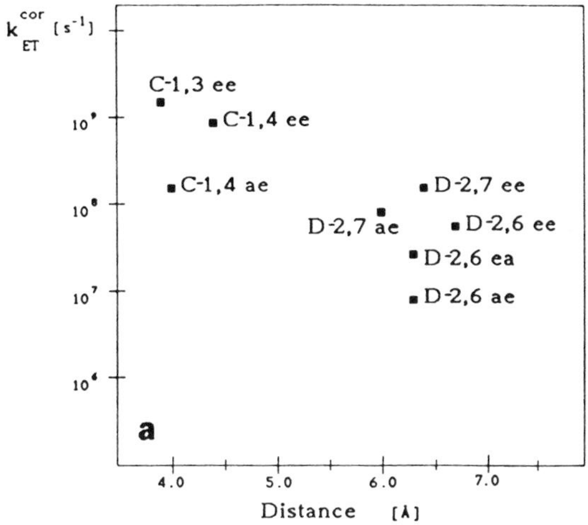

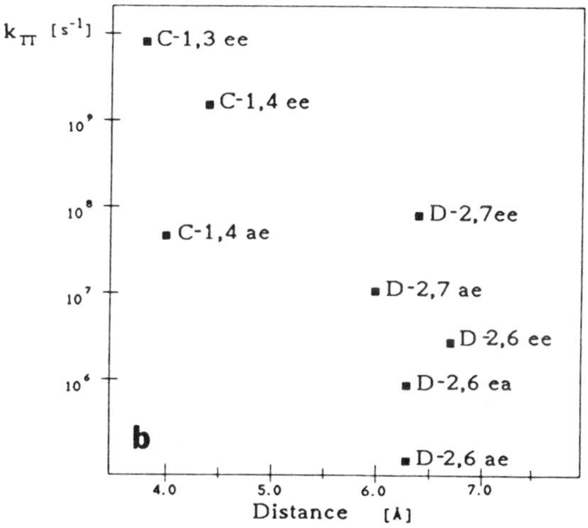

Figure 2. Plot of a) ET and b) TT rate constants against edge to edge distance

Distinguishing the constants describing electron and hole transfer by the indices - and + respectively, we can write the following relationships:

Figure 3 shows the striking confirmation of this prediction. It should be noted that the correlation covers five orders of magnitude on the TT scale!

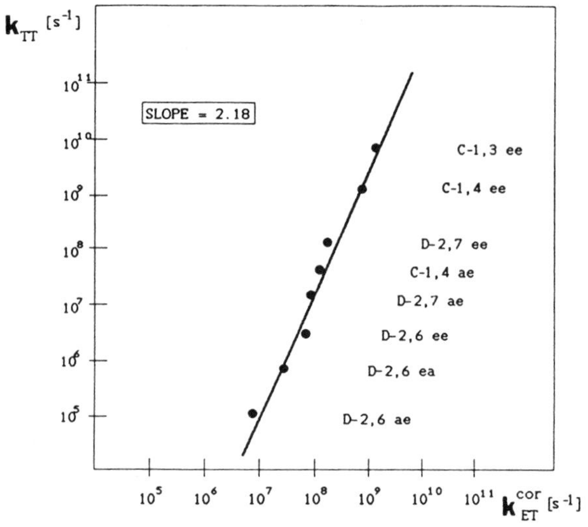

Figure 3. Logarithmic plot of ET vs TT rate constants

The slope of the line in Figure 3 is 2.18, showing the much steeper dependence of the TT transfer on distance and greater sensitivity on stereochemical variations. It also predicts that for the spacers examined, HT should be described by a very similar distance dependence as ET because the slope of the correlation predicts $\beta^{+} = 1.18\ \beta^{-}$. Preliminary data on hole transfer of the equatorial isomers listed in Table I show that HT and ET do indeed have a very similar distance dependence.

The results of this investigation, is a convincing confirmation of the Dexter mechanism, viewing triplet energy transfer as a simultaneous two electron, two site exchange. The quantitative relationship established in this work makes it possible to deduce characteristics of ET from data obtained on TT. Since TT shows little solvent dependence, it is much easier to keep the Franck Condon factors constant throughout a series, simply by using the same donor and acceptor throughout. In that case the relative rates are a fairly direct measure of the electronic coupling matrix elements. With the relationship described here, it is then possible to obtain information on the coupling in ET as well.

Finally, it should be mentioned that energy transfer by the Dexter exchange mechanism is not restricted to the triplet state. Singlet energy transfer may occur by the same mechanism, although often the dipolar coupling mechanism (Förster) predominates. That this is not always so has recently been shown by Verhoeven and collaborators who have reported singlet energy transfer by the exchange mechanism in intramolecular systems [7].

Acknowledgement We are indebted to the many students and postdoctoral fellows who have carried most of the experiments. They are: N.J. Green, L.T. Calcaterra, K.W. Penfield, W.F. Mooney, J.M. McInnis, J. Alfano, and H. Arnold. We thank Professor G.R. Fleming in whose laboratory the picosecond flash photolysis experiments were carried out. The work was generously supported by the National Science Foundation and the Department of Energy, Office of Basic Energy Sciences, Division of Chemical Science.

References

1 Closs G L, Miller J R (1988) Science 240:440, and references cited therein

2 For a recent review and references on ET in photosynthesis see: Budil D E, Gast P, Chang C H, Schiffer M, Norris J R (1987) Annu Rev Phys Chem 38:561

3 Ohta K, Closs G L, Morokuma K, Green N J (1986) J Am Chem Soc 108:1319

4 Closs G L, Calcaterra L T, Green N J, Penfield K W, Miller J R (1986) J Phys Chem 90:3673

5 Closs G L, Piotrowiak P, MacInnes J M, Fleming G R (1988) J Am Chem Soc 110:2652

6 Dexter D L (1953) J Chem Phys 21:836

7 Oevering H, Verhoeven J W, Paddon-Row M N, Cotsaris E, Hush N S (1988) Chem Phys Lett 143:488

DYNAMICS AND ENERGY GAP DEPENDENCES OF PHOTOINDUCED CHARGE SEPARATION AND RECOMBINATION OF THE PRODUCED ION PAIR STATE

NOBORU MATAGA
Department of Chemistry, Faculty of Engineering Science,
Osaka University, Toyonaka Osaka 560, Japan

ABSTRACT

Fundamental aspects of the dynamics and mechanisms of the intra- and intermolecular photoinduced charge separation (CS), charge recombination (CR) of the produced ion pair (IP) and related processes are discussed on the basis of our laser photolysis investigations from nanosecond to subpicosecond regions on the donor (D) and acceptor (A) combined by spacer or directly by single bond, excited state of the charge transfer complexes formed in the ground state, and CS in the fluorescence quenching reactions by encounter between uncombined donor and acceptor. Especially, the dynamics of photoinduced electron transfer (ET) induced by orientation fluctuations of surrounding polar groups in the case of intramolecular exciplex ($D-(CH_2)_n-A$, D-A) systems in liquid solutions as well as polymer films and the energy gap dependences of the rates of photoinduced CS and CR of produced IP in the case of uncombined and combined D, A systems including porphyrins are discussed in detail.

1. INTRODUCTION

The electronic excitation transfer as well as the photoinduced electron transfer (ET) and related phenomena in liquid solutions, rigid matrices, molecular assemblies and biological systems are the most fundamental and important central problems in the photophysical and photochemical primary processes in condensed phase. Especially, the dynamics and mechanisms of the photoinduced charge separation (CS) and charge recombination (CR) of the produced ion pair (IP) state are the most important fundamental problems in the photochemical primary processes [1,2]. The elucidation of factors underlying these processes are of crucial importance and directly related to the mechanisms of the highly efficient ultrafast CS taking place in the biological photosynthetic reaction center.

The elementary processes of inter- and intramolecular photoinduced CS, CR of produced IP and dissociation of IP into free ions, etc. may be summarized as follows.

(a) Intramolecular photoinduced CS and CR of the IP.

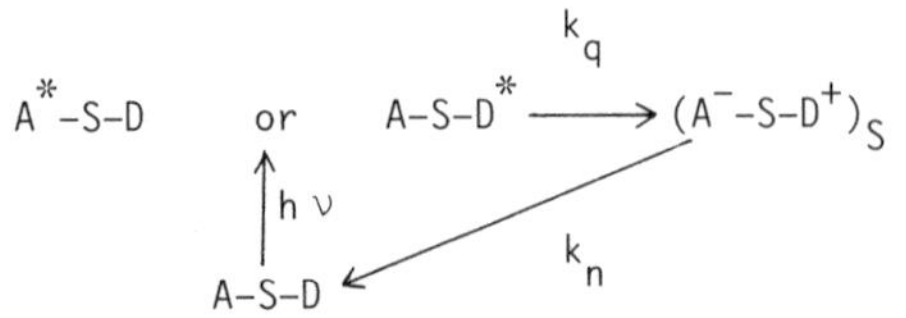

(1)

where -S- represents various spacers and the compound without the spacer includes the so-called TICT system.

(b) Excitation of the ground state CT complex, leading to the formation of

Photochemical Energy Conversion
James R. Norris, Jr. and Dan Meisel, Editors

the geminate IP which undergoes CR and dissociation.

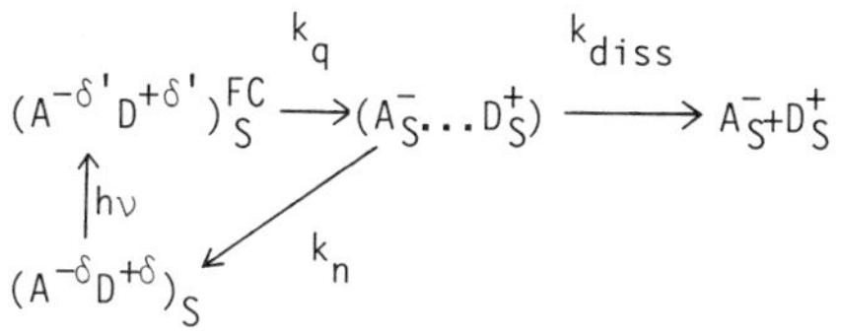

(2)

where $(A^{-\delta'}D^{+\delta'})_S^{FC}$ represents the excited Franck-Condon state.

(c) CS at the encounter between fluorescer and electron donating or accepting quencher leading to the formation of geminate IP which undergoes CR and dissociation.

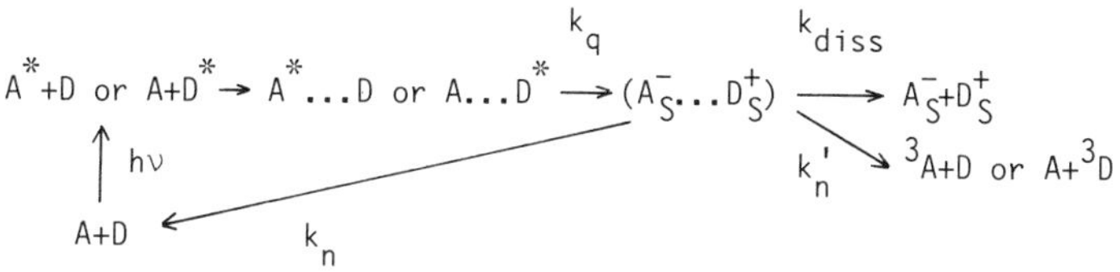

(3)

(d) Slow ET to polar solvent cluster from solute in the relaxed fluorescent state or rapid electron ejection into solvent from higher excited state produced by multiphoton excitation.

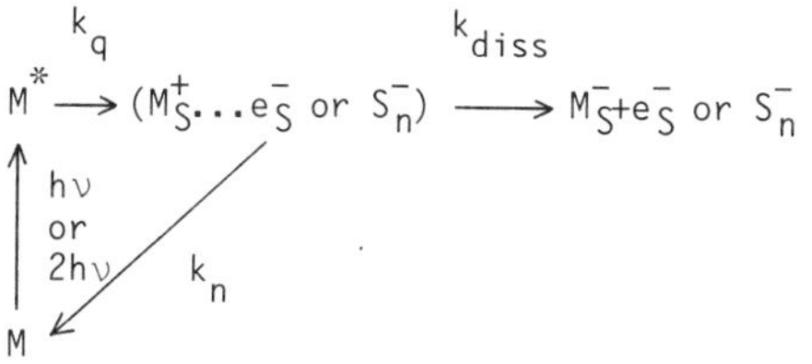

(4)

where S_n^{-} represents solvent aggregate anion.

In this article, we discuss the dynamics and mechanisms of the processes summarized in (a), (b) and (c) above on the basis of results of our ultrafast laser photolysis studies on various systems. Most important problems related to the above reaction processes will be as follows.

(1) Dependences of the rates of photoinduced CS and CR of the produced IP upon the respective energy gaps between the initial and final states of the ET reaction.

(2) Role of solvation of charged species by polar solvents, which affects profoundly the ET processes in photoinduced CS and CR of produced IP, and the dynamics of the ET induced by the solvent orientation fluctuations in the CS process.

(3) Dependence of the CR rate of the geminate IP upon the way of its formation, i.e. the problem of the different CR rate of the IP produced by CS at encounter between fluorescer and quencher from that of the IP produced by excitation of CT complex of the same pair in the same kind of polar solvent. This seems to indicate the different structures of the IP

including the surrounding polar solvent between these two cases.

2. Picosecond Laser Photolysis Studies on the Photoinduced CS and CR of Produced IP State in Some Typical Intramolecular Exciplexes and TICT Systems

It is well-known that the effect of the solvation upon the photoinduced creation of charged species and their annihilation by recombination can be demonstrated clearly by using the typical intramolecular exciplex systems, $A-(CH_2)_n-D$, such as $p-(CH_3)_2N-\phi-(CH_2)_n-$(1-pyrenyl)(Pn) and $p-(CH_3)_2N-\phi-(CH_2)_n-$(9-anthryl)(An) [1,2].

The photoinduced CS of A_3 or P_3 in acetonitrile takes place in a loose configuration without any compact complex formation within ca. 10∿20 ps, while it takes a few ns for the ET state formation in hexane because an extensive conformation change from an extended form to a sandwich form is necessary. In the case of A_3 in acetonitrile, the photoinduced CS occurs within ca. 10 ps and the CR of the intramolecular IP state takes place with time constant of ca. 700 ps. Similarly, the CS of P_3 in acetonitrile occurs within ca. 20 ps and the CR of the IP state takes place with time constant of ca. 1 ns. It is possible that A_S^- and D_S^+ groups can mutually approach a little in the course of the CR process due to the coulombic attraction which will enhance slightly the electronic interaction responsible for ET between two groups. Nevertheless, the rate of ET in CR of IP is much slower than that in photoinduced CS.

We have examined also the photoinduced CS of P_1 and P_2 in acetonitrile by subpicosecond laser photolysis and time-resolved transient absorption spectral measurements [3]. It takes a few ps for P_1 and ca. 8 ps for P_2, which indicates stronger D-A interaction responsible for ET due to the shorter inter-chromophore average distance for the smaller n. However, we have confirmed that it takes ca. 10 ns and 3 ns for the CR of the intramolecular IP of P_1 and P_2, respectively. The much longer time of P_1 for CR may be ascribed to its more restricted geometry which prevents the mutual approach of molecular planes of A and D groups.

In any way, these results can be well comprehended on the basis of the energy gap dependences of the ET rate. The free energy gap for the CS reaction of anthracene*-N,N-dimethylaniline (DMA) pair is $-\Delta G^\circ_{CS}$=0.54 eV and that for pyrene*-DMA pair is 0.48 eV, while that of CR reaction for anthracene$^-$-DMA$^+$ pair is $-\Delta G^\circ_{CR}$=2.74 eV and that for pyrene$^-$-DMA$^+$ pair is 2.86 eV in acetonitrile solution. These ΔG° values indicate that the CR reaction rate is in the inverted region and should be small due to the very large $-\Delta G^\circ_{CR}$ value but the CS reaction is in the normal region with rather large (ET) rate [4,5].

For the compounds A_n and P_n with n=1 and 2, the electronic interaction responsible for the ET between two chromophores becomes stronger and the photoinduced ET in polar solvents becomes much faster compared with the compounds with n=3 as we have discussed already. When the inter-chromophore electronic interaction becomes sufficiently strong and the stabilization due to the inter-chromophore coulombic interaction in the ET state is large owing to some favorable inter-chromophore configuration, the photoinduced ET process will become almost barrierless. Such condition will be realized also in some TICT compoundds in polar solvents. The photoinduced ET in these cases will be induced by a small orientational fluctuation of polar groups in the environment and the longitudinal dielectric relaxation time, $\tau_L=(\varepsilon_\infty/\varepsilon_S)\tau_D$, might be the characteristic relaxation time [7] in the dynamics of the ET in those solutions.

We have examined the dynamics of photoinduced ET of some short chain intramolecular exciplex compounds including A_2 and P_2 fixed in cellulose acetate (CA) film, where the photoinduced ET seems to take place mainly owing to the orientation fluctuations of the surrounding water molecules adsorbed (probably hydrogen bonded) to CA [8]. Only a part of the intramolecular exciplex compounds fixed in the CA film, which seems to have favorable configurations including chromophore pair and surrounding water molecules, undergoes photoinduced ET. This sort of systems will be interesting as a model to study the photoinduced charge separation in the chromophores fixed in high polymers just as those fixed in proteins in biological photosynthetic reaction center.

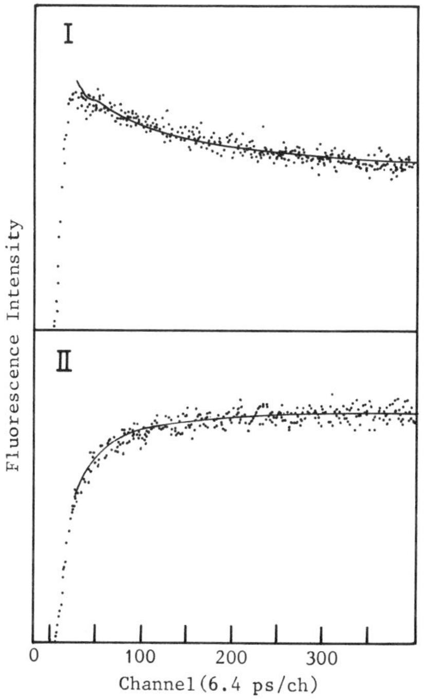

Fig. 1. LE fluorescence decay (I) and CT fluorescence rise (II) curves of P_2 in cellulose acetate film. Full lines are calculated by eqs 6 and 8, respectively, for (I) and (II).

In Fig. 1, the decay curve of pyrene LE fluorescence and the rise curve of CT fluorescence are indicated for P_2 fixed in CA at room temperature. These decay and rise curves cannot be reproduced by the usual exciplex kinetics but the rise of the CT fluorescence is much faster than the decay of the LE fluorescence. Moreover, these decay and rise curves are non-exponential. We can reproduce such experimental results by assuming the time dependent "rate constant" k(t) in the following reaction scheme.

$$\begin{array}{ccc} LE & \xrightarrow{k(t)} & CT \\ \downarrow k_{LE} & & \downarrow k_{CT} \end{array} \quad (5)$$

where k_{LE} is the decay rate constant of LE state in the absence of ET and k_{CT} is the intrinsic decay rate constant of the CT state. Such a time-dependent "rate constant" is a rather common phenomenon as we can see frequently in the case of electronic excitation transfer in rigid state or very rapid fluorescence quenching reactions in general.

The origin of the non-exponential decay of the LE fluorescent state of the anthracene or pyrene part of the bichromophoric compounds undergoing the ET in CA, can be elucidated by analyzing the observed decay and rise curves by means of the solution (eq 6) of the Smoluchowski equation for the barrierless brownian motion on a potential surface and rapid ET quenching at the potential bottom [9].

$$P^{LE}(t) = \exp(-k_{LE}t)\mathrm{erf}(Z(t)) \quad (6)$$

where $P^{LE}(t)$ is the survival probability of LE state, erf represents the error function and,

$$Z(t) = (2\tau_c^0/\tau_c^\omega)^{1/2}\exp(-t/\tau_c^\omega)/[1-\exp(-2t/\tau_c^\omega)]^{1/2} \quad (7)$$

In eq 6, τ_c's are correlation times related to brownian motion.

The rise and decay of the CT state $P^{CT}(t)$ is given by eq 8.

$$P^{CT}(t) = k(t)P^{LE}(t)\exp(-k_{CT}t) \quad (8)$$

where, $k(t)=Z^3(t)\exp(-Z^2(t)/[\pi^{1/2}\tau_c^0\exp(-2t/\tau_c^\omega)\mathrm{erf}(Z(t))]$ (9)

Above equations show that the rate of the LE→CT reaction is actually time dependent (k(t)) as discussed already in relation to eq 5, which leads to the apparently much faster rise of the CT fluorescence compared with the decay of the LE fluorescence as shown in fig. 1. One can see clearly that the observed LE decay and CT rise curves can be well reproduced by the above equations for the barrierless reaction.

Since we are assuming the barrierless model, we have examined the temperature effect upon the LE fluorescence decay and CT fluorescence rise curves. The decay of the LE state and the rise of the CT state do not slow down by lowering temperature, which supports above model assuming barrierless nature of the ET. This result of the photoinduced CS reaction in CA film, which does not show slowing down by temperature lowering, seems to be very interesting from the viewpoint of the ultrafast photoinduced CS in biological photosynthetic reaction center which seems to be barrierless. It should be noted here that by fixing the combined donor acceptor systems in polymer film, only a part of them can take a configuration (including surrounding polar groups) favorable for photoinduced ET and there remains another part of them which cannot undergo ET. Nevertheless, those with configurations favorable for ET can undergo barrierless photoinduced ET which does not show the decrease of the rate at low temperatures.

We have examined also photoinduced ET of A_2 and a directly combined system 2,3,5,6-tetramethyl-4-(9-anthryl)-N,N-dimethylaniline (TM-ADMA) in very viscous alcohol solutions [8]. A_2 and TM-ADMA in ethylene glycol showed behaviors very similar to the case of CA film, which indicates that the bichromophoric solute in favorable configuration including the surrounding alcohols undergoes almost barrierless photoinduced ET which may be induced by a small fluctuational orientation motions of -OH groups of surrounding alcohols.

We have examined the dynamics of the photoinduced ET of these and related bichromophoric compounds also in less viscous alcohols such as butanol and pentanol where the ET process was approximately exponential. However, in many cases, ET rate was not equal to but considerably smaller than the longitudinal dielectric relaxation time τ_L of the solvent [8]. For example, the inverse of the intramolecular photoinduced electron transfer rate of TM-ADMA is 132 ps in 1-pentanol and 99 ps in 1-butanol while that of A_1 is 59 ps in 1-pentanol and 46 ps in 1-butanol. On the other hand, $\tau_L \sim 170$ ps for 1-pentanol and $\tau_L \sim 130$ ps for 1-butanol [8]. Moreover, in the case of $\begin{matrix}\phi \\ (CH_3)\end{matrix}$>N-($CH_2$)-(1-anthryl), the electron transfer times are only 32 ps in 1-pentanol and 27 ps in 1-butanol [8]. It should be noted here that the above compounds, where the initial state of the photoinduced ET reaction is localized in pyrene or anthracene part, should be better system to study the "electron transfer" reaction than such compounds as ANS and N,N-dimethylaminobenzonitrile (DMABN) used previously [7]. It was reported that the photoinduced ET rate of the latter compounds in alcohol solvents was the same as τ_L^{-1} [7]. Solvent reorientation processes in such systems as excited ANS or DMABN in polar solvents may be rather similar to the case of the fluorescence stokes shift due to the interaction of the excited solute dipole with surrounding solvent dipoles, for which the first theoretical formula was given by the present author [10] and Lippert [11] and extended by Bagchi et al. [12] and others to take into account its dynamical aspects. The photoinduced "ET" processes examined in our studies are different from such cases and are affected in a more subtle manner by the initial distribution of the surrounding solvent orientations, electronic interaction matrix elements between D and A chromophores and energy gap related to ET, etc.

3. Picosecond Laser Photolysis Studies on the CR Processes of Geminate IP's of Uncombined Donor Acceptor Systems

As it is pointed out already in 1, the geminate IP formation by photoinduced CS in polar solvents can take place at encounter between excited solute and quencher or when CT complex is excited, although the detailed mechanism of the ET and the IP formation and also the structure of the IP may be different betwen these two cases.

The rate of ET is determined by the electronic interaction responsible for the ET and the energy gap between the initial and final states which is related to the Franck-Condon factor for ET. We have confirmed that the CR deactivation of the IP formed by excitation of the CT complex in polar solvent is much faster than that of the IP formed by CS in the fluorescence quenching reaction between the same pair in the same solvent, which indicates the stronger electronic interaction in the former IP due to the more compact structure [13].

The experimental results on the energy gap dependence of CS in the fluorescence quenching reaction and CR of produced IP are quite limited. Only the result for the normal region has been obtained but no result for the inverted region has been available for the CS in the fluorescence quenching reaction in polar solutions. Contrary to this, only the result for the inverted region but no result for the normal region was available for the CR of geminate IP produced by fluorescence quenching reaction.

We have made some systematic experimental studies on CR processes of geminate IP of various aromatic hydrocarbon - acceptor or donor systems by directly observing the dynamics of the geminate IP with picosecond laser spectroscopy. We have obtained experimental results not only for the inverted region but also for the normal region [2c,13,14] and have proved the bell-shaped energy gap relation for the CR of the geminate IP produced

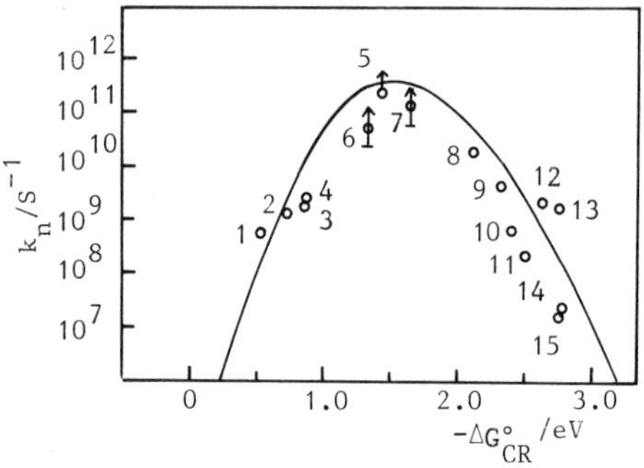

Fig. 2. The energy gap dependence of the CR rate constant k_n of geminate IP produced by fluorescence quenching reaction, $A^*+B \rightarrow A_S^{\mp}+B_S^{\pm}$, in acetonitrile.
o: observed value. 1. Per^+-$TCNE^-$, 2. $BPer^+$-$TCNE^-$, 3. DPA^+-$TCNE^-$, 4. Py^+-$TCNE^-$, 5. EEP^+-TQ^-, 6. Per^+-$PMDA^-$, 7. Per^+-MA^-, 8 Per^+-PA^-, 9. $BPer^+$-PA^-, 10. o-DMT^+-Per^-, 11. DMA^+-Per^-, 12. AN^+-Per^-, 13. DMA^+-c-Py^-, 14. DMA^+-Py^-, 15. Py^+-$DCNB^-$.
Per: perylene, BPer: 1,12-benzoperylene, DPA: 9,10-diphenylanthracene, Py: pyrene, TCNE: tetracyanoethylene, EEP: ethyletiolporphyrin, TQ: toluquinone, PMDA: pyromellitic dianhydride, MA: maleic anhydride, PA: phthalic anhydride, o-DMT: N,N-dimethyl-o-toluidine, AN: aniline, c-Py: 1-cyanopyrene, DCNB: p-dicyanobenzene.
— : calculated (ref. 5).

by ET in the singlet state for the first time [2c,13,14], as shown in fig. 2. Some results of the observations of the geminate IP formation, its decay by CR and dissociation into free ions in acetonitrile solution are indicated in figs. 3 and 4 for the system in the inverted region (perylene anion - N,N-dimethyl-o-toluidine cation) and those in the normal region (perylene cation - TCNE anion), respectively.

It should be noted here that the curve for the normal region in the $k_n \sim -\Delta G_{CR}$ relation as indicated in fig. 2 is quite different from that of the CS reaction obtained for similar aromatic donor-acceptor systems in the same solvent [15]. It seems to be difficult to reproduce those energy gap dependences of both the photoinduced CS and CR of produced IP on the basis of the conventional theories as discussed in the later part of this paper. In view of this result, one should not mix up the data for photoinduced CS and CR of IP to make up a bell-shaped energy gap dependence of ET rate constant.

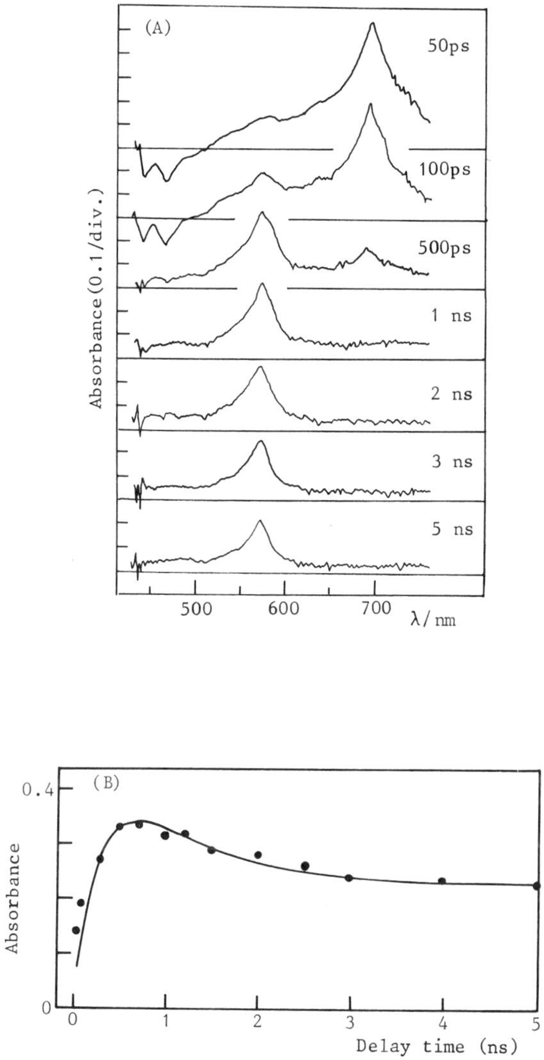

Fig. 3. Picosecond time-resolved transient absorption spectra of Per^{*}–o-DMT in acetonitrile (a system in the inverted region for the CR of IP).
(A) Decay of $S_n \leftarrow S_1$ absorption of Per at 695 nm and rise and decay of Per^{-} absorption at 575 nm.
(B) Time profile of Per^{-} absorbance at 575 nm.

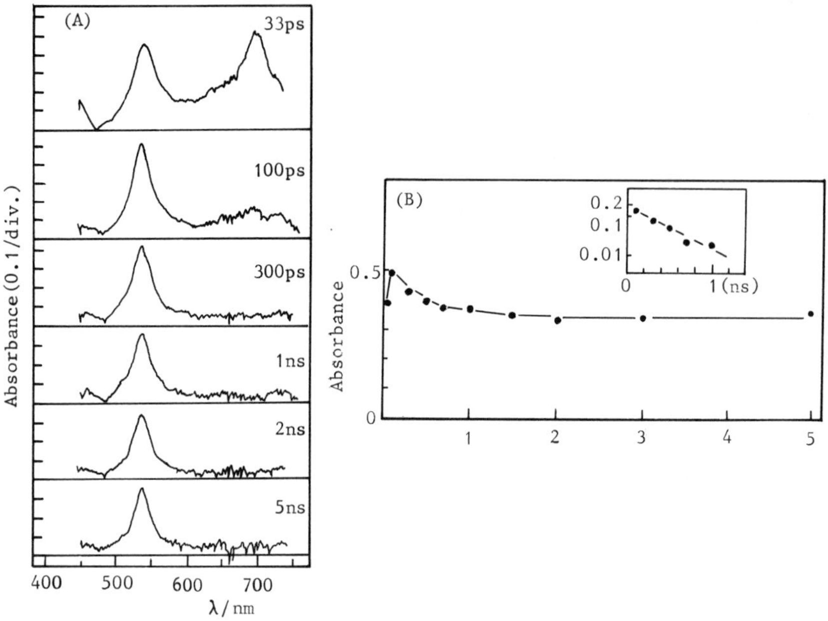

Fig. 4. Picosecond time-resolved transient absorption spectra of Per^{*}-TCNE in acetonitrile (a system in the normal region for the CR of IP).
(A) Decay of $S_n \leftarrow S_1$ absorption of Per^{*} at 695 nm and rise and decay of Per^{+} absorption at 535 nm.
(B) Time profile of Per^{+} absorbance at 535 nm.
Inset: semilogarithmic plot of the absorbance obtained by subtracting the constant value at long delay time from the observed decay curve.

4. Picosecond Laser Photolysis Studies on the CR of $Porphyrin^{+}$-$Quinone^{-}$ State

The above result for the bell-shaped energy gap dependence of the CR rate of geminate IP in polar solvents indicates that the rate of the CR deactivation of $porphyrin^{+}$-$quinone^{-}$ geminate IP is around the top region of the bell-shaped relation in view of its energy gap (- G_{CR} 1.5 eV). and actually we have confirmed its ultrafast CR deactivation in polar solvents [16]. Very similar result has been observed also for the geminate IP, $FTFP^{+}$-$quinone^{-}$, where FTFP is a face to face porphyrin dimer model compound [16b].

We examined also the porphyrin (P)-quinone (Q) combined system, P-$(CH_2)_n$-Q (PnQ), n=2,4,6, to study the more details of the photoinduced CS and CR of the IP state of this pair [17]. The presence of the intervening chain can control the velocity of ET to some extent because the electronic interaction matrix element can be modified by the intervening chain due to the change of the average distance between porphyrin and quinone. The intramolecular photoinduced CS of P^*nQ and CR of the IP state, P^+nQ^-, can be represented by eq 10.

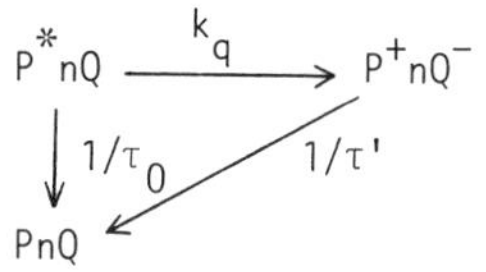

(10)

The lifetime $\tau=(k_q+\tau_0^{-1})^{-1}$ of P^*nQ is determined essentially by k_q for all PnQ with n=2,4,6. From the analysis of the results of picosecond time-resolved transient absorption and fluorescence measurements, the approximate values of the rate constant of photoinduced CS were determined to be, $k_q \sim 10^{11}\ s^{-1}$, $10^{10}\ s^{-1}$ and $10^{9}\ s^{-1}$ for n=2,4 and 6. It has been confirmed that the dependence of $1/\tau'$ upon n was similar to that of k_q. It should be noted here that τ values of PnQ determined by streak camera do not show much dependence upon the solvent polarity in contrast to the case of Pn and An where the photoinduced CS rate is much enhanced by increasing the solvent polarity. This might be ascribed to the more favorable CS energetics in the PnQ than in Pn and An and also weaker solvation of large porphyrin cation ring [17a].

We have examined also the combined system, FTFP-$(CH_2)_n$-Q, n=3,5 [18]. The result was, however, very similar to the case of PnQ. Moreover, owing to the little smaller energy gap between $FTFP^+nQ^-$ and FTFPnQ states compared to the corresponding one of PnQ system, the CR deactivation of the intramolecular IP state appears faster for the FTFPnQ system. In any way, since the CR decay of the intramolecular IP state is essentially governed by the Franck-Condon factor for this pair except that both the photoinduced

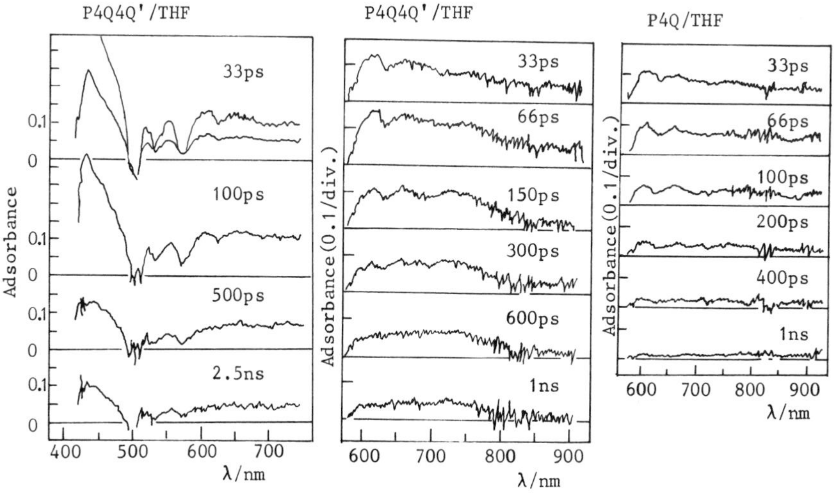

Fig. 5. Picosecond time resolved transient absorption spectra of P4Q4Q' and P4Q in THF.

CS and the CR of the intramolecular IP state become slower with increase of the intervening chain length, the detection of the intramolecular IP state is difficult also in those combined systems. Therefore, not only in the case of uncombined porphyrin-quinone systems but also in the combined systems, the IP state produced from the singlet excited state, especially in polar environment, undergoes very rapid CR deactivation. In this respect, it should be emphasized that we have demonstrated previously [2a,17] that the lifetime of the intramolecular IP state becomes considerably longer in the P4Q4Q' system compared with the P4Q system (where Q' is trichlorobenzoquinone which has higher electron affinity than Q) probably due to the more extensive CS by multistep ET. Fig. 5 shows the picosecond time-resolved transient absorption spectra of P4Q4Q' and P4Q in THF. The broad spectra without structure can be attributed to the IP state. According to the results of our analysis of the experimental results with the method similar to the case of PnQ, the lifetime of IP state of P4Q was shorter than 100 ps while that of P4Q4Q' was ca. 400 ps [17b]. The multistep ET will more or less suppress the rapid CR of IP state.

In any way, on the basis of our observation on the bell-shaped energy gap dependence of CR in the geminate IP produced by CS in fluorescence quenching reaction and the fact that the CR rate of the P^+-Q^- pair is around the top region of this bell-shaped relation, we can understand very well the crucial importance of the multistep ET in the bacterial photosynthetic reaction center, where the following very rapid and highly efficient electron transfer and charge separation processes are realized.

$$\begin{array}{c} (\mathrm{BCh}\ell\text{-b})_2^*(\mathrm{BCh}\ell\text{-b})(\mathrm{BPh\text{-}b})\mathrm{Q}(\mathrm{Fe}^{2+}) \\ \big\downarrow <\text{a few ps} \\ (\mathrm{BCh}\ell\text{-b})_2^+(\mathrm{BCh}\ell\text{-b})(\mathrm{BPh\text{-}b})^-\mathrm{Q}(\mathrm{Fe}^{2+}) \\ \big\downarrow \sim 200\ \text{ps} \\ (\mathrm{BCh}\ell\text{-b})_2^+(\mathrm{BCh}\ell\text{-b})(\mathrm{BPh\text{-}b})\mathrm{Q}^-(\mathrm{Fe}^{2+}) \end{array} \qquad (11)$$

5. Remarks on the Different Energy Gap Dependence of the Photoinduced CS in the Fluorescence Quenching from That of the CR of the Produced Geminate IP

As it is discussed in 2., 3., and 4., we have confirmed directly by means of the ultrafast laser photolysis the bell-shaped energy gap dependence of the CR rate of the geminate IP. It is well-known that, contrary to this result of the CR for the geminate IP, the photoinduced CS in the fluorescence quenching in polar solutions does not show such energy gap dependence. The rate constant of the quenching reaction examined for many systems does not show any decrease even at strongly exothermic (large $-\Delta G^\circ_{CS}$) region.

It is difficult to explain satisfactorily this remarkable difference between the energy gap dependences of photoinduced CS and CR produced IP by the previous classical as well quantum mechanical theoretical consideration [4,19]. The quantum mechanical theory [19b], taking into consideration the contribution of the intramolecular high frequency vibrations of donor and acceptor in the Franck-Condon factor, is the most satisfactory one among those previous treatments. Nevertheless, they predict the same bell-shaped energy gap dependence for both of CS and CR processes.

In this respect, the theoretical treatment proposed by us recently will be important [5,6]. By taking into consideration the fact that the orientational vibration (c-mode) frequency of polar solvent coordinated to a charged solute may be considerably larger than that surrounding neutral solute, we have formulated rate expressions and could give a satisfactory interpretation for his difference between the energy gap dependences of CS and CR reactions. Namely, as indicated schematically in fig. 6 by

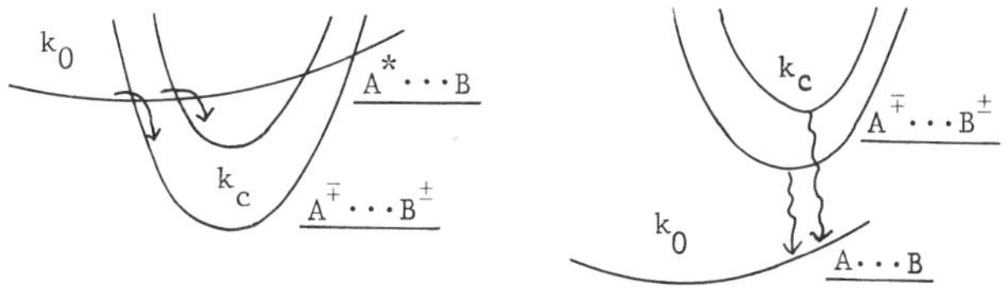

Fig. 6. Schematic potential energy surfaces as functions of the solvent orientational polarization coordinate, taking into consideration that the orientation vibration frequency of the solvent surrounding a charged solute is considerably larger than that surrounding a neutral solute.
(a) photoinduced CS.
(b) CR of IP produced by photoinduced CS.

potential surfaces as functions of solvent orientation (polarization) coordinate, the activation barrier due to this solvent mode does not arise even at very large energy gap region in the photoinduced CS reaction while the solvent mode becomes silent in the CR reaction of the IP, which means the strong energy gap dependence of the CR reaction. Owing to the contribution of this solvent mode to the Franck-Condon factor for the ET, the photoinduced CS reaction will become insensitive to the relevant energy gap at large $-\Delta G^\circ_{CS}$ values while the CR of IP will show a bell-shaped energy gap dependence. We defined a parameter, $\beta=k_0/(k_c-k_0)$, where k_0 and k_c represent force constants for solvent orientation vibration for neutral and charged states respectively. In the case of the potential surfaces as indicated in fig. 6, β should be much smaller than unity.

The ET rate constant can be written by convolution,

$$W(\Delta E) = \int_{-\infty}^{\infty} W_q(\Delta E-\varepsilon)\mathscr{G}(\varepsilon)d\varepsilon \qquad (12)$$

where ΔE is the energy gap between the initial and final states and W_q is the rate constant obtained by considering only the intramolecular vibrational modes as given previously [19b]. $\mathscr{G}(\varepsilon)$ is the Franck-Condon factor due to the solvent c-mode formulated classicaly in analogy to the previous one [19b]. For the CS reaction, $A^{*}\ldots D\rightarrow A_S^{-}\ldots D_S^{+}$,

$$\mathscr{G}(\varepsilon) = \int_{-\infty}^{\infty} D_a(E;E_a^{*})D_d'(E;E_d)dE \qquad (13)$$

where D_a is the electron insertion spectrum of A^{*}, D_d' is the electron removal spectrum of D, E_a^{*} is the energy difference between A^{*} and A_S^{-}, E_d is the energy difference between D_S^{+} and D, and $\varepsilon=(E_a^{*}-E_d)$. Similarly, for the

CR reaction, $A_S^- \ldots D_S^+ \rightarrow A \ldots D$,

$$\mathcal{J}(\varepsilon) = \int_{-\infty}^{\infty} D_a'(E;E_a) D_d(E;E_d) dE \qquad (14)$$

where D_a' is the electron removal spectrum of A_S^- and D_d is the electron insertion spectrum of D_S^+, E_a is the energy difference between A_S^- and A, E_d is the energy difference between D_S^+ and D, and $\varepsilon=(-E_a+E_d)$. More generally speaking, in addition to the solvent c-mode in the first solvent layer around the solute, where the force constant of solvent orientation vibration for charged solute, k_c, is considerably larger than that for neutral solute k_0, there is solvent layer outside the first one, which can contribute to solvation energy but in which $k_0 \sim k_c$. The orientation vibrational mode of the latter solvent layer is called the s-mode. Taking into consideration s-mode, the ET rate constant W can be written by convolution as,

$$W(\Delta E) = \int_{-\infty}^{\infty} d\varepsilon_1 W_q(\varepsilon_1) \int_{-\infty}^{\infty} d\varepsilon_2 \mathcal{J}_c(\Delta E-\varepsilon_2) \mathcal{J}_s(\varepsilon_2-\varepsilon_1) \qquad (15)$$

where $\mathcal{J}_c$ and $\mathcal{J}_s$ are thermally averaged Franck-Condon factors for c and s modes, respectively [5,6]. By means of the above formulations and with β=0.1∿0.3, the observed results of the energy gap dependences of the photoinduced CS and CR of the produced geminate IP can be well interpreted [5,6].

As one can see from the above discussions, the essential point of this theory is the quite different curvature of the potential surface for the solvent mode between the charged and neutral state of solute. We can prove easily that if we assume the linear response in the polarization of solvent due to the charged solute, the curvature of the potential surface for the solvent mode is the same in neutral and charged state of solute. Namely, assuming free energy curves as a function of orientational polarization of

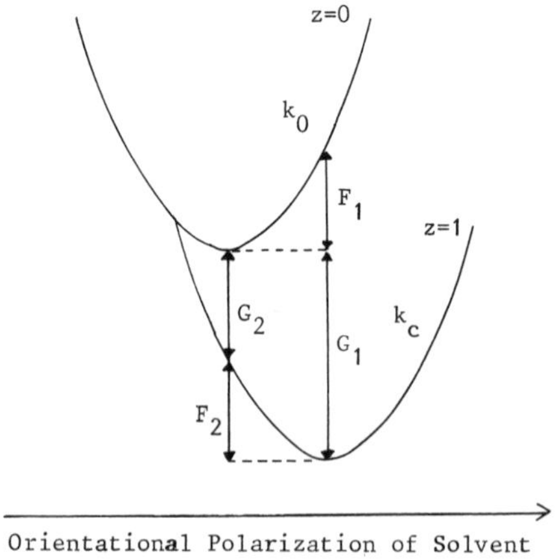

Fig. 7. Schematic diagram of free energy surfaces as functions of orientational polarization of solvent surrounding reactants with charge z=0 and z=1.

solvent surrounding reactants, with charge z=0 and z=1 as indicated in fig. 7, the reorganization energy F_1 is given by Marcus equation as,

$$F_1 = (e^2/2r_0)\left[\frac{1}{\varepsilon_\infty} - \frac{1}{\varepsilon_S}\right] \qquad (16)$$

where r_0 is the radius of solute. The Born solvation energy by charging is given by,

$$G_1 = (e^2/2r_0)\left[1 - \frac{1}{\varepsilon_S}\right],\ G_2 = (e^2/2r_0)\left[1 - \frac{1}{\varepsilon_\infty}\right] \qquad (17)$$

Therefore, $F_2 = G_1 - G_2 = (e^2/2r_0)\left[\frac{1}{\varepsilon_\infty} - \frac{1}{\varepsilon_S}\right] = F_1$,

and $(k_0/k_c) = (F_1/F_2) = 1$. This leads to the parameter $\beta=k_0/(k_c-k_0)=\infty$. Above formulas for reorganization energies and solvation energies are based on the approximation of the linear response of solvent polarization.

Recently, we have made a Monte Carlo simulation for the solvent dipole polarization around solute with z=0 or various z values using a spherical hard-core model for molecules [20]. We have divided the solvent around the solute into thin layers. Our results show that the first thin layer just surrounding a charged solute shows a dielectric saturation, where the polarization P_r divided by the solvent dipole moment μ approaches unity already at z=1. From the population $f^i(P_r/\mu)$ in i-th shell in the region (P_r/μ) to $(P_r/\mu)+(\Delta P_r/\mu)$ in the total Monte Carlo steps $(\Delta P_r/\mu \sim 1/400)$, the relative free energy g^i as a function of (P_r/μ) is given by,

$$g^i(P_r/\mu) \propto -RT\ell n f^i(P_r/\mu) \qquad (18)$$

because $f^i(P_r/\mu)$ is proportional to the probability distribution that (P_r/μ) is realized and should be expressed by the Boltzmann distribution,

$$f^i(P_r/\mu)\ \exp[-\Delta g^i(P_r/\mu)RT] \qquad (19)$$

We have confirmed that the curvature of $\Delta g^1(P_r/\mu)$ as a function of (P_r/μ) increases very much with increase of z. From the ratio of the curvature for z=1 and z=0, the (k_c/k_0) value was obtained to be 10, which leads to the value of $\beta=k_0/(k_c-k_0)\sim 0.1$. This β-value is in the range, $0.1\sim 0.3$, which has been extensively used to reproduce experimental results in terms of the c-mode [5,6].

The above results show clearly the existence of a thin dielectric saturation shell of solvent which contributes to the c-mode, just around the charged solute molecule. This c-mode plays a crucial role in regulating the energy gap laws of photoinduced CS and CR of produced IP. The solvent vibration in the layers outside of the saturation shell constitutes the ordinary polaron mode called the s-mode.

The present work was partly supported by a Grant-in-Aid (No. 62065006) from the Japanese Ministry of Education, Science and Culture to N. M.

REFERENCES

1. N. Mataga and M. Ottolenghi, Molecular Association, Vol. 2, R. Foster Ed., Academic Press, London 1979, p. 1; N. Mataga, Molecular Interactions, Vol. 2, H. Ratajczak and W. J. Orville-Thomas Eds., John-Wiley & Sons, Chichester 1981, p. 509.

2. N. Mataga, (a) Pure & Appl. Chem. 56, 1255 (1984);
(b) J. Mol. Struct. (Theochem.) 135, 279 (1986);
(c) Acta Phys. Polon. A71, 767 (1987), and papers sited there-in.
3. T. Okada and N. Mataga, to be submitted for publication.
4. (a) R. A. Marcus, J. Chem. Phys. 24, 966 (1956),
(b) R. A. Marcus, Ann;. Rev. Phys. Chem. 15, 155 (1964).
5. T. Kakitani and N. Mataga, J. Phys. Chem. 90, 993 (1986).
6. T. Kakitani and N. Mataga, (a) Chem. Phys. 93, 381 (1985),
(b) J. Phys. Chem. 89, 8 (1985); (c) ibid. 89, 4752 (1985);
(d) ibid. 91, 6277 (1987).
7. (a) D. Huppert, H. Kanety and E. M. Kosower, Faraday Discuss, Chem. Soc. 74, 161 (1982).
(b) E. M. Kosower and D. Huppert, Ann. Rev. Phys. Chem.37, 127 (1986).
8. K. Nakatani, T. Okada and N. Mataga, to be submitted for publication.
9. B. Bagchi, G. R. Fleming and D. W. Oxtoby, J. Chem. Phys. 78, 7375, (1983).
10. N. Mataga, Y. Kaifu and M. Koizumi, (a) Bull. Chem. Soc. Jpn. 28, 690 (1955); (b) ibid. 29, 465 (1956).
11. E. Lippert, (a) Z. Naturforsch. 10a, 541 (1955); (b) Ber. Bunsenges. Phys. Chem. 61, 962 (1957).
12. B. Bagchi, D. W. Oxtoby and G. Fleming, Chem. Phys. 86, 257 (1984).
13. (a) N. Mataga, Y. Kanda and T. Okada, J. Phys. Chem. 90, 3880 (1986).
(b) N. Mataga, H. Shioyama and Y. Kanda, ibid. 91, 314 (1987).
14. N. Mataga, T. Asahi and Y. Kanda, to be submitted for publication.
15. D. Rehm and A. Weller, Israel J. Chem. 8, 259 (1970).
16. (a) N. Mataga, A. Karen, T. Okada, S. Nishitani, N. Kurata, Y. Sakata and S.Misumi, J. Am. Chem. Soc. 106, 2442 (1984).
(b) Mataga, A. Karen, T. Okada, S. Nishitani, Y. Sakata and S. Misumi, J. Phys. Chem. 88, 4650 (1984).
17. (a) N. Mataga, A. Karen, T. Okada, S. Nishitani, N. kurata, Y. Saktata, and S. Misumi, J. Phys. Chem. 88, 5138 (1984).
(b) in preparation.
18. (a) Y. Sakata, S. Nishitani, N. Nishimizu, S. Misumi, A. R. McIntosh, J. R. Bolton, Y. kanda, A. Karen, T. Okada and N. Mataga, Tetrahedron Lett. 26, 5207 (1985).
(b) Y. Kanda, A. Karen, T. Okada and N. Mataga, to be submitted for publication.
19. (a) J. J. Hopfield, Proc. Natl. Acad. Sci. 71, 3640 (1974).
(b) J. Jortner, J. Chem. Phys. 64, 4860 (1976).
20. Y. Hatano, M. Saito, T. Kakitani and N. Mataga, J. Phys. Chem. 92, (1988), 92, 1008 (1988).

THERE AND BACK AGAIN: LONG DISTANCE ELECTRON TRANSFER FROM SEMICONDUCTORS, TO PROTEINS, TO SEMICONDUCTORS

G. MCLENDON, K. J. CONKLIN, R. CORVAN, K. JOHANSSON, E. MAGNER, M. O'NEIL, K. PARDUE, J. S. ROGALSKYJ, D. WHITTEN

Department of Chemistry, University of Rochester, Rochester, NY 14627

ABSTRACT

The timeliness of Hopfield's "Light from Distant Pairs" paper is examined on the twenty-fifth anniversary of its publication. Originally addressing unusual luminescence properties in semiconductors, Hopfield's notion of coupling an electron transfer mechanism with a random distribution of trap sites is applicable to electron transfer reactions in systems as far afield as protein-protein couples. Recently, his ideas are being applied again to semiconductors, this time to model electron-hole recombination in cluster whose diameter is smaller than the de Broglie wavelength of pair generating light.

Twenty-fifth anniversaries hold a special significance for many. It has now been twenty-five years since John Hopfield published his insightful analysis of recombination emission from trap sites in semiconductors [1]. In these papers, Hopfield explained the previously bizarre nonexponential emission lineshape, and the complex spectral lineshape, involving only a random distribution of trap sites, which react by a long distance electron tunneling mechanism. It is noteworthy that analogous arguments for electron exchange (Dexter energy transfer) between molecules were simultaneously derived by Inokuti and Hirayama [2]. In this article, we hope to underscore how these simple theories continue to guide research in a variety of superficially unrelated areas including understanding the fundamental aspects of molecular electron transfer, the design of biological redox systems, or even the nonlinear optical properties of "quantum dots".

Despite the elegance of these theories, their significance for understanding electronic effects on the rates of molecular redox reactions remained largely unappreciated for over ten years. This dormancy may stem in part from the spectacular success of the concurrent development, by Marcus [3], of a theory of electron transfer in solution which emphasizes the importance of nuclear reorganization in determining rates, culminating in the famous Marcus equation $\Delta G^{\ddagger} = (\Delta G° - \lambda)^2/4\lambda$ where $\Delta G^{\ddagger}$ is the activation free energy, $\Delta G°$ the overall reaction free energy and λ, the reorganization energy, can be visualized as the Franck Condon energy required to move from the reactant to the product surface without nuclear motion. In its simplest form, Marcus theory assumes the reaction is electronically adiabatic, so that overlap, per se, is not important for the rate. Marcus theory proved spectacularly successful in quantitatively predicting the relative rates of solution redox reactions, over many orders of magnitude in rate. An apparent serious flaw in the theory concerned rates at high $\Delta G°$ ($G° > \lambda$) where rates are expected to *decrease* with increasing ΔG, contrary to almost all results in solution. The ultimate solution to this conundrum came from a reconsideration of Hopfield's and Inokuti's theories to the problem of molecular electron transfer. In a classic series of papers beginning with his thesis, John Miller of Argonne National Labs considered trapping of the solvated electron by molecular acceptors dispersed randomly in glassy matrices [4]. As suggested by

Photochemical Energy Conversion
James R. Norris, Jr. and Dan Meisel, Editors

previous work, the range of possible donor acceptor distances produced a wide range of rates. By studying the electron transfer rate as a function of acceptor concentration, it was possible to show[4] that rate depended exponentially on concentration, much as predicted by theory [1,2,5]. Similar results were subsequently obtained in our lab for photoinduced electron transfer [6] (see Fig. 1).

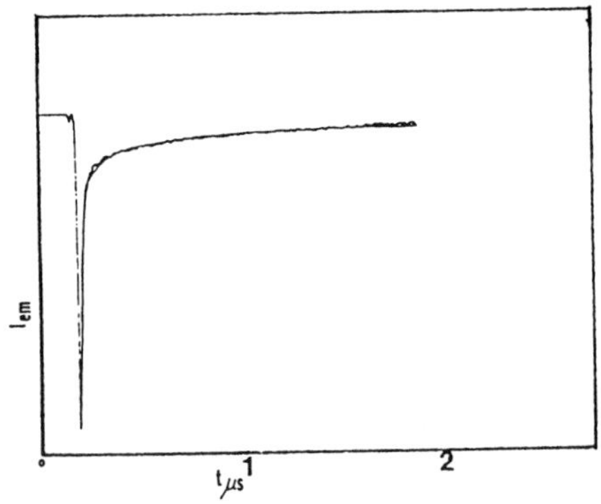

FIG. 1. Luminescence decay of $Ru(me_2phen)_3)Cl_2$/TMPD in lexan. The date are fit assuming $k \approx \exp(-1.2R)$.

FIG. 2. Proposed structure of cyt-c/cyt-b_5 complex based on computer modeling experiments.

Since experimentalists were attuned to considering electron transfer as occurring ubiquitously by close contact, these seminal results were greeted somewhat skeptically. In the past decade, however, elegant work on linked electron donor-acceptor assemblies, by Closs, Miller and others [7], produced overwhelming evidence for long range electron transfer. A number of detailed reviews are available [8], so that details of the theory need not be repeated here. Suffice it to say that for collisionless electron transfer, the rate constant includes an electronic term, the tunneling matrix element, which measures donor-acceptor coupling:

$$(H_{AB}) = V_o \exp(-\alpha R) \quad (1)$$

Rates also depend on the extent of nuclear motion which accompanies electron transfer, as well as the free energy of reaction which accompanies that motion, in the form of a Franck-Condon weighted density of states:

$$FCWD = \sum_{\omega=0}^{\infty} (e^{-S} S^{w}/W!) \exp-(\Delta G^\circ + Es + w\hbar\omega)^2/4Esk_BT \quad (2)$$

Combining these results gives a general expression for electron transfer rate combining the effects of distance, free energy, and nuclear motion:

$$K_{et} = C\, H_{AB}{}^2 FCWD \quad (3)$$

Among the aspects of eq. 3 which have been tested, two are particularly noteworthy. First, the exponential dependence of rate on distance has been clearly verified using a wide variety of chemical reaction systems, media, and techniques. Second, Marcus' free energy relationship predicts that rate increases as ΔG° increases, reaches a maximum when $\Delta G^\circ = \lambda$, and decreases when $\Delta G^\circ > \lambda$. As already mentioned, evidence of "inverted behavior" when $\Delta G^\circ > \Lambda$ was elusive in solution, but has been clearly verified for "long distance" reaction [7]. It is now understood that this effect is largely masked by diffusion limited rates in solution.

With this background, we hope to illustrate how these basic ideas can be applied to understand electron transfer reactions which occur in relatively complex systems, focusing first on electron transfer within protein-protein complexes, and finally returning to where this story began, an examination of trap recombination in semiconductors, but with a present emphasis on small semiconductor clusters, which span the poorly understood region between individual molecules and bulk "materials" which are dominated by collective properties.

1) PROTEIN ELECTRON TRANSFER: THEME AND VARIATIONS

A primary lesson from experiments on protein electron transfer reactions, in which the donor and acceptor sites are held at long distances (10-30 Å) is that the basic rules gleaned from studies of small molecule redox reactions apply equally to larger systems [8a,b]. This general statement is supported by studies of "model" protein-ruthenium conjugates, as well as more biological but also more complicated protein-protein complexes.

For example, detailed studies were carried out of electron transfer within the noncovalent complex formed between cytochrome c and cytochrome b_5 (Fig. 2) [9]. By varying the central metal in the porphyrin active site, it is possible to vary ΔG° over a wide range, without significantly affecting the overall structure of the protein. Cyclic voltammetry using chemically modified electrodes shows that for $Fe^{(II)}$ cytochrome c $E^\circ = 0.26V$, whereas for the Zn(I) substituted derivative, $E^\circ = 0.9V$, (Fig.3), and for the triplet excited state (Zn porphyrin cytochromes) $E^\circ = {}^3E_{00} + E^\circ_{zn} = +0.9V$.

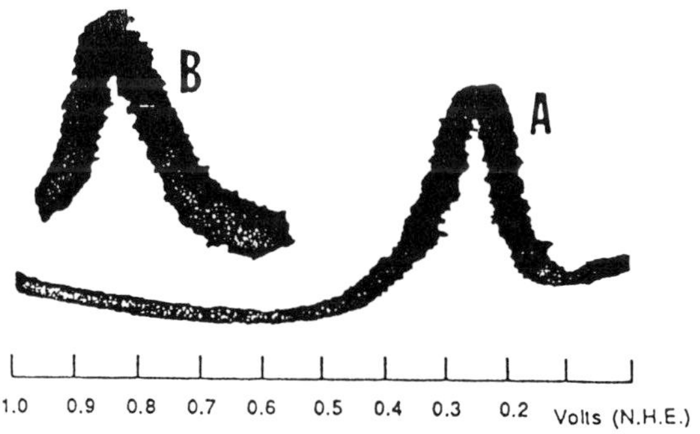

FIG. 3. Differential pulse voltammogram of 3mM Fe (a) and Zn (b) cytochrome c; 3 mm graphite disc electrode; scan rate 2 mV/s; pulse amp. 50 mV.

Associated detailed voltammetric studies under EC conditions establish similar reorganization energies for Zn cytc and Fe cytc. With such data in hand for a variety of metal substituted cytochromes, the dependence of electron transfer rate on $\Delta G°$ for biologically relevant protein-protein complexes could be obtained. As anticipated by Marcus theory, rate increases with increasing $\Delta G°$, reaches a maximum when $\Delta G° = \lambda$, and then decreases with further increase in $\Delta G°$ (Fig. 4a).

The inferred magnitudes of λ, (λ~1V) are significantly larger than had been anticipated for redox proteins [10], perhaps suggesting that for "slow" electron transfer proteins like cytochrome c as distinct from the very rapid reactions which occur in photosynthesis) the protein to protein electron transfer rate may not have been maximized by protein structure, but actually minimized to insure optimal specificity. In the context of specificity, the role of possible specific pathways for electron transfer within proteins is important; that is, what role does specific electronic coupling between the donor and the surrounding matrix play in electron transfer. Following others, we note that such coupling might proceed via "superexchange" [11] involving either the low lying empty orbitals of the medium, or close lying filled orbitals (hole transfer). Recent experiments in our lab and elsewhere suggest that "hole tunneling" may indeed be important. Pioneering studies by Hoffman et al. [12] demonstrated electron transfer occurred over a 20 Å separation in $^3Zn^*/Fe(III)$ hemoglobin hybrids, with $k_{et} = 100\ s^{-1}$. The Fe(II) product in this molecule is

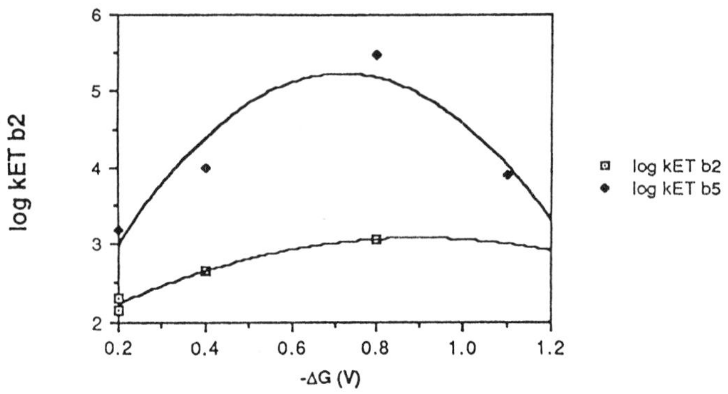

FIG. 4: Rate content vs. $\Delta G°$ for intracomplex electron transfer. Top: cyt c/cyt b_5 complex. Bottom: cyt c/cyt b_5 complex.

partially trapped by a decay (k_d) of the Zn^+ porphyrin cation radical, which occurs competitively with the recombination reaction $Zn^+/Fe(II) \xrightarrow{ke} Zn/Fe(III)$. Drawing on previous work in solar photochemistry of porphyrins [13], we were able to directly prepare the Zn^+ species by irreversibly quenching $^3(Zn)^*$ Hb derivatives with $(Co(III)(NH_3)_5Cl)^{2+}$. In this way, both $k_d = 2000\ s^{-1}$ and $k_r = 4500\ s^{-1}$ can be measured directly (Fig. 5). The surprising result is that $k_r >> k_{et}$, even though the distance and reaction free energy are almost identical for both systems. This result, and other similar findings [14] suggest that "hole transfer" may have a weaker distance dependence than does electron transfer.

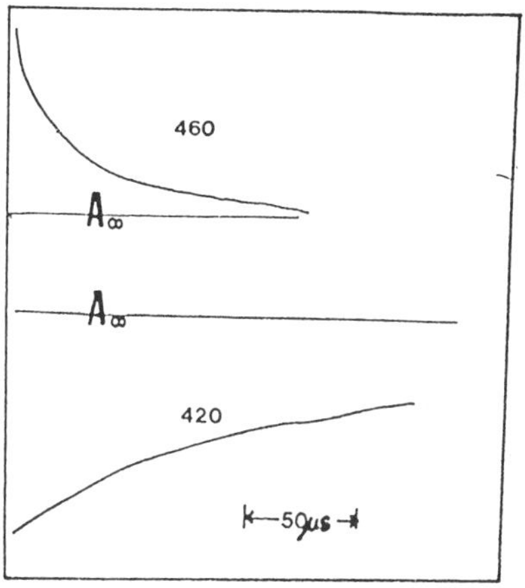

FIG. 5: Rate of decay (k_d) of the Zn porphyrin cation radical. Top: triplet decay: $Co(III) + Zn^* \rightarrow Zn^+ + Co(II)$. Bottom: Soret recovery $Zn+ \rightarrow Zn$. Note: this process is clearly independent of triplet decay per se.

A final caveat in understanding protein redox chemistry arises from the possibility of motion within the protein complex on the time scale of electron transfer. Since rate depends strongly on distance, motion which affects that distance (or angle, or coupling pathway) will modulate the rate of transfer. If motion occurs slowly relative to electron transfer, then an ensemble of complexes may be generated, each with a characteristic reaction rate. If motion is fast, a single rate will be observed, but the characteristic parameters may vary from those expected from static studies of the complex. Finally, motion occurring on the same time scale as electron transfer will produce a complex, time dependent reaction rate: the reaction is "gated" by conformational rearrangements of the proteins [15].

Evidence for conformational heterogeneity in protein complexes which is linked to redox reaction rate is beginning to emerge. For example, the measured equilibrium bonding constant between cytochrome c and cytochrome c peroxidase depends markedly on the method used for measurement: fluorescence titration which measures heme-heme distance, gives $K_{eq} = 10^5\ M^{-1}$, while NMR titrations which are sensitive to the charge around the heme, give $K_{eq} = 10^3 M^{-1}$. Such wide variance shows the complex cannot be characterized by a simple free bound equilibrium, but must involve multiple bound states [16]. Each different complex makes a differently weighted contribution to the observed equilibria, depending on the method used. As suggested above, the individual conformational states of the bound complex may be dynamically interconnected. It may be possible to sense such motion by using energy transfer methods. As derived by Katchalski et al. [17] the lineshape for nonradiative energy transfer dynamics can directly reflect relative motions of the donor with respect to the acceptor which occur during the excited state lifetime [17]. Since energy transfer rates depend strongly on donor-acceptor distance, when the donor and acceptor move with respect to one another during the excited state decay, the rate constant for energy transfer will become time dependent, and a significant deviation from single exponential decay kinetics will be observed. Preliminary experiments have been conducted with the Zn cytochrome c/Fe cytochrome b_5 complex, and the Mg cytochrome c peroxidase/Fe cytochrome c complex. In each complex, motion along the heme-heme coordinate can be sensed by time resolved energy transfer measurements. Although fluorescence decay in the isolated donor is essentially described by a single exponential, energy transfer in the bound complex clearly is not (Fig. 6), suggesting that quite rapid motions along the heme-heme (reaction) coordinate can occur within a bound protein-protein complex.

Finally, if motion within a protein complex can indeed be coupled to rate of electron transfer within this complex, as suggested above, then the possibility exists that the rate of such motion could itself become rate determining for some systems. Evidence for this "conformational gating" kinetic limit is available in studies of the cytochrome c/cytochrome b_2 complex [18]. In this system the dependence of electron transfer rate on $\Delta G°$ is much less than anticipated from theory; indeed, k_{et} is virtually independent of $\Delta G°$, suggesting that another process (conformational change) determines the observed rate of the redox process.

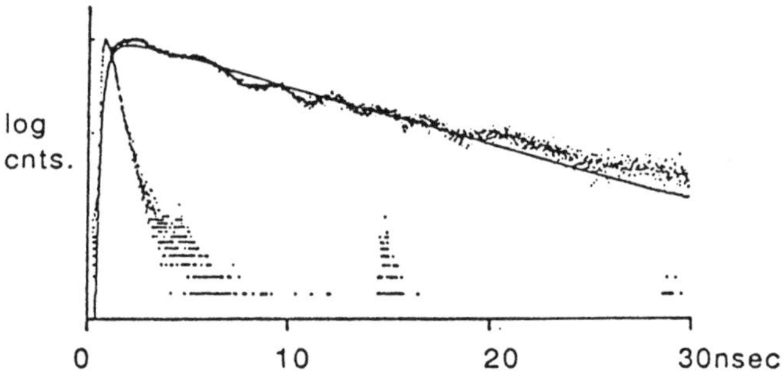

FIG. 6: Fluorescence decay of H_2-cyt c/b_5 complex. Note the nonexponential behavior at long time.

In summary, insights obtained from studies of simple electron transfer reactions proceeding from semiconductors through small molecules have proven equally valuable in understanding the electron transfer reactions which occur in biological systems for energy transduction and metabolism.

In the final section of this paper, we return to the original problem of trap recombination in semiconductors, focusing on the unusual photophysics of "quantum dot" materials which span the gap (literally) between small molecules, and bulk semiconductor materials.

3) BACK AGAIN: PHOTOPHYSICS OF SEMICONDUCTOR CLUSTERS

In an era of cheap oil, solar energy researchers sometimes speak earnestly of "spin off" technologies, for indeed we have much to tell. The bulk of experimental work on long distance electron transfer, for example, springs from an interest in solar conversion. Another interesting spin off arose unexpectedly in the course of experiments by Louis Brus of Bell Labs. Brus, being interested in molecular adsorbates on semiconductor surfaces, turned to studies of high surface area dispersed semiconductor particles as previously studied by Bard, Gratzel, and others [19]. Seeking to maximize signal/noise, he synthesized yet smaller particles using the classical "arrested precipitation" methods of colloid chemistry [20]. During these experiments, Brus noticed [21] that below a critical size (ca 300 Å for CdS) the absorption spectra of microcrystals no longer could be predicted by classical Mie scattering. Instead, the spectra progressively blue shifted with decreasing particle size. He correctly reasoned [21] that this shift was a quantum mechanical effect of exciton confinement, analogous to the "particle in a box" problem for organic polyenes (Fig. 7).

Further work in his and other labs has established the generality of such effects [22-25]. These simple arrested precipitation techniques had thus accessed "diewelt der Verhaltegesst Dimensionen": such clusters represent the missing link between relatively well understood molecular states and equally well characterized bulk semiconductors. Further details of the dependence of electronic structure on cluster size are available elsewhere [21-23], including a review in this volume by Henglein, who has done pioneering work in this area. In the present context we will focus only on the optical properties of these clusters. Understanding these optics lies in characterizing the long distance electron transfer which accompanies electron hole emissive recombination in the clusters. This is, of course, the essential problem with which we began — "light from distant pairs".

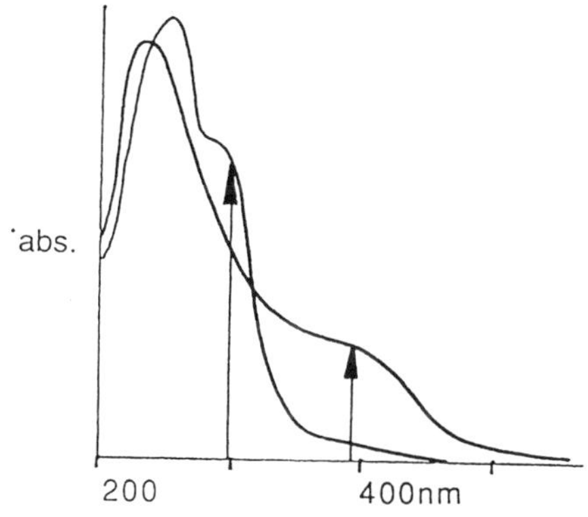

FIG. 7: Absorption spectra of CdS clusters as a function of particle radius. Arrows represent particle in a box type calculations of the band edge with corrections for the effective mass of the electron and hole.

Two optical properties are of particular note: recombination luminescence, and nonlinear optical effects. These properties, as we will show, are strongly interdependent. First, these small clusters show intense recombination luminescence. Both the emission wavelength and quantum efficiency can be tuned by synthetically adjusting the particle radius (which defines the "particle in a box"), or by modifying the surface chemistry of the cluster.

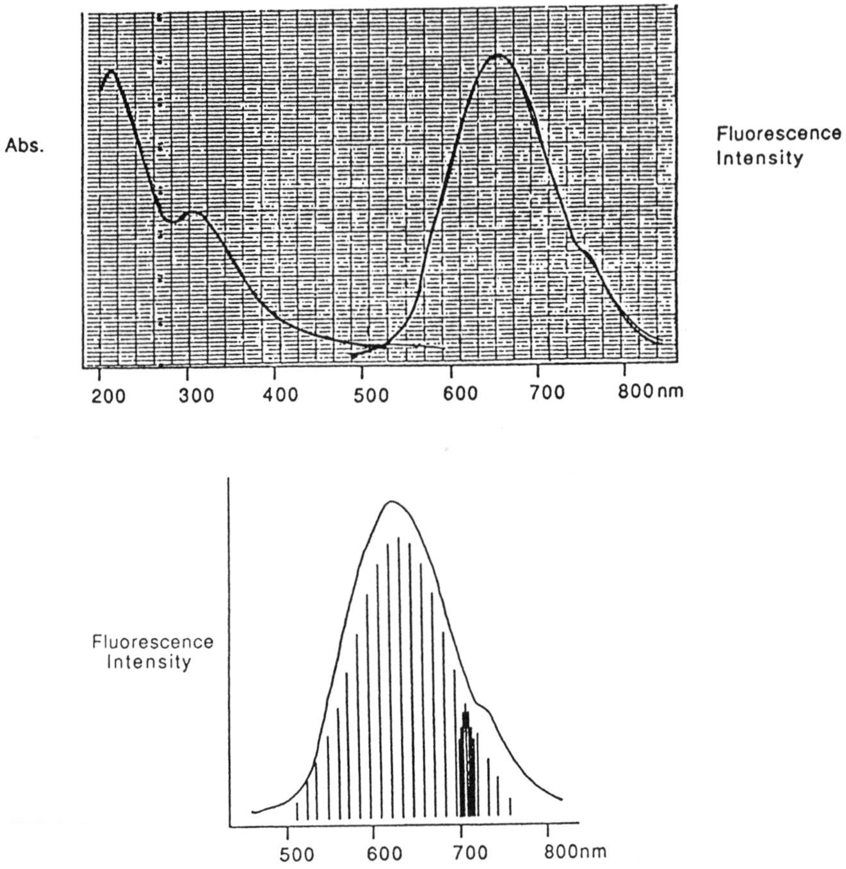

FIG. 8: a) Absorption and emission spectra of 150 Å diameter Cd_3As_2 clusters. Note the large spectral shift between absorption and emission maxima. (b) Coulombic "stick spectrum" of incremental separation of the e^-/h^+ by 10 Å with one representative coupled phonon broadening.

As an example of these effects, we consider the recombination luminescence in Cd_3As_2, as first reported by Henglein [22c], and characterized in some detail in our own lab. A first obvious feature of the luminescence is that the emission onset is shifted to lower energy than the absorption band edge (Fig. 8a). This shift suggests that the initial photogenerated electron hole pair localize at low energy (midband) trap sites. A signature for spatially isolated traps is also shown in the emission band shape, which has the very broad width associated with "light from distant pairs", where the variation in Coulomb energy associated with different e^-/h^+ trap distances provides a corresponding variation in luminescence recombination energy (Fig. 8b). As shown in the figure, the overall lineshape is a convolution of the "stick spectrum" with a coupled phonon broadening.

The shape of the luminescence decay is consistent with trap recombination: close traps, with strong overlap, recombine rapidly while distant traps, with poor overlap recombine slowly, producing a highly nonexponential lineshape which is familiar from the previous discussion of electron transfer reactions among random ensembles of molecular donors and acceptors (Fig. 9).

Finally we note that combining the individual distance dependencies of recombination energy and recombination rate correctly predicts that the higher energy recombination events will generally occur more quickly, as shown in Fig. 9.

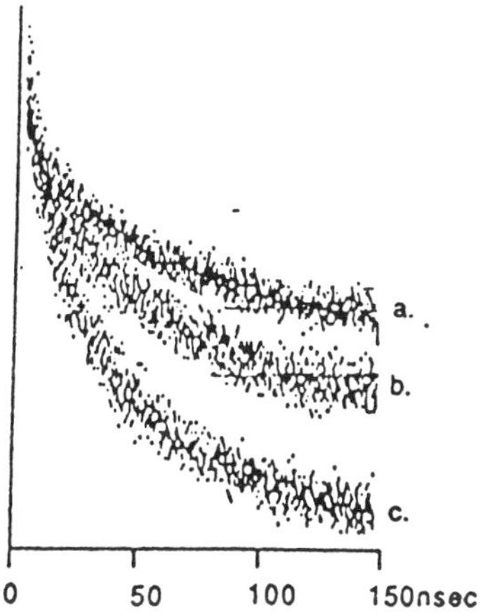

FIG. 9: Fluorescence decay profiles by time correlated single photon counting of three Cd_3As_2 clusters (a) 200 Å dia. (b) 180 Å dia (c) 150 Å dia.

Given that trap sites dominate the luminescence, it remains to ask where the trap sites are found (e.g., surface vs. bulk), and what is their true chemical nature? Strong chemical evidence from our lab and Henglein's favors surface trapping sites [22,25]. Specifically, addition of strong bases like alkylamines, OH^-, and RS results in a blue shift of recombination emission with a corresponding increase in the quantum yield. This effect is particularly dramatic for II-VI materials. For 100 Å CdS clusters, on the addition of OH^- (pH 11), the emission shifts from a weak, red emission at 650 nm to a strong emission (em > 0.8) near the band gap, at ca. 450 nm (Fig. 10).

These effects can be completely reversed by, for example, lowering the pH, or dialyzing away the amine. A simple explanation is that these bases bind directly to the Cd^{2+} ions at the surface and raise the local site energies beyond the band gap. An alternative related explanation is that it is not the ligand binding (Lewis base) properties per se of the additives which is important, but rather their Bronsted basicity to cause, for example, deprotonation of a surface SH atom.

The importance of surface chemistry for the optical properties of semiconductor clusters suggests the need to better control this chemistry. An important step in this direction was taken by Steigerwald et al. in developing synthesis of "capped" clusters [26].

These syntheses proceed in three steps:

First, the initial nucleation clusters are prepared by reacting a metal salt with a chalcogenide source in AOT/H_2O/heptane reverse micelles media. Reverse micelles act to kinetically control the ion redistribution, thereby narrowing the size distribution of the clusters [25-27].

Second, to these initial cluster excess metal ion, equivalent to one monolayer is added to the clusters, producing a reasonably defined reactive surface layer.

Finally, the lattice is terminated by addition of e.g. (TMS)Se(phenyl). The TMS group is readily displaced to form a CdS-phenyl bond. The phenyl group, however, cannot be readily cleaved. Thus capping prevents further lattice growth, and provides a better characterized coordination at the surface. Fortuitously, the phenyl capped compound is sparingly soluble in most organic solvents, and rapidly precipitates from solution. A simple variant on this procedure developed in our own lab uses commercially available thiophenol as the "cap", thereby extending the synthesis to a wider range of media.

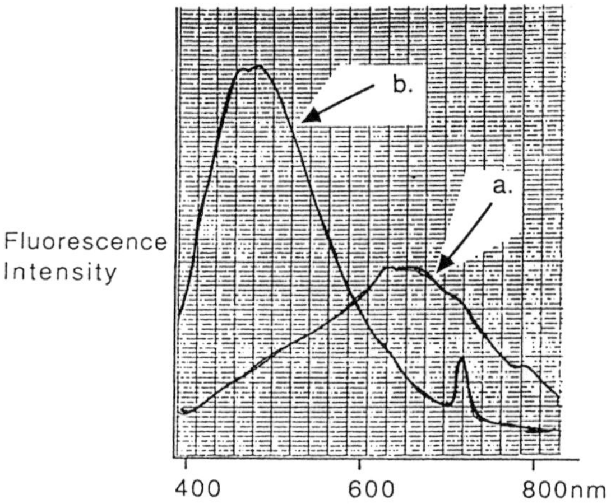

FIG. 10: (a) Fluorescence profile of aqueous 100 Å CdS clusters at pH 7.0 (b) at pH 11.0.

With this background, we conclude by considering a different optical property of these clusters, namely their optical nonlinearity (χ_3). At high applied optical fields many materials exhibit profound nonlinear (i.e.: non "Beer's law") optical behavior. These nonlinear properties are important in a variety of emerging technologies ranging from phase conjugate optics to optical computing [28]. A simple case of χ_3 effects involves resonant excitation in a three level system (Fig. 11).

Intersystem crossing to a triplet produces a molecule with very different absorption properties than the ground state. Thus, the next arriving photon will interact differently with the excited matter than did the first. For this resonant case, there is a necessary trade off between the magnitude of the effect, and the "switching time" between states. For example, organic dyes in a glass matrix show very high χ_3 values, but with switching times approaching one second: far too slow for any use in switching [29]. The "figure of merit" then, involves a tradeoff between magnitude and switching rate. If the transition is collective, as for semiconductors, then larger χ_3 values can be obtained (relative to isolated molecules) without a corresponding loss in switching time. We, and others, reasoned that semiconductor clusters might well have desirable χ_3 properties, and this has indeed been amply demonstrated for a strange class of materials composed of graded CdS/CdSe semiconductor clusters produced advantageously in the production of Corning sharp cut-off filters [30].

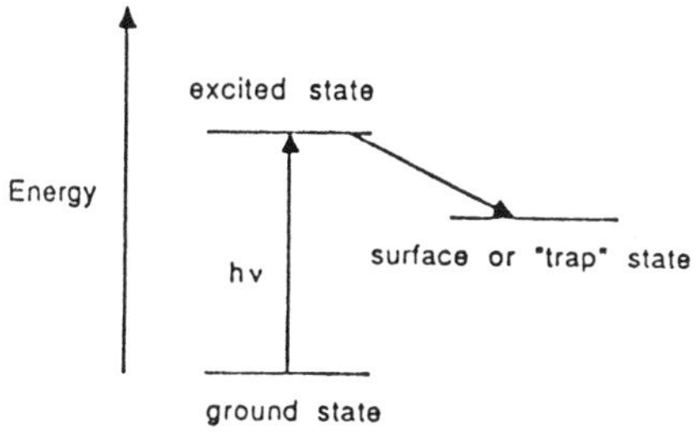

FIG. 11: Schematic representation of resonance excitation of a three level system.

In our own lab, and in parallel work by Ying Wang et al. at DuPont, we have investigated the nonlinear optics of deliberately synthesized semiconductor clusters, using degenerate four wave mixing techniques (Fig. 12).

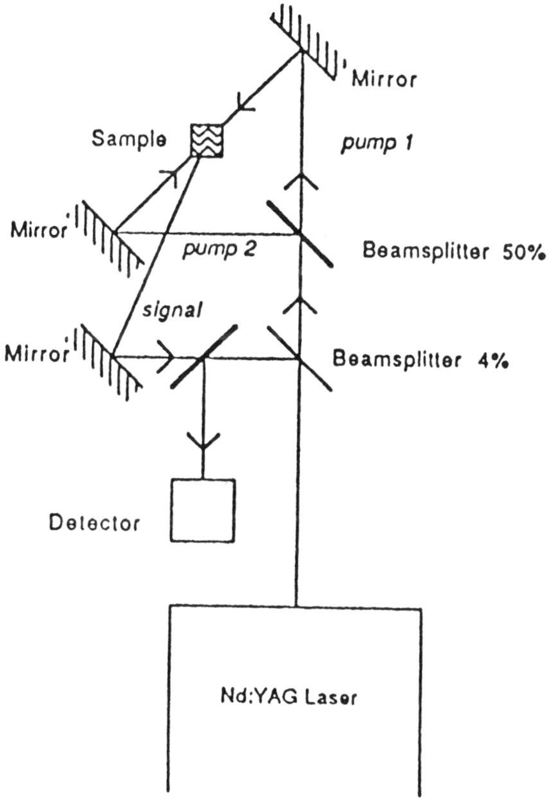

FIG. 12: Typical four wave mixing apparatus. Counterpropagating pump beams interact with the probe beam in the sample. The signal beam is colinear with the probe beam, but propagates in the opposite direction. BS=beamsplitter, M=mirror.

The primary findings are summarized below:

1. Large optical nonlinearities, e.g., $\chi_3 = 10^{-4}$ esu can be obtained, with switching times < 100 ps.

2. These resonant χ_3 effects appear to directly track the trap recombination rates.

Thus, the surface chemistry, which controls the luminescence properties and the photochemistry of these clusters probably controls their nonlinear optical properties as well. By better controlling these surface trap sites, it may become possible to design materials with a widely tunable nonlinear optical response.

REFERENCES

1. J. J. Hopfield, D. Thomas, M. Gershenzon, Phys. Rev. Lett. 10, 162 (1963).
2. M. Inokuti, F. Hiryama, J. Chem. Phys. 43, 1978 (1965).
3. R. Marcus, Discuss. Faraday Soc. 29, 21 (1960).
4. J. Miller, Science 189, 221 (1974).
5. J. Hopfield, Proc. Natl. Acad. Sci. 71, 3640 (1974) and J. Jortner, J. Chem. Phys. 64, 4860 (1976).
6. T. Guarr, M. McGuire and G. McLendon, J. Am. Chem. Soc. 105, 616 (1983).
7. G. Closs, J. Miller and T. Calcaterra, J. Am. Chem. Soc. 106, 3047 (1984). M. Wasielewski and M. Niemczyk, J. Am. Chem. Soc. 106, 5043 (1984).
8. cf. a. R. Marcus and N. Sutin, Biochem. Biophys. Acta 811, 265 (1985).
   b. G. McLendon, Accts. Chem. Res. (in press)
   c. D. Devault, Quant. Rev. Biophys. 13, 387 (1985).
9. G. McLendon and J. R. Miller, J. Am. Chem. Soc. 107, 7811 (1985).
10. G. Moore et al. Discus. Faraday Soc. 74, 311 (1982).
11. J. R. Miller and J. Beitz, J. Chem. Phys. 74, 6746 (1981).
12. J. McGourty, N. Blough and B. Hoffman, J. am. Chem. Soc. 105, 4470 (1983).
13. D. Whitten in: Porphyrins and Metalloporphyrins, I.C. Smith, ed. (Elsevier, Amsterdam).
14. G. McLendon et al. J. Am. Chem. Soc. 107, 739 (1984).
15. B. Hoffman and M. Ratner, J. Am. Chem. Soc. 109, (1987).
16. G. McLendon, J. S. Rogalskyi and F. Sherman, Proc. Nat. Acad. Sci (in press).
17. K. Katchalski et al. in: NATO Workshop in Energy Transfer (Reidel, 1978).
18. G. McLendon, K. Pardue and P. Bak, J. Am. Chem. Soc. 109, 7540 (1987).
19. M. Gratzel, ed., Energy Storage Through Photochemistry and Catalysis (Academic Press, New York, 1983).
20. K. Kalyanasundaram, E. Borgarello, D. Duonhong and M. Gratzel, Angew. Chem. Int. Ed. Engl. 20, 987 (1981).
21. a. R. Rossetti and L. Brus, J. Phys. Chem. 86, 4470 (1982).
    b. L. E. Brus, J. Chem. Phys. 79, 5566 (1983).
    c. R. Rossetti, J. L. Ellison, J. M. Gibson and L. E. Brus, J. Chem. Phys. 80, 4464 (1984).
    d. R. Rossetti, R. Hull, J. M. Gibson and L. E. Brus, J. Chem .Phys. 82, 552 (1985).
    e. N. Chestnoy, R. Hull and L. E. Brus, J. Chem. Phys. 85, 2237 (1986).
22. a. A. Henglein, Ber. Bunsenges. Phys. Chem. 82, 241 (1982).
    b. A. Henglein, J. Phys. Chem. 86, 2291 (1982).
    c. A. Fojtik, H. Weller and A. Henglein, Chem. Phys. Lett. 120, 552 (1985).
    d. S. Baral, A. Fojtik, H. Weller and A. Henglein, J. Am. Chem. Soc. 108, 342 (1986).
    e. L. Spanhel, M. Haase, H. Weller and A. Henglein, J. Am. Chem. Soc. 109, 5649 (1987).
23. a. Y. Wang, A. Suna, W. Mahler and R. Kasowski, J. Chem. Phys. 87, 7315 (1987).

b. Y. Wang and W. Mahler, Opt. Commun. 61, 233 (1987).

24. A. J. Nozik, J. Phys. Chem. 90, 12 (1986).
25. T. Dannhauser, M. O'Neil, K. Johansson, D. Whitten and G. McLendon, J. Phys. Chem. 90, 6074 (1986).
26. M. Steigerwald, A. P. Alivisatos, J. M. Bigson, T. D. Harris, R. Kortan, A. J. Muller, A. M. Thayer, T. M. Duncan, D. C. Douglass and L. E. Brus, J. Am. Chem. Soc. (1988) in press.
27. a. Y. Tricot and J. H. Fendler, J. Am. Chem. Soc. 106, 7359 (1984).
    b. P. Lianos and J. K. Thomas, Chem. Phys. Lett. 125, 209 (1986).
28. J. Hellwarth, Prog. Quant. Elec. 5, 1 (1977).
29. M. A. Kramer, W. R. Tompkin and R. W. Boyd, Phys. Rev. A 34, 2026 (1986).
30. a. J. T. Remillard and D. G. Steel, Opt. Lett. 13, 30 (1988).
    b. S. S. Yao, C. Karaguleff, A. Gabel, R. Fortenberry, C. T. Seton and G. I. Stegeman, Appl. Phys. Lett. 46, 801 (1985).
    c. K. C. Rustagi and C. Flytzanis, Opt. Lett. 9, 344 (1984).
    d. R. K. Jain and R. C. Lind, J. Opt. Soc. Am. 73, 647 (1983).

INTRAMOLECULAR ENERGY AND ELECTRON TRANSFER PROCESSES IN POLYNUCLEAR TRANSITION-METAL COMPLEXES

FRANCO SCANDOLA
Dipartimento di Chimica dell'Università, Centro di Fotochimica CNR,
44100 Ferrara, Italy

INTRODUCTION

Research on solar energy conversion has produced in recent years valuable spin-off's in a variety of fields of fundamental interest [1]. Remarkable examples of this statement are certainly the elegant syntheses of covalently bonded donor-acceptor diads [2], triads [3], and tetrads [4], and the successful attempts to obtain photoinduced charge separation processes within these supermolecules. These studies have actually shed much light on the factors that govern the rates of an elementary chemical process as fundamental as electron transfer.

The charge-separating supermolecules are examples of what can be more generally defined as a molecular photochemical device [5]: a discrete assembly of molecular components capable of performing a specific light induced function. General classes of light induced functions are:

1) Generation and migration of electronic energy;
2) Photoinduced vectorial transport of electric charge;
3) Photoinduced conformational changes.

Examples of specific functions within the above classes are:

1) - Remote energy transfer photosensitization;
   - Remote generation of optical signals;
   - Antenna effect;
   - Light energy up-conversion.
2) - Conversion of light into chemical or electrical energy (charge separation);
   - Photoinduced electron collection;
   - Remote light-induced electron transfer sensitization;
   - Switch on/off of electrical signals.
3) - Conversion of light into chemical energy;
   - Switch on/off of electrical signals;
   - Switch on/off of receptor ability;
   - Switch on/off of access to cavities.

Recently, general criteria for the design of molecular photochemical devices capable of performing such functions have been proposed, analogies with macroscopic or naturally occurring working devices have been pointed out, and examples already present in the literature have been reviewed [5].

The primary act involved in the functioning of any molecular photochemical device is light absorption by a suitable chromophoric unit that may be called the "photosensitizer". The final act will depend on the type of function performed (e.g., light emission, sensitization, heterogeneous electron transfer, complexation, etc.). In between, a number of intramolecular elementary acts must occur for which the thermodynamic and kinetic factors, as well as the spatial organization of the subunits

Photochemical Energy Conversion
James R. Norris, Jr. and Dan Meisel, Editors

involved are likely to be crucial. These intermediate acts are most often *intramolecular* processes of two types: (i) electronic *energy transfer*, (ii) *electron transfer*. It is clear that progress in the understanding of these fundamental intramolecular processes is essential to the design and operation of working molecular photochemical devices.

Transition metal coordination compounds have been studied quite actively in the last decade as reactants in bimolecular energy [6] and electron transfer [7] processes. These systems offer a large degree of flexibility, as several thermodynamic and kinetic parameters of the bimolecular process (excited-state energies and redox potentials, reorganizational barriers and electronic factors) can be controlled by a proper choice of metal and ligands. For these reasons, coordination compounds look as attractive candidates for use as building blocks of molecular photochemical devices. Thus, although synthetic problems may generally be more severe here than for conceptually similar organic systems, efforts in the synthesis of inorganic diads, triads, etc. suitable for the study of intramolecular energy and electron transfer processes seem to be worthwhile.

In the inorganic literature, molecules made up by two or more metal centers, each having a given coordination environment, interconnected by one or more bridging ligands are usually called bi- or *polynuclear complexes*. The prototype is the classic Creutz-Taube ion,

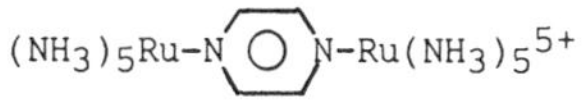

whose synthesis in 1969 [8] initiated a rapid take-off of the chemistry of polynuclear metal complexes. In the last two decades, a large number of combinations of metal centers and bridging ligands have been synthesized and studied [9]. Much of the driving force in this field has been provided by the classical Hush model [10], a simple and powerful theoretical framework that unifies such diverse aspects of the chemistry of polynuclear complexes as mixed-valence behavior, optical properties, and electron transfer kinetics. Thermal intramolecular electron transfer in polynuclear complexes has been extensively studied, with particularly important contributions by Taube [11], Haim [12], and Isied [13]. Light induced intramolecular electron transfer in polynuclear complexes has been often postulated, but directly observed only in relatively few cases, notably by Sutin [14] and Meyer [15]. The possibility of intramolecular energy transfer following light excitation of polynuclear complexes has been investigated by Petersen [16], Endicott [17], and Kane-Maguire [18], with direct evidence [18] for this process being rare.

In this presentation, a number of studies carried out in our laboratory on photoinduced energy and electron transfer processes in polynuclear metal complexes will be reviewed. The polynuclear species investigated consist of $Ru(bpy)_2^{2+}$ (bpy = 2,2'-bipyridine) or related species as the photosensitizer unit, cyanides as bridging ligands, and a variety of metal containing moieties M_1 and M_2 as attached units. The complexes have the following schematic structure:

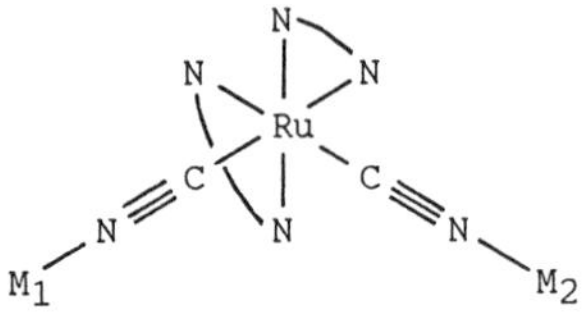

The relationship between the complex optical spectra of these systems and the pathways for photoinduced intramolecular electron transfer relaxation will be pointed out. The use of spectral information in the evaluation of through-bond electronic coupling will be illustrated. Direct observation of intramolecular energy transfer in chromophore-luminophore complexes will also be reported, and the mechanism of such energy transfer processes will be discussed.

PROPERTIES OF THE PHOTOSENSITIZER UNIT. EFFECTS OF BRIDGING LIGANDS AND METALATION.

The $Ru(bpy)_2^{2+}$ unit was chosen as the light-absorbing chromophore, since it is the minimal fragment of the well-known $Ru(bpy)_3^{2+}$ photosensitizer that can be bound to two additional units via bridging ligands. It is known that this unit mantains most of the outstanding properties of the parent $Ru(bpy)_3^{2+}$, among which long MLCT excited-state lifetime and photochemical stability, provided that the two additional ligands do not bring about the presence of low-lying ligand-field states [19] This constitutes a limitation on the choice of bridging ligands to be used for connecting the photosensitizer unit to other molecular subunits, and was the reason for our selection of the strong-field, ambidentate cyanide ligand. Some of the relevant properties of the photosensitizer-bridge combination $Ru(bpy)_2(CN)_2$ are given in Table I. It is seen that the lifetime of the MLCT state is relatively long and that the excited state is expected to be a strong reductant and a mild oxidant, a behavior quite similar to that of the parent $Ru(bpy)_3^{2+}$.

If one wants to bind metal-containing units to the cyanides of $Ru(bpy)_2(CN)_2$ in order to look for intramolecular energy or electron transfer processes, the first question to ask is concerning the effect of metalation, as such, on the properties of the chromophore. The "minimal" effects can be established by looking at polynuclear compounds with metal moieties that do not offer the possibility of intramolecular energy or electron transfer (innocent units). Examples of such compounds are the binuclear and trinuclear complexes containing $Pt(dien)^{2+}$ (Table I) [20]. The most evident effect is the blue shift in emission accompanying the metalation (a parallel and somewhat higher effect is observed in absorption). This reflects the increase in electron withdrawing ability of C-bonded cyanide upon interaction with electron acceptors at the nitrogen end. Quite similar effects have been observed upon metalation with other innocent moieties (Zn^{2+}, Ag^{+}) [21,22] and on protonation or alkylation of the cyanide ligands of $Ru(bpy)_2(CN)_2$ [23,24].

TABLE I. Photophysical and Redox Parameters for Bi- and Trinuclear Complexes of General Formula M_1-NC-Ru(bpy)$_2$-CN-M_2.[a,b,c]

	M_1	M_2	ν_{em} (μm^{-1})	τ (ns)	E_{red} (V)	E_{ox} (V)	$*E_{red}$[d] (V)	$*E_{ox}$[d] (V)
1)	------	------	1.47	205	-1.68	+0.73	+0.37	-1.32
2)	$Pt(dien)^{2+}$	------	1,58	630	-1.62	+1.03	+0.57	-1.16
3)	$Pt(dien)^{2+}$	$Pt(dien)^{2+}$	1.72	90	-1.50	+0.86	+0.81	-1.45
4)	$Ru(NH_3)_5^{2+}$	------	----	---	-1.53	-0.17	----	----
5)	$Ru(NH_3)_5^{3+}$	------	----	---	-0.17	+1.14	----	----
6)	$Ru(NH_3)_5^{2+}$	$Ru(NH_3)_5^{2+}$	----	---	-1.52	-0.17	----	----
7)	$Ru(NH_3)_5^{2+}$	$Ru(NH_3)_5^{3+}$	----	---	-0.17	-0.09	----	----
8)	$Ru(NH_3)_5^{3+}$	$Ru(NH_3)_5^{3+}$	----	---	-0.09	+1.24	----	----
9)	$Ru(NH_3)_4py^{2+}$	------	----	---	-1.54[e]	+0.11[e]	----	----
10)	$Ru(NH_3)_4py^{3+}$	------	----	---	+0.11[e]	+1.12[e]	----	----
11)	$Ru(NH_3)_4py^{2+}$	$Ru(NH_3)_4py^{2+}$	----	---	-1.50[e]	+0.10[e]	----	----
12)	$Ru(NH_3)_4py^{2+}$	$Ru(NH_3)_4py^{3+}$	----	---	+0.10[e]	+0.18[e]	----	----
13)	$Ru(NH_3)_4py^{3+}$	$Ru(NH_3)_4py^{3+}$	----	---	+0.18[e]	+1.35[e]	----	----
14)	$Ru(NH_3)_4py^{2+}$	$Ru(NH_3)_5^{2+}$	----	---	-1.52[e]	-0.11[e]	----	----
15)	$Ru(NH_3)_4py^{2+}$	$Ru(NH_3)_5^{3+}$	----	---	-0.11[e]	+0.16[e]	----	----
16)	$Ru(NH_3)_4py^{3+}$	$Ru(NH_3)_5^{3+}$	----	---	+0.16[e]	+1.25[e]	----	----
17)	$Ru(bpy)_2CN^{+}$	------	1.44	90	-1.53	+0.69	+0.42	-1.26
18)	$Ru(bpy)_2CN^{2+}$	------	----	---	+0.69	+1.31	----	----
19)	$Ru(phen)_2CN^{+}$	------	1.47	400	-1.53	+0.75	+0.47	-1.25
20)	$Ru(phen)_2CN^{2+}$	------	----	---	+0.75	+1.37	----	----
21)	$Ru(bpy)_2CN^{+}$ [f]	$Ru(bpy)_2CN^{+}$ [f]	1.41	50	-1.53	+0.64	+0.29	-1.18
22)	$Ru(bpy)_2CN^{2+}$ [f,g]	$Ru(bpy)_2CN^{+}$ [f,g]	----	---	+0.64	+1.17	----	----
23)	$Cr(CN)_5^{2-}$ [f]	------	1.25	10^6	h	i		
24)	$Cr(CN)_5^{2-}$ [f]	$Cr(CN)_5^{2-}$ [f]	1.25	10^6	h	i		

a) Unless otherwise noted, the central unit contains Ru(II) and C-bonded cyanides;
b) Data from refs. 20,25-28;
c) Unless otherwise noted, redox potentials are in DMF vs. SCE;
d) Excited-state redox potentials estimated using 77K emission energies;
e) In aqueous solution vs. SCE;
f) In these complexes the bridging cyanides are N-bonded to central Ru;
g) In this complex the central Ru is in the +3 oxidation state.
h) Not determined; likely ~ -1.5 V, based on $Cr(CN)_6^{3-}$.
i) Not determined; presumably, immeasurably high.

The lifetime of the MLCT state remains substantial, undergoing even a sizeable increase in the case of the mono-metalated complex. Similar observations have been made for metalation with Zn^{2+} and Ag^{+} [21,22]. This indicates that the energy gap between the upper ligand-field states and the emitting MLCT state is not dramatically decreased upon metalation.

Altogether, these results indicate that metalation as such does not induce quenching of the MLCT state of the chromophore, so that any observation of quenching should be indicative of the opening of intramolecular energy- or electron transfer channels involving the attached units.

ELECTRON TRANSFER. COMPLEXES WITH RUTHENIUM AMMINE SUBUNITS.

For a number of the bi- and trinuclear complexes shown in Table I the redox potentials [25-28] are such that oxidation (complexes n. 5, 7, 8, 10, 12, 13, 15, 16, 18, 20, 22) or reduction (complexes n. 4, 6, 7, 9, 11, 12, 14, 15) of an attached metal moiety by the excited state of the $Ru(bpy)_2^{2+}$ photosensitizer unit is thermodynamically allowed. In these cases, photoinduced intramolecular electron transfer is expected to occur following excitation of the photosensitizer unit. Indeed, for all these complexes the MLCT emission is completely quenched (Table I). These systems exhibit very interesting, complex absorption spectra. It can be shown that the optical transitions are intimately related with the intramolecular electron transfer pathways available within the complex. Some aspects of this interconnection will be briefly discussed in this section.

Optical Electron Transfer.

With respect to the spectra of $Ru(bpy)_2(CN)_2$ and model ruthenium ammine units, the spectra of all of the above-mentioned bi- and trinuclear complexes are far from being additive. In fact, in each case new absorption bands can be detected in the spectrum that can be associated to optical electron transfer between different redox sites in the supermolecule. The variety of electron transfer transitions observed in these systems is exemplified for the case of $M_1 = Ru(NH_3)_4py^{2+}$, $M_2 = Ru(NH_3)_5^{3+}$ (compound n. 15 of Table I) in Figure 1. Spectral assignments in these systems are facilitated by the swithcing on/off of the various types of bands upon oxidation and reduction of the sites involved in the transition. Making reference to Figure 1 and to the complexes of Table I, it is found [25-27] that MLCT transitions of type 1 are exhibited by all the complexes, MLCT transition of type 2 by complexes n. 9, 11, 12, 14, 15, remote MLCT transitions by complexes n. 4, 6, 7, 9, 11, 12, 14, 15, intervalence transfer (IT) transitions by complexes n. 5, 7, 8, 10, 12, 13, 15, 16, 18, 20, 22, and remote IT transitions by complexes n. 7, 12, 15.

The widespread observation of remote MLCT transitions of remarkable intensity (previously unknown) and of sizeable remote IT transitions (previously detected in another trinuclear complex by Taube [29]) is noticeable. Both observations point towards a considerable degree of electronic interaction between the redox centers in the supermolecule. Although for the MLCT transitions this may be somewhat arbitrary, the observed electron transfer bands can be fitted to the Hush model of optical intervalence transfer yielding, among other parameters, the electronic interaction matrix element H_{if}. From the bands of Figure 1, H_{if} values of 8400, 8700, 1500, 1900, and 350 cm^{-1} are obtained for transitions of type 1, 2, 3, 4, and 5, respectively. The values are seen to follow reasonably well either the distance or the number of bonds intervening between the centers involved. This may be related to through-space or through bond mechanisms

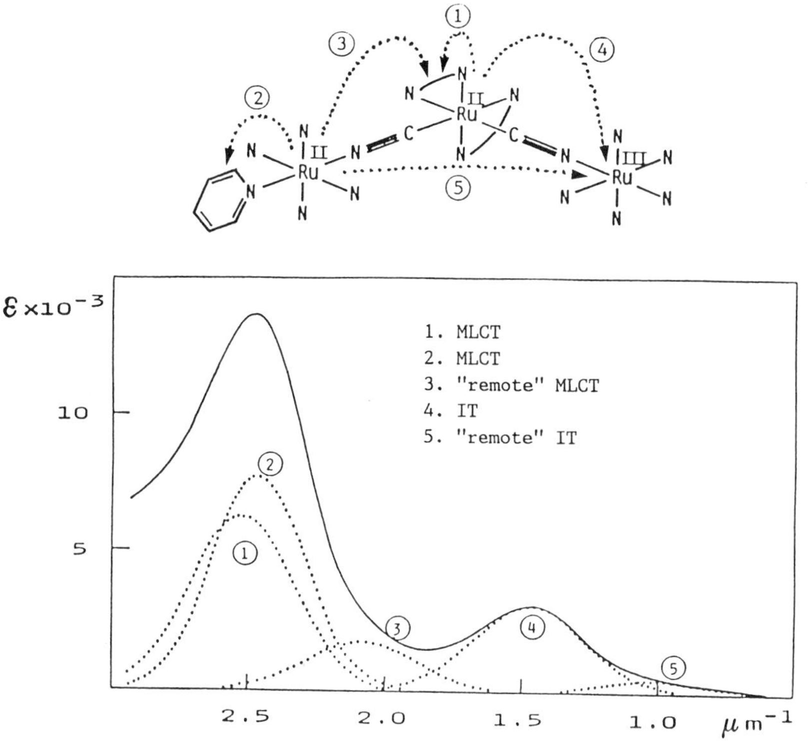

FIG. 1. Absorption spectrum of $py(NH_3)_4Ru\text{-}NC\text{-}Ru(bpy)_2\text{-}CN\text{-}Ru(NH_3)_5^{5+}$ in aqueous solution (lower part) and assignment of the observed bands to site-to-site electronic transitions (upper part).

for site-to-site electronic interaction. This point will be developed for the remote IT in a later paragraph.

Photoinduced Intramolecular Electron Transfer.

The variety of charge transfer and electron transfer states detectable in the absorption spectra of these systems suggests a variety of possible intramolecular electron transfer processes following light absorption by the $Ru(bpy)_2^{2+}$ photosensitizer unit. These processes are shown for the M_1 = $Ru(NH_3)_4py^{2+}$, M_2 = $Ru(NH_3)_5^{3+}$ case in Figure 2, where it is seen that each one of the states that can in principle be reached by intramolecular electron transfer can also be populated by optical transitions. These complexes are indeed good examples to illustrate the concept that thermal electron transfer processes and radiationless transitions are equivalent on a theoretical basis.

Experimentally, in all of the bi- and trinuclear complexes the MLCT

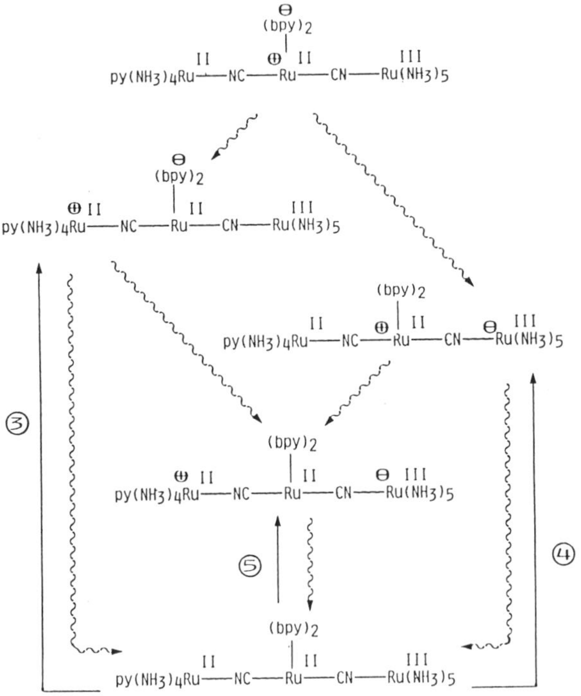

FIG. 2. Relationship between the electron transfer pathways for excited-state deactivation and the optical electron transfer transitions. The trinuclear complex and the numbering of the transitions are the same as in Fig 1.

emission of the $Ru(bpy)_2(CN)_2$ chromophore is completely quenched. This indicates that fast intramolecular electron transfer occurs from the reduced bpy to terminal Ru(III) or from terminal Ru(II) to the oxidized central ruthenium (Figure 2). In no case long-lived transients are observed in laser photolysis, indicating that the subsequent radiationless cascade to the ground state is also very fast. It should be noticed that the remote IT state of Figures 1 and 2 is analogous to the charge-separated state successfully observed in organic triads [3]. The reasons for the lack of long-lived charge separation in these systems seem to lie mainly in the energetics, as discussed in more detail elsewhere [27].

Remote Intervalence Transfer and Through-bond Interactions.

Following the work of the groups of Miller and Closs [2b] and of Hush, Paddon-Row and Verhoeven [2a], it is now widely accepted that through-bond electronic interactions are responsible for long-range electron transfer processes (both thermal and optical) across chemical bridges. This mechanism is related to the "superexchange" mechanism for long-range electron transfer through an intervening medium (e.g., solvent) [30]. A number of theoretical models for the calculation of through-bond electronic coupling matrix elements have been recently developed [31].

In simple qualitative terms, through-bridge interaction can be viewed as an indirect coupling of the initial and final states of the electron transfer process via configuration interaction with excited local charge transfer states involving the bridge. This is schematically shown in Figure 3 for a simple donor (D), bridge (L), and acceptor (A) combination.

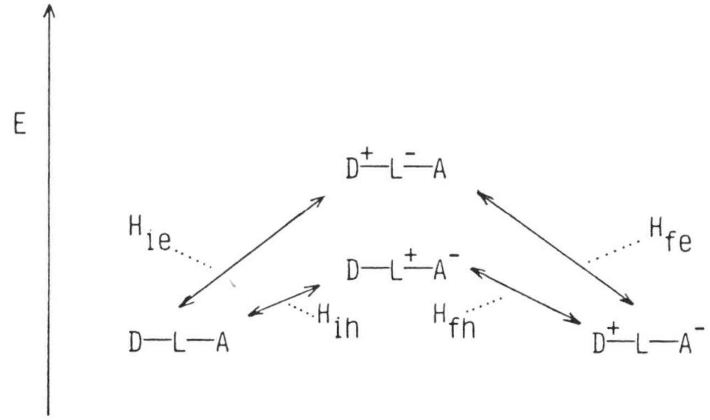

FIG. 3. Schematic representation of the superexchange interaction coupling a donor D and an acceptor A through a bridge L.

Neglecting direct interaction between the initial, ψ_i, and final, ψ_f, configurations (i.e., neglecting through-space D-A orbital overlap), any effective interaction is assumed to come from the admixture of the donor-to-bridge electron transfer, ψ_e, and acceptor-to-bridge hole transfer, ψ_h, excited configurations into ψ_i and ψ_f. Simple second order perturbation theory shows that the effective matrix element between the initial and final states of the electron transfer, H_{if}, is given by

$$H_{if} = \frac{H_{ie} \quad H_{fe}}{\Delta E_{ie}} + \frac{H_{ih} \quad H_{fh}}{\Delta E_{ih}} \qquad (1)$$

where H_{ie} (H_{fe}) and H_{ih} (H_{fh}) are the interactions between the initial (final) states and the electron and hole transfer states, and the ΔE terms are the corresponding energy differences (Figure 3). Therefore, knowledge of the energies and the matrix elements for the virtual electron and hole transfer processes would allow a simple calculation of the through-bridge interaction. Without resorting to theoretical calculations, these quantities could be, in principle, obtained from spectroscopy, provided that the $\psi_i \rightarrow \psi_e$, $\psi_i \rightarrow \psi_h$, $\psi_f \rightarrow \psi_e$, and $\psi_f \rightarrow \psi_h$ transitions can be observed and analyzed. In many organic systems involving saturated bridges, however, the electron and hole transfer states lie at too high energies to be observable (these states are actually often labelled as "virtual states"). In some inorganic cases, (e.g. binuclear complexes with aza-aromatic bridges) some of the relevant transitions (e.q. the electron transfer one) can be observed. Due to the strong metal-ligand interaction, however, such bands can hardly be treated within the Hush model to extract the relevant matrix elements.

It appears that the trinuclear complexes investigated in this laboratory offer a unique opportunity to test the simple configuration-interaction approach on experimental grounds. Using the complex shown in Figure 1 as an example, the end-to-end remote IT transition can be treated as involving $Ru(NH_3)_4py^{2+}$ as the donor D, $Ru(NH_3)_5{}^{3+}$ as the acceptor A, and NC-$Ru(bpy)_2$-CN as the bridge L. It is easily seen that in this case the electron transfer state corresponds to the remote MLCT state, while the hole transfer state corresponds to the IT state. In this case, therefore, both the electron and the hole transfer states are observed spectroscopically (Figure 1), and the H_{ie}, E_{ie} and H_{ih}, E_{ih} parameters can be easily obtained from the intensity, energy, and halfwidth of the corresponding bands. The H_{fe}, E_{fe} and H_{fh}, E_{fh} parameters can be similarly taken from the corresponding bands of the trinuclear complex containing two pentammine moieties. Use of these experimentally determined parameters in eq 1 leads to the prediction of H_{if} = 360 cm^{-1} for the end-to-end IT transition. This is to be compared with the experimentally determined value of 350 cm^{-1}. Given the approximations used in the spectral analysis, the agreement appears to be quite satisfactory, showing clearly that most of the experimental intensity of the end-to-end transition comes from through-bridge interaction. Of the two terms in eq 1, that involving the hole transfer (IT) state accounts for the greater part (ca. 70%) of the intensity of the remote IT transition, as this state is both the lowest and the more heavily coupled of the two.

In these systems the "bridge" connecting the D and A units of the remote IT process is itself a three-component subsystem made of a central metal-containing moiety and two bridging ligands. It should be pointed out that composite bridges of this type may prove useful to achieve relatively independent changes in the various factors affecting the donor-acceptor electronic coupling. In fact, with bridges of this type and keeping the D and the A units constant, the H_{ih}, and H_{fh} interactions are primarily determined by the bridging ligands, while the energy of the hole-transfer state is primarily governed by the redox properties of the central metal containing moiety. An interesting perspective would be the possibility to modulate the rates of end-to-end electron transfer by simply tuning the redox properties of the central metal component of the bridge. Studies in this direction are in progress in this laboratory.

ENERGY TRANSFER. COMPLEXES WITH CHROMIUM CYANIDE SUBUNITS.

All the polynuclear complexes that do not have intramolecular electron transfer channels available for excited-state deactivation (complexes n. 2, 3, 17, 19, 21, 23, 24 in Table I) exhibit emission. As discussed previously, in complexes that do not have low-lying excited states on the attached subunits (compounds n. 2 and 3) the emission is from the $Ru(bpy)_2^{2+}$ photosensitizer unit.

The analysis of the behavior of the polynuclear complexes containing $Ru(bpy)_2CN^+$ as the attached fragments (compounds n. 17, 19, 21) is complicated by the fact that they contain two or three $Ru(bpy)_2^{2+}$ subunits that differ in excited-state energy by small amounts depending on the C- or N-bonded orientation of the bridging cyanide(s). Therefore, absorptions by the various subunits overlap considerably, and the identification of the emitting unit relies on quantitative arguments. As discussed in detail elsewhere [27], however, efficient intramolecular energy transfer can be demonstrated to occur from C-bonded to N-bonded $Ru(bpy)_2^{2+}$ subunits in these complexes.

Clear-cut evidence for intramolecular energy transfer has been recently obtained with complexes in which one or two $Cr(CN)_5^{2-}$ units are bound to $Ru(bpy)_2(CN)_2$. As shown schematically in Figure 4 for the trinuclear

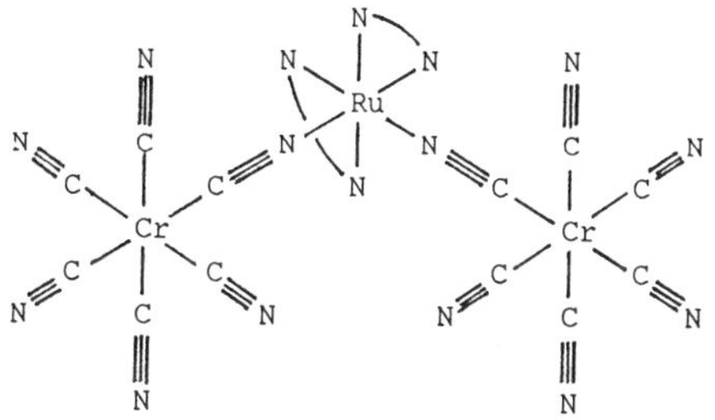

FIG. 4. Schematic structure of the $Ru(bpy)_2[Cr(CN)_6]_2^{4-}$ complex.

species, the bridging cyanides in these complexes are N-bonded to Ru. The photophysical behavior expected for N-bonded $Ru(bpy)_2(CN)_2$ can be deduced from that of complex n. 21 in Table I: emission at slightly lower energy with respect to the C-bonded isomer with reasonably long lifetime. The photophysical behavior of free $Cr(CN)_6^{3-}$ in DMF has been studied in detail by Wasgestian [32] and Balzani [33], and is summarized in Figure 5. In practice, the intraconfigurational doublet state can be easily detected from its sharp, long-lived phosphorescent emission, while the non-emitting quartet excited state is responsible for an efficient photosolvation reaction. Absorption, emission and excitation spectra for free $Cr(CN)_6^{3-}$ and the trinuclear complex are shown in Figure 6. Absorption in the visible by the trinuclear complex is exclusively due to the $Ru(bpy)_2^{2+}$ photosensitizer unit, as the chromium complex is totally transparent in this region. In the trinuclear complex, the emission of the sensitizer unit (expected at $\sim$ 700 nm) is totally quenched, while the typical Cr(III) phosphorescence is obtained under visible excitation (Figure 6).

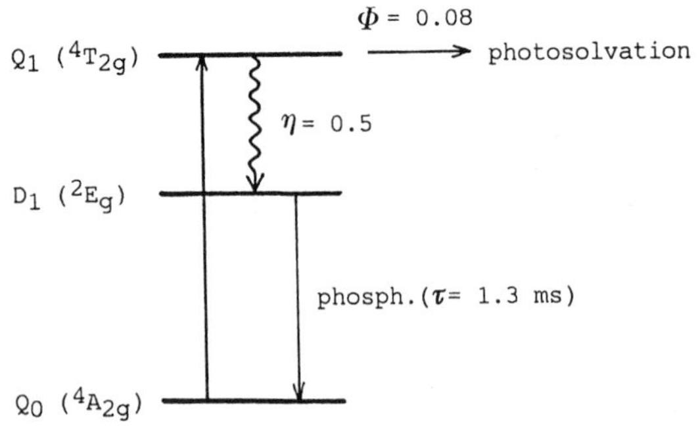

FIG. 5. Photochemical and photophysical behavior of $Cr(CN)_6^{3-}$ in DMF.

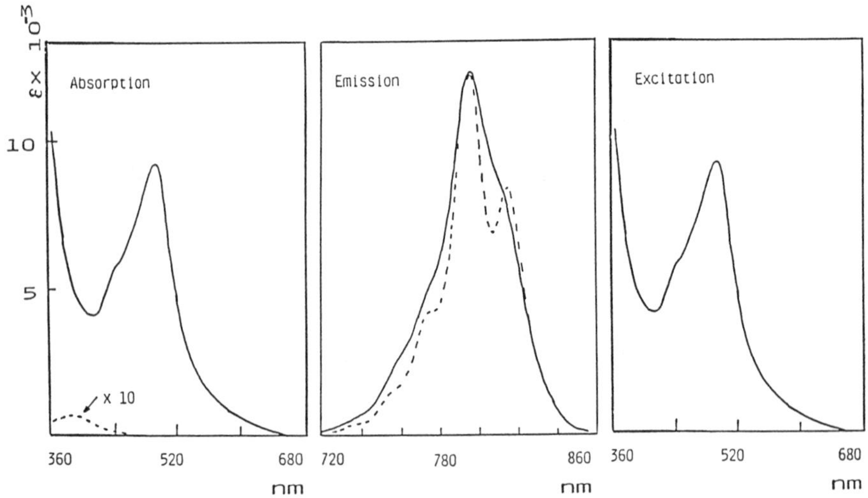

FIG. 6. Absorption (left), emission, (center), and excitation (right) spectra of $Cr(CN)_6^{3-}$ (broken line) and $Ru(bpy)_2[Cr(CN)_6]_2^{4-}$ (full line) in DMF.

This is convincing evidence for the occurrence of intramolecular energy transfer form the sensitizer unit to the chromium cyanide moieties. The energy transfer process is further defined by the observation that sensitization of the phosphorescence is not accompanied by photosolvation, indicating direct transfer from the triplet MLCT state of the $Ru(bpy)_2^{2+}$ unit to the doublet state of $Cr(CN)_6^{3-}$, bypassing the reactive quartet state. Indeed, a comparison between the absolute intensities of the emissions of free $Cr(CN)_6^{3-}$ and the trinuclear complex (allowing for the difference in the lifetimes of the two emissions) indicates an essentially unitary efficiency for the energy transfer step. The behavior of the trinuclear complex is summarized in Figure 7. Very similar behavior is exhibited by the binuclear complex.

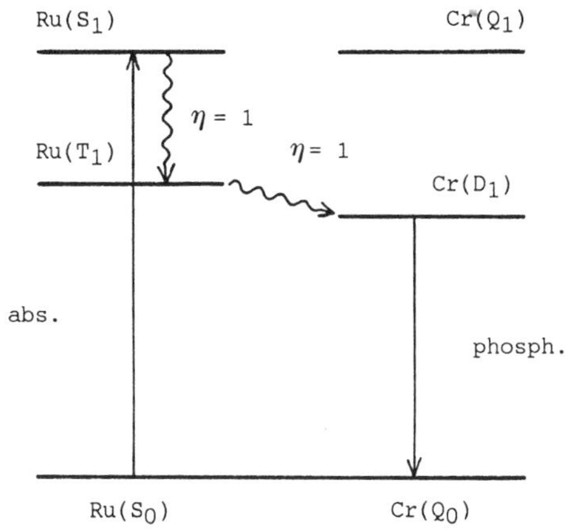

FIG. 7. Photophysical behavior of $Ru(bpy)_2[Cr(CN)_6]_2^{4-}$ in DMF.

The intramolecular energy transfer process is found to be complete in less than 10 ns. It is to be remarked that in systems of this kind the energy transfer must occur via an exchange mechanism, as both the virtual transitions within the donor and the acceptor are spin forbidden. With such a mechanism, the transfer rates should be very sensitive to the nature of the bridge. In the present case, the transfer process likely benefits from the high electronic coupling provided by cyanide (see the above discussed large electronic factors for IT across cyanide bridges). Studies on intramolecular exchange energy transfer in other bi- and polynuclear complexes are now in progress to assess the role of the nature of the bridging ligands and of the driving force, with the aim of arriving at stepwise energy transfer over several molecular units.

An interesting possibility offered by the trinuclear complex is that of achieving excitation of both Cr(III) centers in the complex by means of two successive absorption-transfer sequences. This possibility should be favored by the large extinction coefficient of the $Ru(bpy)_2^{2+}$ unit and by the long lifetime of the excited Cr(III) centers. It is expected that a doubly excited trinuclear complex may give rise to some interesting physics (e.g. up conversion processes) and chemistry (e.g., intramolecular annihilation, sensitized biphotonic processes). In this perspective, it is interesting to notice that bimolecular doublet-doublet annihilation processes can be readily observed with the bi- and trinuclear complexes under laser excitation.

These bi- and trinuclear complexes can be used to illustrate some of the advantages of coupling a chromophore and a luminophore into a supramolecular structure. In the present case, the performance of the $Cr(CN)_6^{3-}$ luminophoric unit is improved by the presence of the $Ru(bpy)_2^{2+}$ chromophore in several ways:

(i) the overall cross-section for light absorption is increased by orders of magnitude;

(ii) spectral sensitization in the visible region is achieved;

(iii) energy wasting through the quartet state is avoided by direct access to the emitting state from the excited chromophore;

(iv) the luminophore is protected towards photodecomposition by the chromophore acting at the same time as an inner filter and as a selective luminescence sensitizer.

Some of these features recall the roles played by specific chromophores in natural photosynthetic systems. Based on points (i), (ii), and (iii) the $Ru(bpy)_2^{2+}$ fragment can be considered as an efficient antenna attached to the luminophore. Point (iv) is reminiscent, despite the difference in mechanism, of the protective role played by carotenoids in photosynthetic membranes [34].

CONCLUSIONS.

The polynuclear complexes described here are an example of the variety of situations that can be created by combining together transition-metal-based molecular fragments into a supramolecular structure, even by remaining within a relatively narrow class of compounds (the $Ru(bpy)_2^{2+}$ unit as the photosensitizer and cyanides as bridging ligands). The field is still at an early stage. Many other classes of molecular units and bridges deserve attention, and extensive experimental studies are needed to obtain a general understanding of the role of structural, thermodynamic, and kinetic factors in determining the efficiency of intramolecular energy and electron transfer in polynuclear transition metal complexes. Right now, however, it can be said that transition metal complexes look as versatile molecular components for the study of intramolecular energy and electron transfer and, in perspective, as attractive building blocks for the construction of molecular photochemical devices.

ACKNOWLEDGMENT.

I am particularly grateful to C.A. Bignozzi, who devised and carried out all of the the hard synthetic work that is hidden behind the results presented here. The contribution of M. T. Indelli to the photophysical part of the work is also gratefully acknowledged. This work was supported by the Consiglio Nazionale delle Ricerche (C.N.R.) and by the Ministero della Pubblica Istruzione.

REFERENCES

1. J-M. Lehn, Nouv. J. Chim., 11, 77, (1987).
2a. H. Oevering, M. N. Paddon-Row, M. Heppener, A. M. Oliver, E. Cotsaris, J. W. Verhoeven, and N. S. Hush, J. Am. Chem. Soc., 109, 3258 (1987).
b. G. L. Closs, L.T. Calcaterra, N. J. Green, K. W. Penfield, and J. R. Miller, J. Phys. Chem, 90, 3673 (1986).
c. M. R. Wasielewski, M. P. Niemczik, W. A. Svec, and E. B. Pewitt, J. Am. Chem. Soc, 107, 1080 (1985).
d. A. D. Joran, B. A. Leland, G. G. Geller, J. J. Hopfield, and P. B. Dervan, J. Am. Chem. Soc., 106, 6090 (1984).
e. J. R. Bolton, T.-F. Ho, S. Liauw, A. Semiarczuk, C. S. K. Wan, and A, C. Weedon, J. Chem. Soc. Chem. Commun. 559 (1985).
3a. M. R. Wasielewski, M. P. Niemczik, W. A. Svec, and E. B. Pewitt, J. Am. Chem. Soc., 107, 5562 (1985)
b. D. Gust, T. A. Moore, P. A. Liddell, G. A. Nemeth, L. R. Makings, A. L. Moore, D, Barrett, P. J. Pessiki, R. V. Bensasson, M. Rougée, C. Chachaty, F. C. de Schryver, M. Van der Auweraer, A. R. Holzwarth, and J. S. Connolly, J. Am. Chem. Soc., 109, 846 (1987).
4. D. Gust, T. A. Moore, A. L. Moore, D. Barrett, L. O. Harding, L. R. Makings, P. A. Liddell, F. C. De Schryver, M. Van der Auweraer, R. V. Bensasson, and M. Rougée, J. Am. Chem. Soc., 110, 321 (1988).
5. V. Balzani, L. Moggi, and F. Scandola in: Supramolecular Photochemistry, V. Balzani, ed. (Reidel, Dordrecht,1987), p. 1.
6a. F. Scandola and V. Balzani, J. Chem. Educ. 60, 814 (1983).
b. F. Wilkinson and S. L. Collins in: Supramolecular Photochemistry, V. Balzani, ed. (Reidel, Dordrecht 1987) p. 225.
c. J. F. Endicott, Coord. Chem. Rev., 61, 193 (1985).
7a. V. Balzani, F. Bolletta, M. T. Gandolfi, and M. Maestri, Top. Curr. Chem., 75, 1 (1978)
b. N. Sutin and C. Creutz, Pure Appl. Chem., 56, 2717 (1980).
c. N. Sutin and C. Creutz, J. Chem. Educ., 60, 809 (1983).
d. V. Balzani and F. Scandola in: Energy Resources through Photochemistry and Catalysis, M. Gratzel, ed. (Academic Press, New York 1981) p. 1.
8. C. Creutz and H. Taube, J. Am. Chem. Soc., 91, 3988 (1969).
9. C. Creutz, Prog. Inorg. Chem., 30, 1 (1983).
10. N. S. Hush, Prog. Inorg. Chem., 8, 391 (1967).
11. H. Taube in: Tunneling in Biological Systems, B. Chance, et al.,eds (Academic Press, New York 1979, p. 173.
12. A. Haim, Prog. Inorg. Chem, 30, 273 (1983).
13. S. S. Isied, Prog. Inorg. Chem., 32, 443 (1984).
14. C. Creutz, P. Kroger, T. Matsubara, T. L. Netzel, and N. Sutin, J. Am. Chem. Soc., 101, 5442 (1979).
15. J. C. Curtis, J. S. Bernstein, and T. J. Meyer, Inorg. Chem., 24, 385 (1985).
16. J. D. Petersen in: Supramolecular Photochemistry, V. Balzani, ed. (Reidel, Dordrecht 1987) p. 135.
17. J. F. Endicott, R. B. Lessard, Y. Lei, and C. K. Ryu in: Supramolecular Photochemistry, V. Balzani, ed. (Reidel, Dordrecht 1987) p. 167.
18. N. A. P. Kane-Maguire, M. M. Allen, J. M. Vaught, J. S. Hallcock, and A. L. Heatherington, Inorg. Chem., 22, 3851 (1983).

19. J. V. Caspar and T. J. Meyer, J. Phys. Chem., 87, 952 (1983).
20. C. A. Bignozzi and F. Scandola, Inorg. Chem., 23, 1540 (1984).
21. V. Balzani, N. Sabbatini and F. Scandola, Chem. Rev., 86, 319 (1986).
22. M. G. Kinnaird and D. G. Whitten, Chem Phys. Lett.,88, 275 (1982).
23. M. T. Indelli, C. A. Bignozzi, A. Marconi, and F. Scandola in Photochemistry and Photophysics of Coordination Compounds, H. Yersin and A. Vogler, eds. (Springer-Verlag, Berlin 1987) p. 159.
24. M. T. Indelli, C. A. Bignozzi, A. Marconi, and F. Scandola, submitted for publication
25. C. A. Bignozzi, S. Roffia, and F. Scandola, J. Am. Chem. Soc., 107, 1644 (1985).
26. C. A. Bignozzi, C. Paradisi, S. Roffia, and F. Scandola, Inorg. Chem., 27, 408 (1988).
27. F. Scandola and C. A. Bignozzi in: Supramolecular Photochemistry, V. Balzani, ed. (Reidel, Dordrecht 1987) p. 121.
28. S. Roffia, C. Paradisi, and C. A. Bignozzi, J. Electroanal. Chem. Interfacial Electrochem., 200, 105 (1986).
29. A. von Kameke and H. Taube, Inorg. Chem., 17, 1790 (1978).
30. J. R. Miller, Nouv. J. Chim., 11, 83 (1987).
31. N. S. Hush in: Supramolecular Photochemistry, V. Balzani, ed. (Reidel, Dordrecht 1987) p. 53, and references therein..
32. H. F. Wasgestian, J. Phys. Chem., 76, 1947 (1972).
33. N. Sabbatini, M. A. Scandola, and V. Balzani, J. Phys. Chem., 78, 541 (1974).
34. P. A. Liddell, D. Barrett, L. R. Makings, P. J. Pessiki, D. Gust, and T. A. Moore, J. Am. Chem. Soc., 108, 5350 (1986).

ENERGY CONVERSION AT THE MOLECULAR LEVEL

Thomas J. Meyer
Department of Chemistry
The University of North Carolina
Chapel Hill, NC 27599-3290

INTRODUCTION

Many small molecule reactions that are candidates for light to chemical energy conversion such as water splitting,

$$2H_2O \longrightarrow 2H_2 + O_2$$

or reactions involving carbon dioxide,

$$2H_2O + 2CO_2 \longrightarrow 2HCO_2H + O_2$$

are oxidation-reduction reactions. As such, they can be broken up into their component half reactions,

$$2H^+ + 2e \longrightarrow H_2 \qquad\qquad 2H_2O \longrightarrow 4H^+ + 4e + O_2$$

which immediately suggests the possibility of using excited state electron transfer chemistry to drive them.[1-5]

An approach based on molecular systems is suggested by Scheme 1.

Scheme 1.

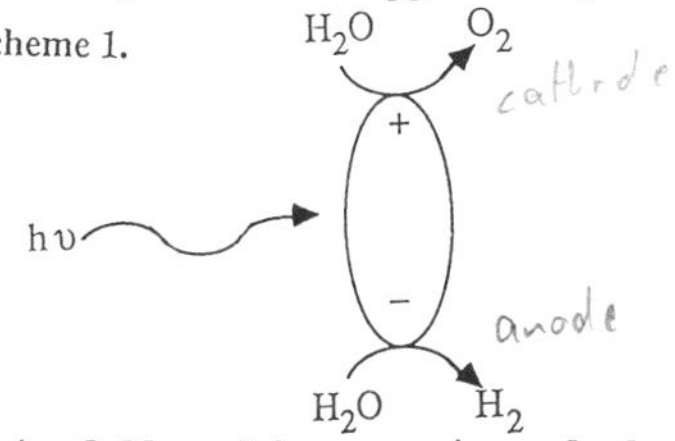

In Scheme 1 light absorption is followed by a series of electron transfer events which lead to spatially separated oxidative and reductive equivalents. Those equivalents, in turn, are utilized to drive separate catalysts for the oxidation or reduction of water.

The optical excitation-charge splitting sequence suggested in Scheme 1 is, of course, analogous to what happens at the reaction center in photosynthesis.[6-8] The principles are available and processes are known at the molecular level which are required to build artifical systems which accomplish the same end. In a sequential order the processes that must occur are molecular light absorption, excited state electron transfer, electron transfer over

Published 1989 by Elsevier Science Publishing Co., Inc.
Photochemical Energy Conversion
James R. Norris, Jr. and Dan Meisel, Editors

long distances, and electron transfer induced activation of molecular catalysts. The catalysts must have the ability to carry out complex multiple electron transfer reactions.

The theme here is to point to the fundamental principles that play an important role in the design of devices for optical excitation-charge splitting at the molecular level. The specific cases chosen as illustrations will come largely from our own work and the work of others based on metal to ligand charge transfer (MLCT) excited states. It is possible to find in these systems examples that nicely illustrate the interplay between fundamental principles and the design of molecular systems.

Excited State Decay. For the metal to ligand charge transfer (MLCT) excited states of complexes like $[Ru(bpy)_3]^{2+}$ (bpy is 2,2'-bipyridine) $[(bpy)Re(CO)_3Cl]$, or $[Os(phen)_3]^{2+}$ (phen is 1,10-phenanthroline) the change in

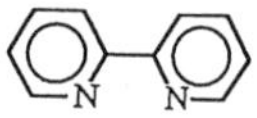

(bpy)

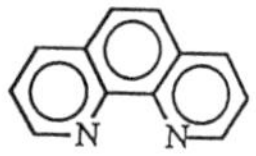

(phen)

electronic configuration between the ground and excited states is,

$$(d_{\pi})^6 \xrightarrow{h\nu} (d_{\pi})^5(\pi^*)^1$$

The acceptor orbital is a vacant π^* orbital on the polypyridyl ligand. Temperature dependent lifetime and emission spectra measurements have revealed much about the properties of these excited states.[9,10] They are typically weak emitters in fluid solution at room temperature, their lifetimes are dominated by nonradiative decay processes. In nonradiative decay the energy that is released by back electron transfer,

$$(d_{\pi})^5(\pi^*)^1 \longrightarrow (d_{\pi})^6$$

appears largely (> 90%) in polypyridyl based ring stretching modes and, to a lesser extent, in low frequency modes, including metal ligand stretching vibrations, and in the solvent.[9,11,12]

From time dependent perturbation theory, the nonradiative decay rate constant, k_{nr}, is predicted to be proportional to the product of a vibrationally induced electronic coupling term and a "Franck-Condon" factor.[12,13] To zero order the excited and ground state wave functions are orthogonal and can not mix. Vibrations of the appropriate symmetry, the promoting modes, can increase orbital overlap which causes the states to mix.

The Franck-Condon factor is a mathematical function which describes how the energy released when electron transfer occurs is distributed amongst the various vibrational modes of the system.

An important development has been the utilization of the results of resonance Raman and emission spectral fitting to acquire the parameters required for the calculation of the Franck-Condon factors for the decay of MLCT excited states.[11d,12,13g] The agreement between experiment and theory in this area shows that it is possible to learn how to control excited state properties systematically by taking advantage of physical insight. For example, theory predicts that k_{nr} should vary as the energy gap, E, as in eq. 1.[12]

$$k_{nr} \propto \exp-(\gamma E/\hbar\omega) \qquad (1)$$

In eq. 1 $\hbar\omega$ is an averaged vibrational quantum spacing for the series of polypyridyl ring stretching modes which act as energy acceptors. γ is defined as, $\gamma = (\ln \frac{E}{S\hbar\omega})-1$

S is the electron-vibrational coupling constant (Huang-Rhys factor). It is related to the difference in equilibrium displacement in the averaged ν(bpy) mode between the ground and excited states. Experimentally, it is possible to vary k_{nr} systematically by using changes in the ligands in complexes such as $[(bpy)Os(L)_4]^{2+}$ (L = py, PR_3, CH_3CN, ...) to vary the energy gap.[12,14]

The photochemistry of MLCT excited states is complicated by the appearance of other states such as dd states and higher lying MLCT states.[14,15] However, the decay properties of the lowest MLCT states are very much in line with predictions based on fundamental physical principles: 1) The extent of the structural changes at the polypyridyl ligands in the excited states increases as the energy gap between states is increased. This is understandable since as the energy gap increases, mixing between the ground and excited states decreases. As the extent of mixing decreases the extent of charge transfer increases which causes the greater distortion in the excited state. 2) The same pattern of polypyridyl acceptor vibrations appears to dominate excited state decay where the acceptor ligand is of the type 4,4'-$(X)_2$-2,2'-bipyridine (X = NH_2, CH_3, H, C(O)OEt, ...).[16] However, each ligand has its own individual nuances with S varying differently with the energy gap. 3) For complexes of Ru and Os where there are equivalent Franck-Condon factors, non-radiative decay is faster for Os by a factor of ~3 because greater spin-orbit coupling at Os (λ_{Os} ~3000 cm^{-1} vs. λ_{Ru} ~1100 cm^{-1}) increases the magnitude of the vibrationally induced electronic coupling between states.

Spin effects play a role in both radiative and nonradiative decay. The lowest lying MLCT excited states are largely triplet $^3[(d\pi)^5(\pi^*)^1]$ in character and the ground state is a $(d\pi)^6$ singlet. The operators that mix them, the electric dipole moment operator for radiative decay ($\langle\psi_e|e\bar{r}|\psi_g\rangle$) and the promoting mode ($\langle\psi_e|\partial/\partial_q|\psi_g\rangle$) for nonradiative decay, do not include spin. When spin-orbit coupling is included in the analysis, the "triplet" excited states mix with low lying singlet excited states. The matrix elements then become non-zero and provide the dynamic basis for excited state decay.

Excited State Electron Transfer. In an outer sphere excited state quenching scheme, there are a series of microscopic steps that occur before the redox products appear in solution. They are preassociation of the reactants, electron transfer quenching, and separation of the redox products. Separation is in competition with back electron transfer, Scheme 2.

SCHEME 2

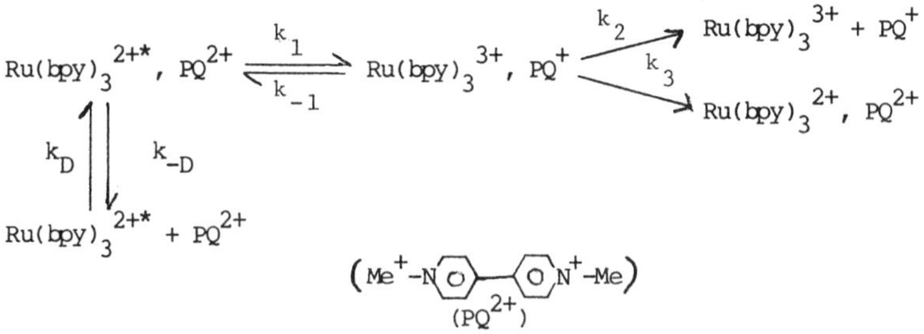

The role of free energy change (ΔG) in the electron transfer quenching step (k_1) has been well established and is in agreement with the theoretical predictions of Hush and Marcus.[17] In the classical limit, the maximum electron transfer rate is predicted to occur when $-\Delta G = \lambda$. λ is the sum of the intramolecular (λ_i) and solvent dipole orientational (λ_o) barriers to electron transfer. The yields of separated redox products depend upon the relative magnitudes of k_2 and k_3 in Scheme 2. k_2 is strongly influenced by electrostatic effects with the repulsion between like charged ions greatly favoring separation. The energy released in the back electron transfer step in Scheme 2 is considerable (~-1.7 eV). In fact, the back electron transfer reaction is in the "inverted region" where $-\Delta G > \lambda$.[18] In this region it is predicted theoretically that k_3 should decrease as the free energy stored in the quenching step is increased. In the limit that the energy released is much larger than λ, the electron transfer rate constant is predicted to have the same dependence on the energy gap as for nonradiative decay in eq. 1. Such effects play an important role in these systems where quantum efficiencies for the separation of redox products as high as unity have been achieved.[19]

Spin effects can also influence the magnitudes of separation efficiencies. An apparent example of this occurs in the excited state quenching of $[Ru(bpy)_3]^{2+*}$ by PQ^{2+}. Under conditions of complete quenching, less than 20% of the excitation events lead to separated redox products.[20] By contrast, an indirect mechanism in which $[Ru(bpy)_3]^{2+*}$ is initially quenched by energy transfer to an anthracene derivative, reaction 2, followed by electron transfer quenching of the anthracene, reaction 3, occurs with nearly unit efficiency.[19b]

$$Ru(bpy)_3^{2+*} + 9\text{-MeAn} \longrightarrow Ru(bpy)_3^{2+} + {}^3(9\text{-MeAn})^* \qquad (2)$$

(9-MeAn is 9-methylanthracene)

$${}^3(9\text{-MeAn})^* + PQ^{2+} \longrightarrow (9\text{-MeAn})^+ + PQ^+ \qquad (3)$$

For the organic example in reaction 2 there is a spin prohibition to back electron transfer. The spin prohibition arises because of the differences in spin character between the triplet excited state and the singlet products,

$${}^3(9\text{-MeAn})^*, PQ^{2+} \longrightarrow {}^3[(9\text{-Me(An)}^+, PQ^+] \longrightarrow {}^1[9\text{-MeAn}, PQ^{2+}]$$

The sequence k_1 followed by k_3 in Scheme 2 is closely related.

However, spin-orbit coupling at Ru is significant. The low lying "triplet" MLCT excited state has sufficient singlet character to break down the spin prohibition to electron transfer.[21]

In an attempt to combine the virtues of the Ru-bpy visible light absorber and the spin advantage of electron transfer based on triplet anthracene, we are studying complexes of the type $[Ru(4\text{-}Me\text{-}4'\text{-}(9\text{-}CH_2OCH_2An)\text{-}2,2'\text{-}bpy)_3]^{2+}$.[22]

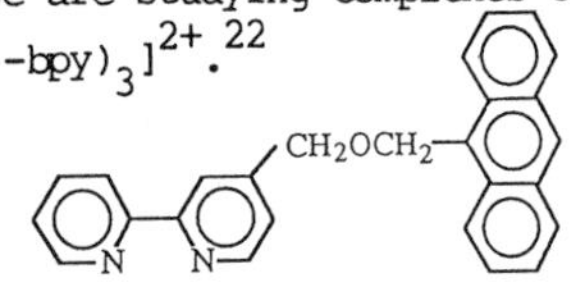

[4-Me-4'-(9-CH_2OCH_2 An)-2,2'-bpy]

In this complex MLCT emission is essentially completely quenched. Triplet anthracene appears following MLCT excitation as shown by transient laser experiments.

Intramolecular analogs of the pyridinium quenching in Scheme 2 have also been prepared. In these complexes the effect of intramolecular structural changes can play a major role in rates of intramolecular electron transfer and can even dictate whether or not intramolecular electron transfer occurs. An example is the chromophore-quencher complex $[(bpy)Re(CO)_3(MQ^+)]^{2+}$.[23,24]

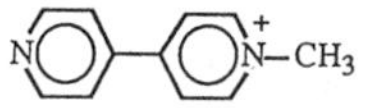

(MQ^+)

An x-ray structural study of the PF_6^- salt of this complex has shown that the angle between the pyridyl rings of the pyridinium ligand is approximately 45°.[25] Spectroscopic studies suggest that an angle close to that is maintained in solution as well. The results of transient absorbance experiments show that following Re → bpy excitation of the complex, intramolecular electron transfer occurs to the pyridinium ligand, reaction 4

$$\xrightarrow{h\nu} [(\overset{\frown}{bpy})Re^{II}(CO)_3(N\text{py}\text{—}\text{py}N^+\text{-}Me)]^{2+*} \longrightarrow [(bpy)Re^{II}(CO)_3(N\text{py}\text{—}\text{py}N^{\cdot}\text{-}Me)]^{2+*} \quad (4)$$

In the quenched product there is a flattening of the angle between the rings as shown by the appearance of typical reduced viologen absorption bands in the transient absorbance experiments.[23b] The driving force for flattening is the delocalization of the added electron over both rings. Spectroscopic studies show that the Re-bpy and Re-MQ^+ based MLCT excited states lie close in energy. Upon vertical excitation the bpy-based state lies lowest. When the rings flatten, the MQ^+-based state is stabilized by ~ 300 mV and it becomes the lowest excited state in fluid solution.[25] In a frozen 4:1 ethanol-methanol (V:V) glass at 77 K the vertical ordering of the states is maintained because of the inhibition to ring rotation in the glass. Emission is observed from the Re --> bpy MLCT state even following Re --> MQ^+ excitation because the inversion in state ordering leads to reverse electron transfer,

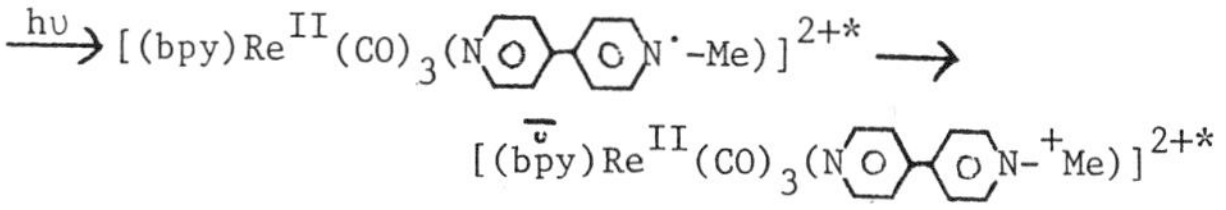

(5)

Whether or not intramolecular quenching occurs in these complexes can be manipulated by substituent effects.[26] In the sequence $[(4,4'-(X)_2-bpy)Re(CO)_3(MQ^+)]^{2+}$ (X = C(O)OEt, H, NH_2), the ester substituents lower the π*-bpy acceptor level and the state ordering is MLCT(Re-bpy) < MLCT(Re-MQ^+). Intramolecular quenching does not occur even in fluid solution. For X = H the levels are closely matched. For X = NH_2 the energy of the π*(bpy) acceptor level is increased and intramolecular quenching occurs even in an EtOH-MeOH glass at 77K.

A closely related effect has been observed in the 4,4'-bipyridine (4,4'-bpy) bridged complex $[(bpy)_2(CO)Os(4,4'-bpy)OsCl(phen)($<u>cis</u>-dppene$)]^{3+}$ (phen is 1,10-phenanthroline).[27]

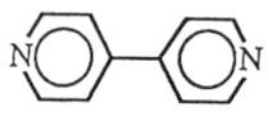

(4,4'-bpy)

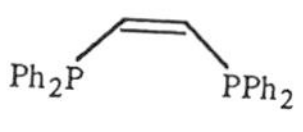

(<u>cis</u>-dppene)

For this complex Os → phen MLCT excitation in the glass at 77 K leads to emission only from the corresponding Os-phen MLCT excited states. However, transient absorbance experiments show that in fluid solution at room temperature initial MLCT excitation is followed by intramolecular electron transfer to the 4,4'-bpy bridge. This was shown by the appearance of a reduced viologen like spectrum.

$$\xrightarrow{h\nu} [(bpy)_2(CO)Os(4,4'\text{-}bpy)Os^{III}Cl(\overset{\bullet}{\overline{phen}})(dppene)]^{3+*} \longrightarrow \quad (6)$$

$$[(bpy)_2(CO)Os(4,4'\text{-}\overset{\bullet}{\overline{bpy}})Os^{III}Cl(phen)(dppene)]^{3+*}$$

By adjusting the redox levels of appropriate donors and acceptors ΔG effects can be used to dictate the sequence of steps that occur in multiple electron transfer processes induced by optical excitation. In the bpy subsituted complex shown below

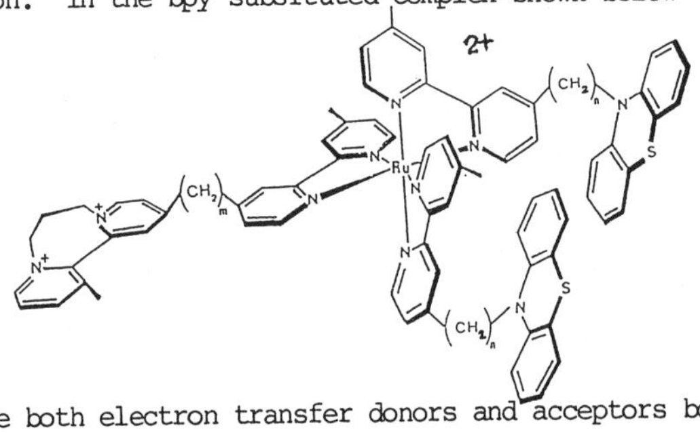

there are both electron transfer donors and acceptors bound to the MLCT chromophore.[28] The series of electron transfer events that lead to the charge separated state with reductive and oxidative equivalents stored on separate bpy ligands could be initiated by either initial reductive or oxidative electron transfer. Kinetic studies on model systems containing only the electron transfer donor or the electron transfer acceptor show that at room temperature both processes are very rapid. However, by adjusting the relative redox levels of donors and acceptors it is possible to utilize only one of the two processes to trigger the net electron transfer sequence. For example, in the complex,

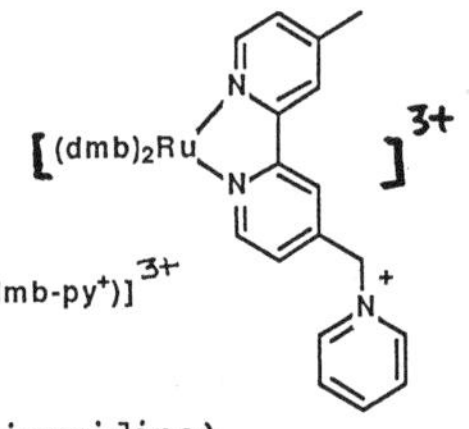

$[Ru(dmb)_2(dmb\text{-}py^+)]^{3+}$

(dmb is $4,4'\text{-}(CH_3)_2\text{-}2,2'$-bipyridine)

the chemically attached pyridinium ligand has a relatively low potential as an electron acceptor, $E_{p,c}(Mepy^{-+/o}) = -1.1$ V in 0.1 M $[N(n\text{-}Bu)_4](PF_6)\text{-}CH_3CN$ vs. SSCE.[29] Optical excitation of the MLCT chromophore leads to negligible quenching. However, with added 10-methylphenothiazine,

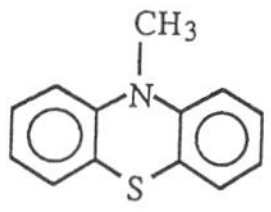

(10-MePTZ)

reductive electron transfer quenching occurs,

$$[Ru^{III}(dmb)_2(\dot{\overline{dmb}}\text{-}py^+)]^{2+*} + 10\text{-MePTZ} \longrightarrow [Ru^{II}(dmb)_2(\dot{\overline{dmb}}\text{-}py^+)]^+ + (10\text{-MePTZ}^+) \qquad (7)$$

to give the bpy-localized reduction product. The bpy reduced product is enhanced as a reductant, -1.3 V compared to -0.8V for the excited state. The enhanced reducing power triggers a now spontaneous intramolecular electron transfer to the pyridinium group, reaction 8.

$$[Ru^{II}(dmb)_2(\dot{\overline{dmb}}\text{-}py^+)]^+ \longrightarrow [Ru^{II}(dmb)_2(dmb\text{-}\dot{py})]^+ \qquad (8)$$

Medium Effects. In normal electron transfer reactions, $-\Delta G < \lambda$. In this domain in the classical limit, the rate constant for intramolecular electron transfer varies with ΔG and λ as shown in eq. 9.

$$k = \nu_{et} \exp-[(\lambda+\Delta G)^2/4\lambda RT] \qquad (9)$$

In eq. 9 the solvent plays a double role. It contributes to ΔG and to λ via the outer-sphere reorganization energy, λ_o. It can also help to dictate the pre-exponential term, ν_{et}, by solvent dynamics.[30]

The effects on λ_o and ΔG lie in the change in electronic distribution between the initial and final states. In solution the surrounding solvent dipoles adopt the average equilibrium orientations appropriate to the electronic distributions of the two states. Differing degrees of interaction lead to the solvent dependence of ΔG. For the transition between states to occur there is a requirement that solvent dipoles reorient. The reorientation is the origin of λ_o. The motions involved are analogous to rotations in the gas phase but are necessarily collective in nature in the liquid state. They are referred to as librations and can be treated, at least conceptually, by analogy with phonons in the solid state. Librational modes are characteristically of low frequency, 1-10 cm^{-1}, and can be treated as classical motions of the system at all but the lowest of temperatures.

Electron transfer in the normal region occurs from reacting partners which have those nonequilibrium solvent dipole and vibrational distributions appropriate for electron transfer to occur. There is a temperature dependence to electron transfer arising from the requirement to populate these nonequilibrium distributions.

The situation is quite different in the inverted region where $-\Delta G > \lambda$. If the energy released is far greater than λ, there is a changed role for the solvent. The factors that influence rates of electron transfer are the same as those for nonradiative excited state decay. The key is the ability of the system to release energy into the surrounding molecular vibrations when electron transfer occurs. A prediction of time dependent perturbation theory and quantum mechanics is that the energy release will be dominated by those modes of highest frequency which have significant changes in equilibrium displacement between the initial and final states.[12,13] It is for those modes that maximum vibrational overlap exists between the initial and final states and they provide the major channel for the energy release associated with the excited to ground state transition. Typically, solvent modes do participate but only as a minor partner because of their classical nature. Rather, the solvent can play a major role by its effect on ΔG.[31] As in excited state decay, the energy gap between the initial and final states is important since it determines the extent of vibrational overlap for the higher frequency modes.

These predictions are borne out and play an important role in light induced electron transfer. The intramolecular quenching reaction in 4 provides an example. In fluid solution, Re $\rightarrow$ bpy excitation is followed by sub nanosecond intramolecular electron transfer. In a 1:1 ethanol/methanol (V:V) glass at 77 K, the Re $\rightarrow$ bpy emission is not quenched.

In addition to the flattening of the two rings of the pyridinium acceptor, the loss of solvent dipole mobility in the glass plays a major role. Compared to the ground state, there is a greater charge transfer distance in the Re-MQ^+ based state in reaction 4 compared to the initial Re-bpy MLCT state. This increases the contribution to λ_o for this state and contributes to the inversion in the ordering of the Re-bpy and Re-MQ^+ states in the glass. The solvent dipoles can not reorient following Re $\rightarrow$ bpy excitation because they are frozen. Even Re $\rightarrow$ MQ^+ excitation leads to Re-bpy based emission following the reverse electron electron transfer step in reaction 5.

Intramolecular quenching can occur in the glass if the reaction is sufficiently spontaneous. This has been shown by Re-bpy MLCT quenching and the appearance of Re-MQ^+ based emission in the bpy-substituted complex $[(4,4'-(NH_2)_2\text{-bpy})Re(CO)_3(MQ^+)]^{2+}$. The important criterion that must be met in order for quenching to occur in the glass is that $-\Delta(\Delta G) > \lambda_o$.[26] $\Delta(\Delta G)$ is the difference in excited state energies between the final (Re^{II}-$\dot{MQ}$) and initial ($\bar{bpy}Re^{II}$) excited states. $\Delta\lambda_o$ is the difference in solvent reorganizational energies between the Re-MQ^+ based state and the ground state and between the Re-bpy based state and the ground state.

Frozen solvent dipole effects also appear in nonradiative decay but they have a different origin. It is characteristic of MLCT excited states that their emissions shift to lower energies and nonradiative lifetimes shorten in the glass to fluid transition region.[32] In a frozen glass, solvent dipole orientations are frozen. There is no way for reorientation to occur following excitation of the ground state. Emission is at high energy because the solvent dipoles have configurations appropriate to the ground state rather than the MLCT excited state. In fluid solution the dipoles can reorient following excitation and the emission is shifted to lower energy. Dynamically, the effect of solvent is indirect. It influences k_{nr} by its effect on the energy gap through eq. 1.[33] In fluid solution, rapid dipole relaxation occurs, the energy gap is lower, and the nonradiative decay rate increases.

Intermediate States. Multiple Electron Transfers. The lifetimes of the states which are the immediate products of intramolecular electron transfer quenching are themselves subject to the usual rules which dictate electron transfer and/or nonradiative decay. However, different kinds of states can be reached which differ in a fundamental way depending on the extent of electronic interaction between the electron transfer donor and acceptor sites. For example, intramolecular MQ^+-based quenching in $[(bpy)Re(CO)_3(MQ^+)]^{2+}$ involves an intramolecular electron transfer from levels which are largely $\pi^*(bpy)$ in character to levels which are largely $\pi^*(MQ^+)$, reaction 4. Here, intramolecular electron transfer leads to an interconversion between excited states. The MQ^+-based state which is reached is also a metal to ligand charge transfer excited state of the system. It emits in fluid solution at room temperature.[23b] Its decay to the ground state by electron transfer from $MQ^{\cdot}$ to Re(II) is subject to the normal rules of nonradiative decay. They include the use of "promoting modes" to mix the excited and ground state electronic wave functions.[12,13]

The situation is different if electronic coupling is weak. An example is shown in reaction 10. In reaction 10

$$\xrightarrow{h\nu} [(\bar{\dot{b}}py)Re^{II}(CO)_3(py\text{-}PTZ)]^+ \longrightarrow [(\bar{\dot{b}}py)Re^{I}(CO)_3(py\text{-}PTZ^+)]^+ \quad (10)$$

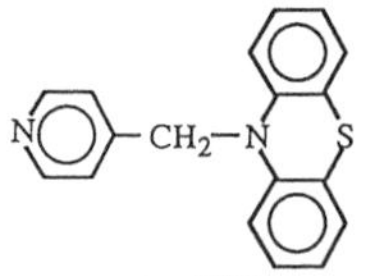

(py-PTZ)

initial Re $\longrightarrow$ bpy excitation is followed by electron transfer from the -PTZ electron transfer donor to Re(II). There is no orbital basis for strong electronic coupling between the photoproduced $-PTZ^+$ and $\bar{\dot{b}}py$ sites. Back electron transfer between them,

$$[(\bar{\dot{b}}py)Re^{I}(CO)_3(py\text{-}PTZ^+)]^{+*} \longrightarrow [(bpy)Re^{I}(CO)_3(py\text{-}PTZ)]^+ \quad (11)$$

and excited state decay of $[(bpy)Re^{II}(CO)_3(MQ^{\cdot})]^{2+*}$ are related processes in that ~2eV are released when back electron transfer occurs. In both cases the large energy release must appear in the surrounding molecular vibrations. However, decay of the $\overset{-}{bpy}$-$\overset{+}{PTZ}$ based state is an electron transfer process between sites that are weakly coupled electronically. The $\overset{-}{bpy}$-$\overset{+}{PTZ}$ state is not a true excited state of the system in that emission is not expected to occur. The matrix element that mixes the charge separated state with the ground state is the usual electrostatic operator of normal electron transfer.

In the conversion of light energy to chemicals, most of the attractive energy storage reactions, such as water splitting, involve multiple electron changes. Invariably, in these small molecule reactions, single electron steps lead to high energy intermediates such OH or H which are highly reactive and energetically unacceptable. This creates a problem for excited state-electron transfer based schemes. In all such schemes one photon produces one oxidizing equivalent and one reducing equivalent. There is, therefore, a requirement in a reactivity sense that the excitation-electron transfer apparatus be cycled more than once with the redox equivalents being stored for the following multielectron steps.

In Photosystem II a series of monophotonic events at the reaction center leads in a sequential fashion to the production and storage of multiple oxidative equivalents. The oxidative equivalents are transferred by electron transfer chains to the catalytic site where the oxidation of water occurs after four equivalents are accumulated.

In an artificial system the necessity for collecting and storing multiple redox equivalents also exists. One approach to the accumulation problem is to utilize ligand-bridged metal complexes or multiple redox sites bound to polymers. At first glance a fatal flaw appears to exist in such strategies. The flaw is illustrated for a polymer containing the multiple chromophoric sites C in Scheme 4.

SCHEME 4

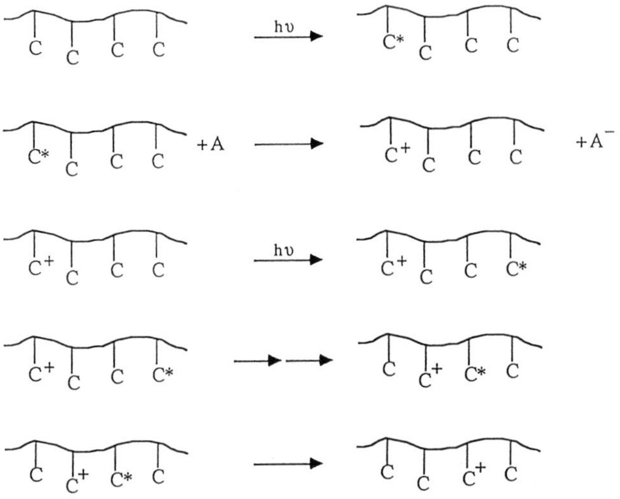

Excitation and quenching, in this case to an external acceptor A, would lead to a multichromophoric array which contains an oxidized site. Rapid electron transport amongst these sites via electron hopping, reaction 12,

(12)

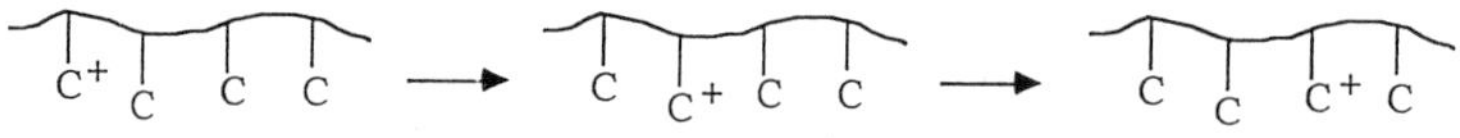

is a required property if redox equivalents are to be transported to a catalyst site. If electron transfer, or energy transfer, reaction 13,

(13)

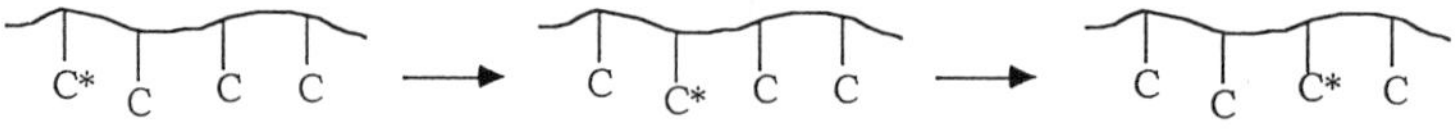

between sites is facile, a basis exists for self-quenching by electron transfer as shown by the final reaction in Scheme 4.

Typically such self-quenching reactions are highly favored thermodynamically. For example, the oxidative quenching of $[Ru(bpy)_3]^{2+*}$ by $[Ru(bpy)_3]^{3+}$ is favored by 2.1 eV. Because of the large energy release, self-quenching need not necessarily inhibit the buildup and storage of multiple redox equivalents in a single large molecule. The same energy gap law which helps to dictate rates of nonradiative decay and electron transfer in the inverted region has an important role to play in self-quenching. This has been known for some time from studies on ligand-bridged mixed-valence complexes. For the complexes $[(bpy)_2(CO)Os^{II}(4,4'\text{-}bpy)Os^{III}Cl(phen)(\underline{cis}\text{-}dppene)]^{4+}$ [27] or $[(bpy)_2ClRu^{II}(4,4'\text{-}bpy)Ru^{III}(NH_3)_5]^{4+}$, [35] Os ⟶ bpy or Ru ⟶ bpy MLCT excitation is <u>not</u> followed by facile intramolecular electron transfer as in,

$$\xrightarrow{h\nu} (bpy)(\overset{-}{bpy})(CO)Os^{III}(4,4'\text{-}bpy)Os^{III}Cl(phen)(cis\text{-}dppene)^{4+*}$$

$$(bpy)_2(CO)Os^{III}(4,4'\text{-}bpy)Os^{II}Cl(phen)(dppene)^{4+}$$

For the mixed-valence Os complex, intramolecular quenching is favored by > 2eV but it is slow. The quenching is slow because of the large energy release and the relatively weak coupling between $\overset{-}{bpy}$ and the remote M(III) site in the initial excited state.

We have observed the same phenomenon in soluble polymers. In this work, we have attached Ru-bpy and Os-bpy chromophores to soluble 1:1 polystyrene/chloromethylpolystyrene polymers by nucleophilic displacement of the chloro groups.[36] Samples have been prepared in which complete substitution leads to as many as 30 Ru(II) or Os(II) complexes chemically attached to a single polymeric strand. The repeating structural unit is shown below,

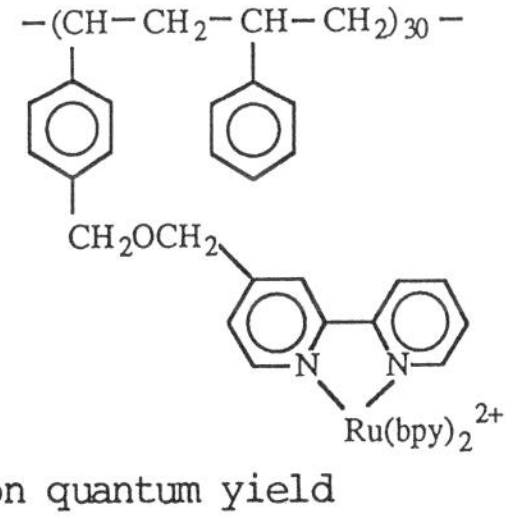

We have carried out lifetime and emission quantum yield measurements on these multiple light absorbers as a function of the relative content of M(II) and M(III). The degree of mixed-valence character can be adjusted by adding controlled amounts of chemical oxidants such as Ce(IV). In the pure M(II) containing polymers, multiple excitation and energy transfer does lead to a self-quenching pathway apparently via,

However, for the mixed-valence polymers we find that self-quenching of M(II)* by M(III) does not play an important role.[37] The introduction of the potentially quenching M(III) sites has, at best, a subtle effect on the excited state properties of the remaining M(II) sites. The absence of self-quenching has allowed us to create and store as many as 30 oxidative equivalents on a single polymeric strand by sequential excitation/oxidative quenching steps by using the irreversible oxidative quencher $[Co(NH_3)_5Br]^{2+}$ in acidic solution.[37]

Perspectives. The processes of importance in excited state electron transfer schemes depend upon the fundamental principles of chemistry and physics. From those principles, equations can be derived, which give quantitative or semiquantitative insight into quantum yields and rates. In turn, those equations are couched in terms of the molecular properties of the system. Between the underlying principles and an appropriate manipulation of properties, it does appear that it should be possible to dictate rates and quantum yields of charge separation in molecular systems in a systematic way.

In Table 1 an attempt has been made to summarize how such properties as energy gap, spin, and the solvent influence the various excited state processes which determine the efficiency with which MLCT excited states function in energy converson schemes.

TABLE 1. RELATIONSHIPS BETWEEN EXCITED STATE PROCESSES AND MOLECULAR PROPERTIES FOR MLCT EXCITED STATES

Process	Role of Molecular Properties
nonradiative decay	energy gap - $k \propto \exp -(\gamma E/\hbar\omega)$. In this limit k decreases as the energy stored increases k also increases as the change in equilibrium displacement between the ground and excited states increases for the dominant acceptor modes. spin - $k \propto \langle\psi_e\|\partial/\partial q\|\psi_g\rangle^2$. Spin-orbit coupling mixes singlet character into the excited state. solvent - The solvent affects k indirectly through the energy gap, E, and slightly as a direct energy acceptor. Dynamic effects occur in the glass to fluid transition region.
radiative decay	energy gap - $k \propto E^3$. k is predicted to increase with E^3 and the square of the transition dipole moment. spin - $k \propto \langle\psi_e\|e\bar{r}\|\psi_g\rangle^2$. The transition dipole moment can be evaluated from the integrated absorption band connecting the states. Spin-orbit coupling mixes singlet character into the excited state.
electron transfer quenching	energy gap - $k \propto \exp -[(\lambda+\Delta G)^2/4\lambda RT]$ in the clasical limit. k increases with $-\Delta G$. The reaction is barrierless at $-\Delta G = \lambda$. The inverted region occurs for $-\Delta G > \lambda$ at which point k decreases with $-\Delta G$. solvent - Influences k via ΔG and λ_o ($\lambda = \lambda_i+\lambda_o$). Solvent dynamic effects may also contribute to ν_{et}. Electron transfer quenching in a frozen solution is inhibited unless $-\Delta(\Delta G_{ES}) > -\Delta\lambda_o$ and the reaction occurs in the inverted region.
separation of redox equivalents following quenching	energy gap - If $-\Delta G > \lambda$, back electron transfer occurs in the inverted region and $k \propto$

	$\exp-(\gamma E/\hbar\omega)$. The greater the energy stored, the greater the efficiency of charge separation. spin - If there is a spin change between the initial excited state and the final redox products, back electron transfer may be inhibited. electrostatics - Coulombic repulsion between like charged quenching products increases the rate by which they separate which increases the efficiency of charge separation.
decay of intermediate states	energy gap - With sufficient energy stored, $-\Delta G > \lambda$ for excited state decay and $k \propto \exp-(\gamma E/\hbar\omega)$. strong electronic coupling - $k_{nr} \propto (\langle\psi_e \| \partial/\partial_q \| \psi_g\rangle)$. Emission may occur. weak electronic coupling - $k \propto \nu_{et}$. No emission. by electron transfer - $k \propto \exp-[(\lambda+\Delta G)^2/4\lambda RT]$.
accumulation and storage redox equivalents	energy gap - Self-quenching by of adjacent oxidative or reductive sites may be slow if $-\Delta G > \lambda$; $k \propto \exp-(\gamma E/\hbar\omega)$.

Acknowledgements are made to the Department of Energy under grant #DE-FG05-86ER13633, US Army Research Office under grant #DAAG29-85-K-0121 and to the National Science Foundation under grant #CHE-8503092 for support of this research.

REFERENCES

1. a) Balzani, V.; Bolletta, F.; Gandolfi, M. F.; Maestri, M., Top.Curr. Chem., 1978, 75, 1.
   b) Kalyanasundaram, K., Coord. Chem. Rev., 1982, 46, 159.

2. a) Meyer, T. J., Acc. Chem. Res., 1978, 11, 94.
   b) Meyer, T. J., Prog. Inorg. Chem., 1983, 30, 389.
   c) Whitten, D. G., Acct. Chem. Res., 1980, 13, 83.

3. a) Sutin, N.; Creutz, C., Pure App. Chem., 1980, 52, 2717.
   b) Sutin, N., J. Photochem., 1979, 10, 19.

4. Balzani, V.; Scandola, F., in "Energy Resources Through Photochemistry and Catalysis", Graetzel, M., Ed., Academic Press, 1983, 1.

5. Gratzel, M., "Energy Resources Through Photochemistry and Catalysis", Academic Press, New York, 1983.

6. a) Youvan, D. C.; Marrs, B. L., Sci. Amer., 1987, 256, 42.
   b) Feher, G.; Okamura, M. Y. In, "The Photosynthetic Bateria", Clayton, R. K.; Sistram, W. R., Eds.; Plenum Press: New York, 1978, 349.
   c) Parson, W. W.; Woodbury, N. W. T.; Becker, M.; Kirmaier, C.; Holten, D. in "Anntennas and Reaction Centers of Photosynthetic Bacteria", Michel-Beyerle, M. E., Ed.; Springer-Verlag, Berlin, 1985, 278.

7. Gunner, M. R.; Robertson, D. E.; Dutton, L., J. Phys. Chem., 1986, 90, 3783.

8. a) Won, Y.; Friesner, R. A., Proc. Nat. Acad. Sci. USA, 1987, 84, 5511.
   b) Creighton, S.; Hwang, J. -K.; Warshel, A.; Parson, W. W.; Norris, J., Biochem., 1988, 27, 774.

9. Meyer, T. J., Pure and App. Chem., 1986, 58, 1193.

10. a) Crosby, G. A., Acc. Chem. Res., 1975, 8, 231.
    b) DeArmond, M. K. and Carlin, C. M., Coord. Chem. Rev., 1981, 36, 325.
    c) Kemp, T. J., Prog. React. Kinetics, 1980, 10, 301.
    d) Ferguson, J.; Herren, F.; Krausz, E. R.; Maeder, M.; Vrbancich, J., Coord. Chem. Rev., 1985, 64, 21.

11. a) Bradley, P. G.; Kress, N.; Hornberger, B. A.; Dallinger, R. F.; Woodruff, W. H., J. Am. Chem. Soc., 1981, 103, 7441.
    b) Forster, M.; Hester, R. E., Chem. Phys. Lett., 1981, 81, 42.
    c) Smothers, W. K.; Wrighton, M. S., J. Am. Chem. Soc., 1983, 105, 1067.
    d) Caspar, J. V.; Westmoreland, T. D.; Allen, G. H.; Bradley, P. G.; Meyer, T. J.; Woodruff, W. H., J. Am. Chem. Soc., 1984,

106, 3492.
e) Poizet, O.; Sorisseau, C., J. Phys. Chem., 1984, 88, 3007.

12. Kober, E. M.; Caspar, J. V.; Lumpkin, R. S.; Meyer, T. J., J. Phys. Chem., 1986, 90, 3722.

13. a) Henry, B. R.; Siebrand, W., Organic Molecular Photophysics, (Vol. 1), Chap. 4, J. B. Birks, London, 1973.
b) Fong, F. K., "Radiationless Processes", Top. Appl. Phys, (Vol. 15), Springer-Verlag, New York, 1976.
c) Avouris, P.; Gelbart, W. M.; El-Sayed, M. A., Chem. Rev., 1977, 77, 793.
d) Freed, K. F., Acc. Chem. Res., 1978, 11, 74.
e) Lim, E. C., Excited States, Academic Press, New York, 1979.
f) Lin, S. H., Radiationless Transitions, Academic Press, New York, 1980.
g) Heller, E. J.; Brown, R. C., J. Chem. Phys., 1983, 79, 3336.

14. Caspar, J. V.; Meyer, T. J., Inorg. Chem., 1983, 22, 2444.

15. a) Van Houten, J.; Watts, R. J., J. Am. Chem. Soc., 1978, 17, 3381; Inorg. Chem., 1978, 17, 3381.
b) Durham, B.; Caspar, J. V.; Nagle, J. K.; Meyer, T. J., J. Am. Chem. Soc., 1982, 104, 4803.
c) Wacholtz, W. F.; Auerbach, R. A.; Schmehl, R. H., Inorg. Chem., 1986, 25, 227.

16. Barqawi, K.; Llobet, A.; Meyer, T. J., J. Am. Chem. Soc., in press.

17. a) Bock, C. R.; Connor, J. A.; Gutierrez, A. R.; Meyer, T. J.; Whitten, D. G.; Sullivan, B. P.; Nagle, J. K., J. Am. Chem. Soc., 1979, 101, 4815; Chem. Phys. Lett., 1979, 61, 522.
b) Ballardini, R.; Varani, G.; Indelli, M. T.; Scandola, F.; Balzani, V., J. Am. Chem. Soc., 1978, 100, 7219.

18. Marcus, R. A., J. Chem. Phys., 1965, 43, 1261; J. Chem. Phys., 1963, 39, 1734.

19. a) Shioyama, H.; Masuhara, H.; Mataga, N., Chem. Phys. Lett., 1982, 88, 161.
b) Mau, A. W.-H.; Johansen, O.; Sasse, W. H. F., Photochem. Photobiol., 1985, 41, 503.

20. Hoffman, M. Z.; Prasad, D. R. in "Supramolecular Photochemistry"; Balzani, V. Ed., NATO, ASI Ser-C, 1987, 214, 153, D. Reidel Pub. Co., Dordrecht.

21. Olmsted, J., III; Meyer, T. J., J. Phys. Chem., 1987, 91, 1649.

22. Boyde, S.; Jones, W., work in progress.

23. a) Westmoreland, T. D.; LeBozec, H.; Murray, R. W.; Meyer, T. J., J. Am. Chem. Soc., 1983, 105, 5952.
b) Chen, P.; Danielson, E.; Meyer, T. J., manuscript in preparation.

24. Meyer, T. J. in "Supramolecular Photochemistry"; Balzani, V. Ed., D. Reidel Pub. Co., Dordrecht, Holland, 1987, 103.

25. Chen, P. Y.; Curry, M. E.; Meyer, T. J., submitted.

26. Chen, P.; Danielson, E.; Meyer, T. J., J. Phys. Chem., in press.

27. Schanze, K. S.; Neyhart, G. A.; Meyer, T. J., J. Phys. Chem., 1986, 90, 2182.

28. Danielson, E.; Elliott, C. M.; Merkert, J. W.; Meyer, T. J., J. Am. Chem. Soc., 1987, 109, 2519.

29. Boyde, S., work in progress.

30. a) Kosower, E. M., J. Am. Chem. Soc., 1985, 107, 1114.
b) Calef, D. F.; Wolynes, P. J., J. Chem. Phys., 1983, 78, 470.
c) Zusman, L. D., Chem. Phys., 1980, 49, 295.
d) Weaver, M. J.; Gennett, T., Chem. Phys. Lett., 1985, 113, 213.
e) Su, S.-G.; Suman, J. D., J. Phys. Chem., 1987, 91, 2693.

31. a) Caspar, J. V.; Meyer, T. J., J. Am. Chem. Soc., 1983, 105, 5583.
b) Caspar, J. V.; Sullivan, B. P.; Kober, E. M.; Meyer, T. J., Chem. Phys. Lett., 1982, 91, 91.

32. a) Danielson, E.; Lumpkin, R. S.; Meyer, T. J., J. Phys. Chem., 1987, 91, 1305.
b) Kim, H.-B.; Kitamura, N.; Tazuke, S., Chem. Phys. Lett., 1988, 91, 1305.
c) Ferguson, J.; Krausz, E., Chem. Phys. Lett., 1986, 90, 5307.

33. Lumpkin, R. S.; Meyer, T. J., J. Phys. Chem., 1986, 90, 5307.

34. Chen, P.; Westmoreland, T. D.; Danielson, E.; Schanze, K. S.; Anthon, D.; Neveux, Jr. P. E.; Meyer, T. J., Inorg. Chem., 1987, 26, 1116.

35. Curtis, J. C.; Bernstein, J. S.; Meyer, T. J., Inorg. Chem., 1985, 24, 385.

36. Younathan, J. N.; McClanahan, S. F.; Meyer, T. J., Macromolec., in press.

37. Worl, L., work in progress.

FACTORS DETERMINING THE EFFICIENCY OF ELECTRON TRANSFER SENSITIZATION WITH EMPHASIS ON RUTHENIUM(II) COMPLEXES.

SHIEGEO TAZUKE, NOBORU KITAMURA, AND HAENG-BOO KIM
Research Laboratory of Resources Utilization, Tokyo Institute of Technology, 4259 Nagatsuta, Midori-ku, Yokohama 227 Japan

ABSTRACT

As part of a basic study of achieving efficient electron transfer sensitization, various factors determining the absorption wavelength, oxidation-reduction potentials, excited state lifetime and mode of relaxation of ruthenium(II) complexes were studied. The structure of ligands, solvent and temperature are all important in determining the ground and excited state properties of the complexes. The kinetics of electron transfer quenching of the excited Ru(II) complexes are then discussed with special reference to the thermodynamic parameters of the rate processes. It is concluded that the electrostatic interactions after primary electron transfer are the most important parameter determining the participation of back electron transfer to the excited state when the electron transfer process is slower than diffusion controlled rate. The discussion is extended to other electron transfer systems. Several examples of utilizing photochemically produced ion radicals for energy up-conversion are also presented.

1) INTRODUCTION

Achieving highly efficient photoredox charge separation is the ultimate goal of chemical conversion and storage of light energy. The conditions to meet this requirement are as follows: (a) strong absorption over a broad wavelength range by the sensitizer for efficient uptake of incident energy; (b) long excited state lifetime of the sensitizer to allow high probability of encounter with the reactants; (c) appropriate matching of oxidation-reduction potentials of the excited state with the reactants to accomplish highly efficient forward electron transfer; (d) prevention of back electron transfer either to the excited state or to the ground state in order to achieve effective charge separation, and finally, (e) clever uses of the active species after charge separation.

As part of a basic study for the construction of highly efficient photoredox systems, we chose ruthenium(II) complexes as representative photosensitizers and are trying to determine systematically the various factors relevant to the conditions mentioned above. Although ruthenium complexes may not be the best of all electron transfer sensitizers, their photophysical and photochemical properties have been best characterized among metal complexes. Furthermore, in designing efficient systems, transition metal complexes as sensitizers seem more advantageous than organic sensitizers due to the tunability of their redox and excited state properties by the variation of ligand structures and due to their superior turnover number as sensitizers.

While the study on photophysical properties of ruthenium complexes is specific to these complexes, kinetic and mechanistic studies on electron transfer reactions are more general, and the results can be applied to organic systems as well.

Photochemical Energy Conversion
James R. Norris, Jr. and Dan Meisel, Editors

2) DESIGN OF REDOX AND EXCITED STATE PROPERTIES OF Ru(II) COMPLEXES

2a) Redox Potentials and Absorption Bands

Twelve Ru(II) tris-chelates were studied systematically by means of spectroscopic and electrochemical technique [1-4]. Figure 1 shows the structures of the ligands (L) studied.

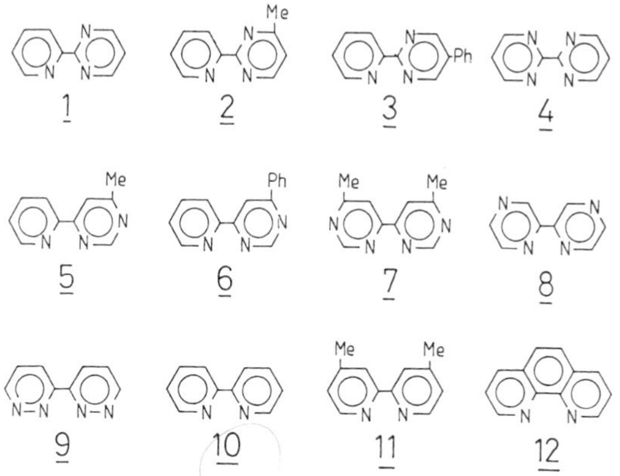

FIG. 1. The structure of ligands (L).

All Ru(II)L_3 complexes exhibited broad metal-to-ligand charge transfer (MLCT) absorption and emission bands around 400-500 nm and 600-700 nm, respectively, at room temperature. For the presently studied Ru(II) complexes, the lowest unoccupied molecular orbital (LUMO) and the highest occupied molecular orbital (HOMO) belong to a ligand π^* and a metal t_{2g} orbital and consequently, electron transfer to and from Ru(II) are expected to be ligand centered and metal centered, respectively.

As shown in Fig. 2a, a fairly good linear correlation between the reduction potentials of Ru(II)L_3 and those of L was obtained with a slope of about unity. The slope indicates that the reduction potentials of Ru(II)L_3 are exclusively determined by the reduction potentials of L, i.e., the π-acceptor strength of the ligands. On the other hand, the σ-donor strength of L, reflected in pK_a of the ligating nitrogens of L, is the main factor modulating the metal t_{2g} orbital energies. Weaker σ-donating ability to the central metal ion results in a higher formal charge of the metal ion and consequently, the stabilization of the metal t_{2g} orbitals. A good correlation between the oxidation potentials of Ru(II)L_3 and the pK_a values of L was obtained as shown in Figure 2b.

The energy difference between the metal t_{2g} HOMO and the ligand π^* LUMO orbitals should correspond to the MLCT transition energy. Indeed, the MLCT absorption energies of Ru(II)L_3 are correlated linearly with the difference between the oxidation (HOMO) and reduction potentials (LUMO) of Ru(II)L_3, $\Delta E_{1/2}$. A linear correlation was also obtained between the MLCT emission energies and $\Delta E_{1/2}$. Since the oxidation and reduction potentials of Ru(II)L_3 were shown to be related to the pK_a and the reduction potential of L, respectively, fine tuning of the redox potentials as well as of the MLCT transition energies of Ru(II)L_3 is possible by modulating the ligand properties. As far as photoreduction of methyl viologen is concerned, a nearly quantitative photoreduction is achieved by employing $Ru(\underline{5})_3^{2+}$ or $Ru(\underline{6})_3^{2+}$ as a sensitizer with triethanolamine as a sacrificial donor [5].

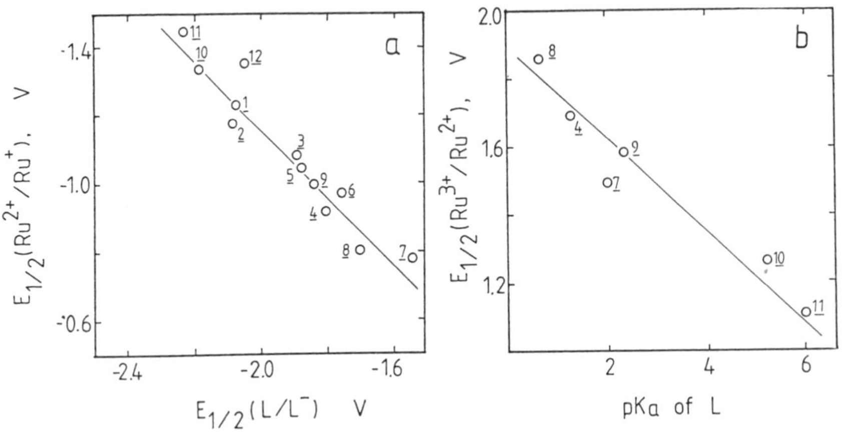

FIG. 2. Dependence of electrochemical redox potentials of Ru(II)L_3 on ligand properties. a) Reduction potential vs. those of L; b) Oxidation potential vs. pK_a of L.

2b) Excited State Lifetime

Controlling the excited state lifetime or emission quantum yield is a more difficult task. The difficulty arises principally from uncertain decay processes of the excited state, in particular the complicated temperature dependence [6-8]. To understand further the effects of the ligand structures on the excited state properties of RuL_3^{2+}, we studied temperature dependence of the emission lifetime of RuL_3^{2+} in acetonitrile.

Although the emission lifetimes, τ, of all RuL_3^{2+} decrease with increasing temperature, the temperature dependence of τ is strongly dependent on L as shown in Figure 3 and analyzed by eq (1) [7,9].

$$\tau^{-1} = k' + \nu \exp(-\Delta E/RT) = k_r + k_{nr} + \nu \exp(-\Delta E/RT) \qquad (1)$$

where k_r and k_{nr} are the temperature independent radiative and non-radiative decay rate constants, respectively. Provided the decay of the d-d* excited state is very fast, ν and ΔE are the frequency factor and the activation energy for the thermal activation from the lowest emitting MLCT excited state (^{3}MLCT*) to the lowest non-emitting d-d excited state (d-d*), respectively. When ^{3}MLCT* and d-d* are in equilibrium, ν and ΔE are the rate of (^{3}MLCT → d-d*) excitation and the energy difference between the two states. The best-fit simulation of the data in Figure 3 using eq. (1) gave k' = (k_r + k_{nr}), ν, and ΔE. With the measured emission quantum yield of RuL_3^{2+}, k_r and k_{nr} can be calculated according to eqs. (2) and (3). Results are given in Table I.

$$k_r = \frac{\phi^{em}}{\tau} \tag{2}$$

$$k_{nr} = k' - \frac{\phi^{em}}{\tau} \tag{3}$$

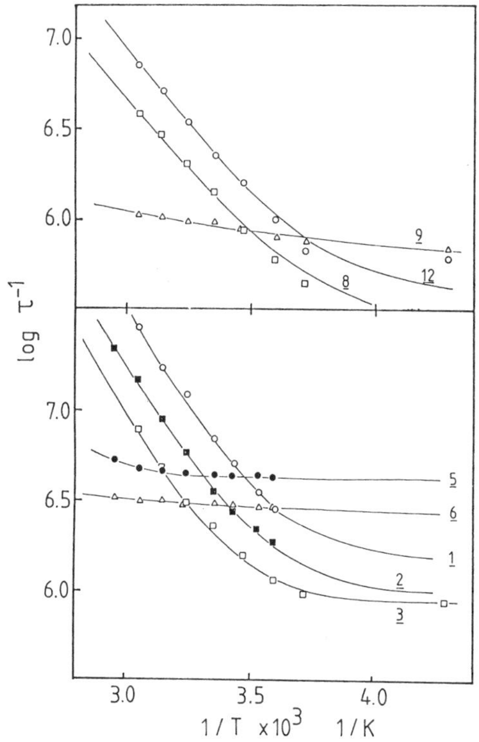

FIG. 3. Temperature dependence of the emission lifetime of RuL_3^{2+} in acetonitrile. For identification of numbers, see FIG. 1.

For RuL_3^{2+} with 1-3, 8, 10 or 12 as ligands, τ is strongly dependent on temperature, similar to $Ru(bpy)_3^{2+}$, and the observed ν (10^{12} - 10^{14} s^{-1}) and ΔE (2700 ~ 4000 cm^{-1}) values are comparable to those reported for $Ru(bpy)_3^{2+}$ ($\nu = 3 \times 10^{12} \sim 5 \times 10^{13}$ s^{-1} and ΔE = 3200 ~ 3800 cm^{-1}, depending on solvent) [7]. The results manifest that the ^{3}MLCT* state of RuL_3^{2+} (L = 1-3, 8, 10, and 12) deactivate via the d-d* state as concluded for $Ru(bpy)_3^{2+}$ [7,9]. In marked contrast to these complexes, $Ru(6)_3^{2+}$ and $Ru(9)_3^{2+}$ showed small temperature dependence of τ and ν and ΔE were calculated to be ~10^7 s^{-1} and 1000 ~ 1200 cm^{-1}, respectively. For $Ru(5)_3^{2+}$, the computer simulation of the data in Figure 3 by eq. (1)

gave ν = 5.2 x 10^{15} s^{-1} and ΔE = 5270 cm^{-1}. However, the emission lifetime of $Ru(\underline{5})_3^{2+}$ is essentially temperature independent similar to that of $Ru(\underline{6})_3^{2+}$ or $Ru(\underline{9})_3^{2+}$. The ν and ΔE values for $Ru(\underline{5})_3^{2+}$, estimated by Arrhenius equation, were 1 x 10^7 s^{-1} and 190 cm^{-1}, respectively.

When one compares ν and ΔE for RuL_3^{2+} (L = 5, 6, and 9) with those for $Ru(bpy)_3^{2+}$, the values for the former complexes are apparently too small to be ascribed to thermal activation to the d-d* state and thus, the excited state decay model involving ^{3}MLCT* and d-d* is not appropriate for RuL_3^{2+} with 5, 6, or 9 as ligands.

TABLE I. Temperature dependence of the emission lifetime of RuL_3^{2+} in acetonitrile

L RuL_3^{2+}	k' $x10^5 s^{-1}$	k_r $x10^4 s^{-1}$	k_{nr} $x10^6 s^{-1}$	ν s^{-1}	ΔE cm^{-1}
1	16	0.64	1.6	$3.1x10^{14}$	3700
2	9.5	3.4	0.92	$9.2x10^{13}$	3600
3	9.1	5.6	0.85	$2.8x10^{14}$	3990
4		5.4			
5	44	6.9	4.3	$5.2x10^{15}$	5270
				$(1.1x10^7)$a)	(190)a)
6	28	8.1	2.7	$8.3x10^7$	1190
				$(5.5x10^6)$a)	(120)a)
7		4.5			
8	2.6	3.8	0.22	$5.9x10^{11}$	2730
9	6.0	4.2	0.56	$3.3x10^7$	970
				$(3.3x10^7)$a)	(250)a)
10b)	5.8	7.7	0.48	$5.8x10^{13}$	3800
11		7.3			
12	4.0	5.0	0.35	$2.5x10^{12}$	2920

a) Calculated based on Arrhenius equation.b) Data compiled from [7]. Error limits are as follows: k_r = ±15%, k_{nr} = ±5%, ν = ±20~30%, ΔE = ±100~200 cm^{-1}.

Participation of the d-d* excited state in the decay process is indicated also in the yields for photodecomposition (photoanation) of RuL_3^{2+} in the presence of 0.1 M of KSCN in acetonitrile. This reaction supposedly proceeds via the d-d* excited state. The efficient photodecomposition of $Ru(\underline{2})_3^{2+}$, comparable to that of $Ru(bpy)_3^{2+}$, is ascribed to the participation of the d-d* state in the excited state decay. This conclusion agrees very well with the large temperature dependence of τ; ν = 9 x $10^{13} s^{-1}$ and ΔE = 3600 cm^{-1}. On the other hand, relatively small ν (1~8 x $10^7 s^{-1}$) and ΔE (200~1200 cm^{-1}) of RuL_3^{2+} (L = 5, 6, and 9) correspond to higher stabilities against photoanation. The decay of these complexes probably does not involve the d-d* state. The yields of the photodecomposition for RuL_3^{2+} ((L = 5, 6, and 9) were only ~20% of that for $Ru(bpy)_3^{2+}$. Several other Ru complexes with mixed ligands (L', L = 4, 8, 11, 4,4'-dicarboxylic-bpy, 4,4'-dicarboethoxy-bpy, and CN^-) are known to exhibit small temperature dependence of τ as well.

There is a broad correlation showing that RuL_3^{2+} exhibiting small temperature dependence of τ (RuL_3^{2+}, L -= 5, 6, and 9) possess relatively low energy MLCT excited states as compared with those showing large temperature dependence of τ (ν = 10^{12} ~ $10^{14} s^{-1}$ and ΔE = 3000~4000 cm^{-1}). This relation is also applicable to the $RuL_nL'_{3-n}$ mixed complexes mentioned above. The emission energies of these complexes were all in the range of 14.1 ~ 15.2 x $10^3 cm^{-1}$ [10-12], which are considerably lower in energy relative to that of $Ru(bpy)_3^{2+}$ (16.1 x $10^3 cm^{-1}$ in acetonitrile). It is very probable that the energy of the d-d* state is affected by the present series of ligands to a lesser degree than the ^{3}MLCT* state and then the energy difference between the emitting ^{3}MCLT* state and the non-emitting d-d* state (ΔE (^{3}MCLT* - d-d*) increases with lowering the energy of the ^{3}MLCT* state. The ΔE(^{3}MLCT*-d-d*) for $Ru(bpy)_3^{2+}$ has been reported to be ~3600 cm^{-1}[6], those for $Ru(5)_3^{2+}$, $Ru(6)_3^{2+}$, and $Ru(9)_3^{2+}$ may be as large as 4200 ~ 5200 cm^{-1}, as estimated from the emission energies at 77 K. Thermal activation to the d-d* state will not be possible for these complexes. Instead, by analogy to $RuL_nL'_{3-n}$, we propose that the ^{3}MCLT* state deactivates through the fourth MLCT excited state, MLCT'. For $Ru(bpy)_3^{2+}$, the existence of MLCT' has been predicted theoretically to locate 600 ~ 1000 cm^{-1} above ^{3}MLCT* [10]. This process is only discernible when excitation to the d-d* state does not occur due to the large activation energy.

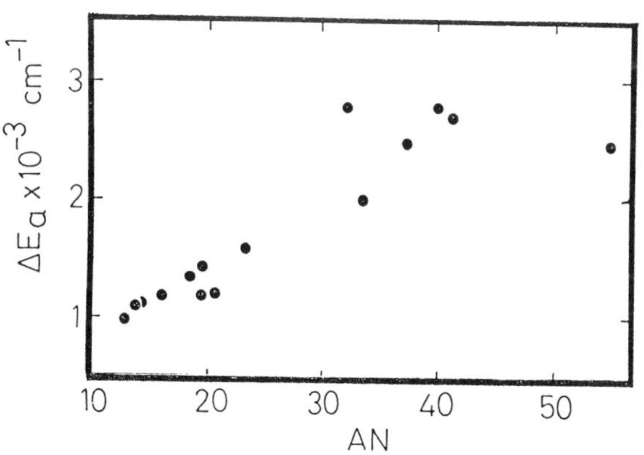

FIG. 4. Temperature dependence of τ for cis-$Ru(phen)_2(CN)_2$ vs. solvent acceptor number.

Further support for the present explanation was obtained from the study of solvent effects [13]. The temperature dependent emission lifetime of <u>cis</u>-$Ru(phen)_2(CN)_2$ showed that ν and ΔE increased from 10^7 to $10^{12} s^{-1}$ and from 1000 to 2800 cm^{-1}, respectively, with increasing the Gutmann's solvent acceptor number (AN) as shown in Figure 4 [13]. Since the emission energy (E^{em}) of the cyanide complex shifts to higher energy with AN, the solvent effects unequivocally indicate that the decrease in $\Delta E(^3MLCT^*\text{-}d\text{-}d^*)$ brings about a larger temperature dependence of τ.

The non-radiative decay constant (k_{nr}) is much larger (~10^2) than the radiative decay constant (k_r in Table I). If the vibrational overlap between the ground and excited states (e.g., Franck-Condon factor) determines k_{nr}, ln k_{nr} should be linearly correlated with E^{em} as predicted by the energy gap law [14]. While the structure of ligands is considerably different, the energy gap law is satisfactorily applicable to the present systems (correlation coefficient: -0.94) excluding the complexes with 8 and 9 ligands.

From the above discussion, a long excited state lifetime is expected if we prepare a ruthenium tris-diimine complex in which the lowest d-d* excited state is at high energy, whereas the $^3MLCT^*$ state is about 5,000-6,000 cm^{-1} lower than the d-d* state so that both the decay process via the d-d* excited state and that governed by the energy gap law are suppressed.

3) KINETICS OF PHOTOINDUCED ELECTRON TRANSFER REACTIONS

3a) Redox Quenching of Tris(2,2'-bipyridine) Ruthenium(II)

This is one of the best studied electron transfer systems. However, the kinetic mechanistic details, including the interpretation of thermodynamic parameters have not been investigated until recently [15,16]. In Figure 5, the results of quenching experiments are summarized as a function of the overall free energy change (ΔG_{23}). While both reductive and oxidative quenching rate constants (k_q) fall on the same curve (Figure 5a), dependence of their activation enthalpies ($\Delta H^‡$) and entropies ($\Delta S^‡$) on ΔG_{23} are quite different for reductive and oxidative quenching.

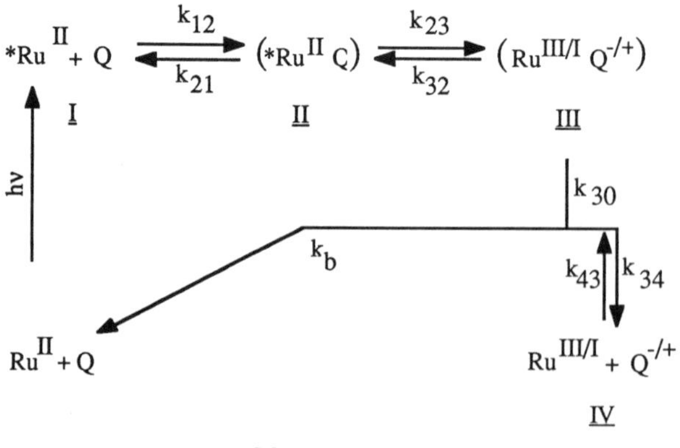

SCHEME I

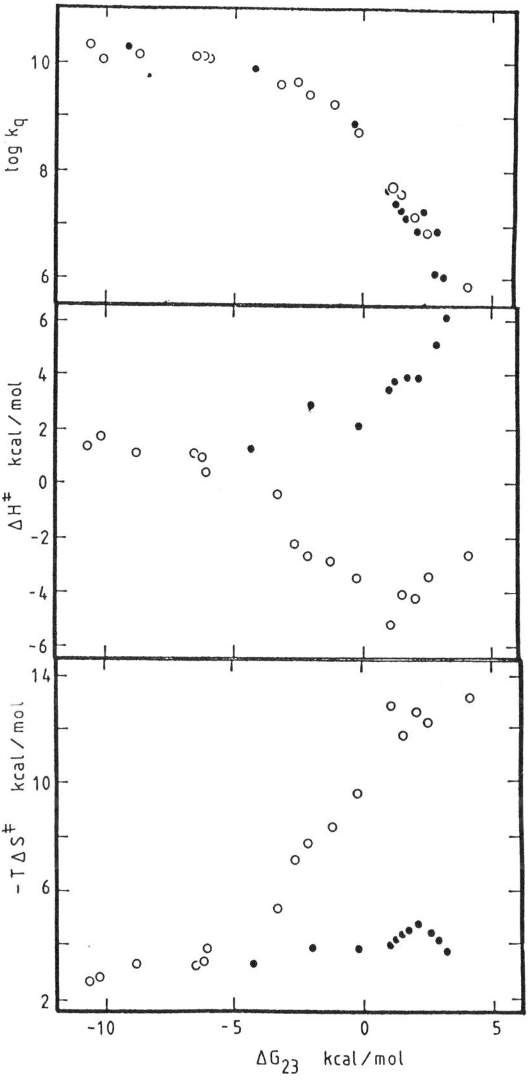

FIG. 5. Free Energy vs. log. k_q, $\Delta H^{\ddagger}$ and $\Delta S^{\ddagger}$ for reductive (•) and oxidative (O), quenching of $^*Ru(bpy)_3^{2+}$ in acetonitrile.

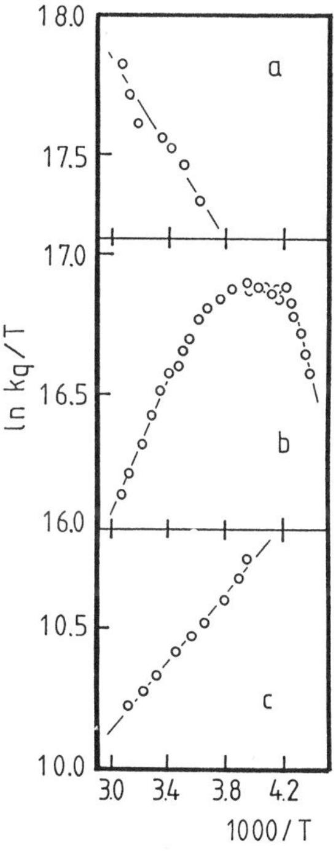

FIG. 6. Eyring plots for quenching of $^*Ru(bpy)_3^{2+}$ by 1,4-naphthoquinone (a), duroquinone (b), and methyl m-benzoate (c), in acetonitrile

The most striking feature is the appearance of a "bell-shaped" Eyring plot (Figure 6) which recalls the shape of the Stevens-Ban plot [17] applicable to excimer/exciplex formation. The results are explainable by Scheme I and eq. 4.

$$k_q = K_{12}\, k_{23} \frac{k_{30}}{k_{30} + k_{32}} \tag{4}$$

where $K_{12} = k_{12}/k_{21}$. Depending of the relative magnitude of k_{32} to k_{30}, k_q in eq. (4) can be simplified into eqs. (5) and (6) [13]:

$$\text{Case I: } k_{30} >> k_{32} \quad k_q = K_{12}\, k_{23} \tag{5}$$

$$\text{Case II: } k_{32} >> k_{30} \quad k_q = K_{12}K_{23}k_{30} \tag{6}$$

where $K_{23} = k_{23}/k_{32}$. Applying transition state theory to the k_{23}, k_{32}, and k_{30} processes, we obtain eq. (7) and (8) for case I and II, respectively.

Case I $\quad k_q = K_{12}k_{23}$

$$\Delta H^{\ddagger} = \Delta H_{23}^{\ddagger} \qquad \Delta S^{\ddagger} = \Delta S_{23}^{\ddagger} \tag{7}$$

Case II $\quad k_q = K_{12}K_{23}k_{30}$

$$\Delta H^{\ddagger} = \Delta H_{23} + \Delta H_{30}^{\ddagger} \qquad \Delta S^{\ddagger} = \Delta S_{23} + \Delta S_{30}^{\ddagger} \tag{8}$$

It is apparent that the negative temperature dependence corresponds to Case II when ΔH_{23} is sufficiently negative as compared with $\Delta H_{30}^{\ddagger}$.

The results in FIG. 6 are a reflection of the shift of reaction mechanism from Case I to Case II with increasing temperature. From the analysis of the "bell-shaped" Eyring plot, together with electrochemical data summarized in Table II, the following may be concluded:

i) The results are perfectly normal when ΔH_{23} and ΔS_{23} are taken into account. $\Delta H_{23}^{\ddagger}$ and $\Delta S_{23}^{\ddagger}$ indicate that the electron transfer from $^*Ru(bpy)_3^{2+}$ to an electron acceptor proceeds via an enthalpy controlled reaction path similar to the reductive quenching systems. As far as the <u>II</u> → <u>III</u> process in Scheme I is concerned, there is no evidence to support a mechanistic difference between oxidative and reductive quenching of $^*Ru(bpy)_3^{2+}$, while the orbitals involved are different for each mode of quenching.

ii) The observed large and negative $\Delta S^{\ddagger}$ (-26 ~ -44 e.u.) in oxidative quenching is due to the unfavorable ΔS_{23} values (-19 ~ -38 e.u.). The major contribution to ΔS_{23} is the change in solvation before and after electron transfer. Calculation of solvation entropy according to the Born model clearly indicates that the II → III process producing oppositely charged ions (i.e., $Ru(bpy)_3^{3+}$ + A^-) is entropically unfavorable while that leading to electrostatically repulsive ions (i.e., $Ru(bpy)_3^{+}$ + D^+) is entropically advantageous [18]. The situation is reversed for the back electron transfer to the excited reactant pair. k_{32} is entropically favorable for oxidative quenching, rendering the Case II mechanism to dominate.

iii) The activation enthalpy and entropy calculated from the Marcus expression using ΔH_{23} and ΔS_{23} run parallel to those observed. Qualitatively, both the II → III and III → II processes of Scheme I are explainable within the framework of the Marcus theory. More quantitative investigation is now underway.

3b) The Importance of Electrostatic Interactions Following Forward Electron Transfer

The participation of back electron transfer to the excited state is observed only in the electron transfer systems where attractive ion pairs are produced after the forward electron transfer reaction. The importance of electrostatic interactions was proven by ionic strength effect studies on the quenching constant of $Ru(bpy)_3^{2+}$/neutral donors and acceptors systems [1,3,19]. We have also demonstrated that the quenching of singlet excited state of pyrene by organic donors and acceptors [19] and that of excited $Ru(phen)_2(CN)_2$ by neutral donors and acceptors [18] are subjected to negative temperature dependence. All these combinations produce a uni-positive — uni-negative pair both in oxidative and reductive quenching. $Ru(phen)_2(CN)_2$ is of particular interest since both oxidative and reductive quenchings can be compared under identical electrostatic conditions. In the case of reduction, an electron is donated to the ligand LUMO, while in oxidation an electron is transferred from the central metal d orbitals (pseudo t_{2g} orbital in octahedral approximation). Such differences in the participating orbitals might bring about different kinetic behavior depending upon the mode of quenching. The results are summarized in Table II and equations (9) - (12).

TABLE II. Quenching of excited cis-$Ru(phen)_2(CN)_2$ and $Ru(bpy)_3^{2+}$ in acetonitrile.

Reaction: Parameter	$*Ru(phen)_2(CN)_2$ +A	$*Ru(phen)_2(CN)_2$ +D	$*Ru(bpy)_3^{2+}$ +A	$Ru(bpy)_3^{2+}$ +D
w_p	-0.85	-0.85	-2.3	+0.78
ΔH_{23}	-10 ~ -1.0	-10 ~ -4.5	-17 ~ -6.0	+0.8 ~ +5.0
ΔS_{23}	-13 ~ -2.0	-27 ~ -22	-38 ~ -19	+3.4 ~ +14.4
ΔG_{32}^{es}	-0.95	-0.95	-1.8	-0.10
ΔH_{32}^{es}	+0.48	+0.48	+0.89	+0.05
ΔS_{32}^{es}	+4.8	+4.8	+8.9	+0.52
$k_{34}(s^{-1})$	3.6×10^9	3.6×10^9	5.3×10^8	1.1×10^{10}
$K_{34}(M^{-1})$	0.094	0.094	0.0059	1.1

w_p, ΔH: in kcal/mol, ΔS: in e. u.

$$*RuCN + A \longrightarrow \boxed{-} + \boxed{+} \quad (w_p = -0.85 \text{ kcal/mol}) \tag{9}$$

$$*RuCN + D \longrightarrow \boxed{+} + \boxed{-} \quad (w_p = -0.85 \text{ kcal/mol}) \tag{10}$$

$$*Ru(bpy)_3^{2+} + A \longrightarrow \boxed{3+} + \boxed{-} \quad (w_p = -2.34 \text{ kcal/mol}) \tag{11}$$

$$*Ru(bpy)_3^{2+} + D \longrightarrow \boxed{+} + \boxed{+} \quad (w_p = +0.78 \text{ kcal/mol}) \tag{12}$$

Reaction enthalpies (ΔH_{23}) and reaction entropies (ΔS_{23}) were calculated from the dependence of oxidation-reduction potentials as well as dielectric constants in the work terms on temperature. When ΔH_{23} is negative, $\Delta H^{\ddagger}$ is also negative, consistent with the kinetic expression for Case II. It is apparent that electrostatic interactions are largely responsible for the difference in the thermodynamic parameters.

The electrostatic interactions control the dissociation process, $\underline{III} \rightarrow \underline{IV}$ in Scheme I as well. When process $\underline{III} \rightarrow \underline{IV}$ is a simple diffusion process, k_{34} can be calculated by eq. (13).

$$k_{34} = \frac{2k_BT}{\pi d^3 \eta} \cdot \frac{w_p/RT}{1-\exp(-w_p/RT)} \tag{13}$$

As a matter of course, the magnitude of k_{34} is directly related to w_p. As shown in Table II, k_{34} is as large as $1.12 \times 10^{10} s^{-1}$ for reductive quenching of $*Ru(bpy)_3^{2+}$,indicating that the rate determining step is not the dissociation process. On the other hand, when the produced ions are electrostatically attractive, k_{34} decreases to 5.28×10^8 s^{-1} in the case of oxidative quenching of $*Ru(bpy)_3^{2+}$. This value is smaller than or comparable to the overall quenching constant, indicating that the dissociation process could be the bottle-neck of the overall process. The cyanide complex again falls in between the two extreme cases.

The sign and magnitude of w_p also reflect on the equilibrium constant, $K_{34} = k_{34}/k_{43}$ expressed by eq. (14).

$$K_{34} = 3000/(4\pi N_O d^3 \exp(-w_p/RT)) \tag{14}$$

Although a quantitative interpretation on the basis of purely electrostatic interaction is not possible, the kinetic mechanistic characteristics of photoinduced electron transfer reactions of Ru(II) complexes are largely decided by the difference in Coulombic interactions before and after the forward electron transfer. The nature of interacting orbitals does not affect the mechanism in agreement with the assumption of outer sphere electron transfer mechanism. As a consequence, the thermodynamic parameters can be interpreted by the Marcus theory.

Previously, Baggott [20] observed negative temperature dependence in fluorescence quenching of pyrene by nitrile compounds in ethanol. Such an abnormal temperature dependence is not because of a specific combination of a quencher and solvent but is a general feature for the combination of neutral fluorescers with neutral electron transfer quenchers.

4) ENHANCEMENT OF CHARGE SEPARATION IN PHOTOREDOX REACTIONS

The preceeding discussion is partially specific to the ruthenium complexes discussed. Other features are relevant to electron transfer in photoredox reactions in general. A useful conclusion is the importance of the Coulombic work term in suppressing the back electron transfer either to the excited state or to the ground state and in facilitating charge separation. In addition to this conclusion, back electron transfer processes can also be minimized if the primary reaction products are removed as stable reaction products immediately after charge separation. Several examples from our results are presented below.

4a) Photoreduction of Methyl Viologen

In the well-known combination of RuL_3^{2+}/methyl viologen (MV^{2+})/triethanolamine (TEA) [5,21], the photoredox quantum yield (ϕ_{redox}) depends very much on the ligand properties as shown in Table III. In the case of reductive quenching of $*RuL_3^{2+}$ by TEA (Mechanism I), TEA is irreversibly oxidized, and the overall yield is high.

TABLE III. Photoreduction of MV^{2+} by RuL_3^{2+} in Acetonitrile.[a)]

L	Q_e (ΔG, kcal/mol)[b)] MV^{2+}	TEA	ϕMV^{2+}	$\phi redox$[c)]	Quenching Mechanism
1	0.11(-5.5)	0 (+0.2)	0.031	0.28	II
2	0.25(-6.2)	0 (- 0.2)	0.089	0.36	II
3	0.26(-4.4)	0.14 (- 2.3)	0.29	0.73	I + II
5	0.18(-3.7)	0.16 (-0.19)	0.16	0.47	I + II
6	0.34(-2.5)	0.18 (-1.6)	0.26	0.50	I + II
8	0 (+3.2)	1.00 (-10.6)	1.7	1.7	I
9	0.14(-0.9)	0.40 (-3.7)	0.42	0.78	I + II
10	0.75(-9.9)	0 (+2.3)	0.33	0.44	II
12	0.62(-11.3)	0 (+1.2)	0.25	0.40	II

a) Determined in the presence of 0.15 M of TEA under deaerated conditions. b) Free energy change of the quenching. $\Delta G(MV^{2+}) = E_{1/2}(Ru^{3+}/{*}Ru^{2+}) - E_{1/2}(MV^{2+}/MV^{+})$. $\Delta G(TEA) = E_{1/2}(TEA^{ox}/TEA) - E_{1/2}(+Ru^{2+}/Ru^{+})$. $E_{1/2}(MV^{2+}/MV^{+})$ and $E_{1/2}(TEA^{ox}/TEA)$ are -0.44 and 0.90 V (vs. SCE), respectively. c) $\phi_{redox} = \phi_{MV^+}/Q_e$. Error limits of ϕ_{MV^+} and ϕ_{redox} are ±5%.

Mechanism I (reductive quenching):

$$*RuL_3^{2+} + TEA \longrightarrow RuL_3^{+} + TEA^{ox} \quad // \quad > RuL_3^{2+} + TEA$$

$$RuL_3^{+} + MV^{2+} \longrightarrow RuL_3^{2+} + MV^{+}$$

Mechanism II (oxidative quenching):

$$*RuL_3^{2+} + MV^{2+} \longrightarrow RuL_3^{3+} + MV^{+} \longrightarrow RuL_3^{2+} + MV^{2+}$$

$$RuL_3^{3+} + TEA \longrightarrow RuL_3^{2+} + TEA^{ox}$$

In the photoreduction of MV^{2+} by phenothiazine derivatives bearing positive, negative or no charge, Coulombic effects [22] as well as neutral salt effect [23] were observed. The unexpected salt effects were interpreted by the suppression of the back electron transfer reaction process within the geminate ion pair.

4b) Reductive Photocarboxylation of Aromatic Hydrocarbon

Irradiation of aromatic hydrocarbons, in particular phenanthrene, in the presence of tertiary amines such as N,N-dimethylaniline or N,N-dimethyltoluidine in CO_2 atmosphere and in polar aprotic solvents is known to bring about reductive carboxylation of the aromatic hydrocarbon [24,25]. With phenanthrene, the main product is 9,10-dihydrophenanthrene-9-carboxylic acid with a quantum yield of 0.1 - 0.2. There seems to be a number of equilibrium processes, as shown in Scheme II, involved in this reaction.

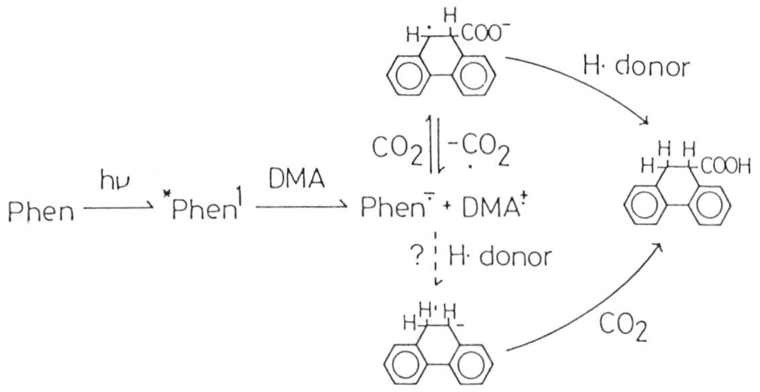

SCHEME II

Addition of compounds with labile hydrogen atoms such as cumene, decalin, tetraline, and 9,10-dihydrophenanthrene improves the quantum yield appreciably. These hydrogen atom donors will stabilize the reactive CO_2^- - anion radical adduct to give the final product.

More recently, we observed photocarboxylation of phenanthrene in the presence of 1,4-dicyanobenzene and carbon dioxide in rigorously dried acetonitrile. Although the yield is low (<1% after prolonged irradiation), the reaction seems to proceed via electron transfer from the acceptor anion radical to CO_2 [26]. The CO_2^- anion radicals either couple with phenanthrene cation radicals or attack phenanthrene. Such processes are inevitably inefficient since the back electron transfer process is likely to predominate. Interestingly, addition of tetrabutylammonium chloride (0.1 M) doubles the yield of photocarboxylation, indicating that the charge separation process is facilitated by the addition of the neutral salt.

5) CONCLUDING REMARKS

We attempted to present general principles to construct efficient photoredox systems in homogeneous solutions. It is possible to achieve efficient charge separation by a number of ways including the uses of Coulombic effect, neutral salt effect, and additive effects to remove the reactive intermediates participating in the back electron transfer from the system or to stabilize the reaction products. In comparison with the complicated processes of the back electron transfer to the ground state and the subsequent side reactions, the primary electron transfer process (emission quenching process) is kinetically as well as mechanistically simpler. Efficient quenching is a necessary condition but seldom determines the overall yield.

REFERENCES

1. S. Tazuke, Reports of Special Project Research on Energy under Grant-in-Aid of Scientific Research. The Ministry of Education, Science and Culture. October 1987, pp. 99-104.
2. Y. Kawanishi, N. Kitamura, Y. Kim and S. Tazuke, Riken Sci. Papers 78, 212 (1984).
3. S. Tazuke, N. Kitamura and Y. Kawanishi, J. Photochem. 29, 123 (1985).
4. N. Kitamura, Y. Kawanishi and S. Tazuke, Chem. Phys. Lett. 97, 103 (1983).
5. N. Kitamura, Y. Kawanishi and S. Tazuke, Chem. Lett. 1185 (1983).
6. V. Van Houten and R. J. Watts, J. Am. Chem. Soc. 98, 4853 (1976).
7. J. V. Caspar and T. J. Meyer, J. Am. Chem. Soc. 105, 5583 (1983).
8. D. P. Rillema, G. Allen and T. J. Meyer, Inorg. Chem. 22, 1617 (1983).
9. B. Durham, J. V. Caspar, J. K. Nagle and T. J. Meyer, J. Am. Chem. Soc. 104, 4803 (1982).
10. G. H. Allen, R. P. White, D. P. Rillema and T. J. Meyer, J. Am. Chem. Soc. 106, 2613 (1986).
11. W. F. Wacholtz, R. A. Auerbach and R. H. Schmehl, Inorg. Chem. 25, 227 (1986).
12. L. J. Henderson, Jr. and W. R. Cherry, J. Photochem. 28, 143 (1985).
13. N. Kitamura, M. Sato, H.-B. Kim, R. Obata and S. Tazuke, Inorg. Chem. 27, 651 (1988).
14. T. J. Meyer, Pure Appl. Chem. 58, 1193 (1986) and references therein.
15. N. Kitamura, S. Okano and S. Tazuke, Chem. Phys. Lett. 90, 13 (1982).
16. H.-B. Kim, N. Kitamura, Y. Kawanishi and S. Tazuke, J. Am. Chem. Soc., 109, 2033 (1987).
17. J. B. Birks, Photophysics of Aromatic Molecules, Wiley-Interscience, London, 1970.
18. N. Kitamura, R. Obata, H.-B. Kim and S. Tazuke, J. Phys. Chem. 91, 2033 (1987).
19. S. Tazuke, N. Kitamura and H.-B. Kim, Supramolecular Photochemistry, V. Balzani, ed. (Reidel, Dordrecht 1987) NATO ASI Series C: 214, pp. 87-102.
20. J. E. Baggott, J. Phys. Chem. 87,, 5223 (1983).
21. Y. Kawanishi, N. Kitamura and S. Tazuke, to be published.

22. Y. Kawanishi, N. Kitamura and S. Tazuke, J. Phys. Chem., 90, 2469 (1986).
23. Y. Kawanishi, N. Kitamura and S. Tazuke, J. Phys. Chem., 91, 6034 (1986).
24. S. Tazuke and H. Ozawa, J. Chem. Soc., Chem. Commun., 237 (1975)
25. S. Tazuke, S. Kazama and N. Kitamura, J. Org. Chem. 51, 4548 (1986).
26. S. Tazuke and N. Kitamura, Nature, 275, 301 (1978).

EXCIPLEX FORMATION AND ELECTRON TRANSFER IN POLYCHROMOPHORIC SYSTEMS

NIEN-CHU C. YANG,* DAVID W. MINSEK,* DOUGLAS G. JOHNSON,** AND MICHAEL R. WASIELEWSKI**
*Department of Chemistry, University of Chicago, Chicago, IL 60637;
**Chemistry Division, Argonne National Laboratory, Argonne, IL 60439

ABSTRACT

The rates of excited anthracene decay and intramolecular exciplex formation from bichromophoric molecules containing an anthryl group and an amine donor vary with the length of the chain link, the nature of the amine donor and the viscosity of the medium. The results indicate that the intramolecular exciplex formation may proceed via more than one pathway. Our recent experimental results suggest that electron transfer from the amino donor to the excited anthryl group may play a role in the exciplex formation in viscous alkanes.

INTRODUCTION

Through the classical contribution of Weller and his coworkers [1], it has been demonstrated that excited aromatic hydrocarbons may interact with amines to undergo either exciplex formation or electron transfer depending on the experimental conditions. Exciplexes are usually formed in non-polar media, while electron transfers usually occur in polar media. Our current interest concerns the possibility of electron transfer in such systems in non-polar media, and we have demonstrated that such a process does occur intramolecularly in acyclic polychromophoric systems [2]. We have also shown that excited anthracene may be quenched by aromatic amines at a rate considerably higher than the diffusion-controlled rate in viscous alkanes [3]. In this investigation, we studied the excited state behavior of several bichromophoric molecules linked with an aliphatic chain containing either 7, 1a and 1b, or 3 carbon atoms, 2a and 2b, in a variety of solvents of different polarity and viscosity. The chain length was selected on the basis of previous work which shows that the 3-atom link allows facile intramolecular excimer formation, while the 7-atom link is particularly unfavorable in an analogous system [4,5]. Our results suggest that this electron transfer between the amino group and the excited anthryl group may preceed exciplex formation in these molecules.

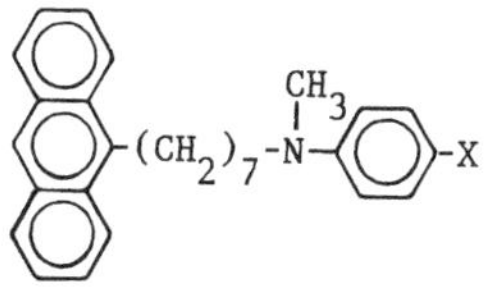

1a, X = H; 1b, X = OCH_3

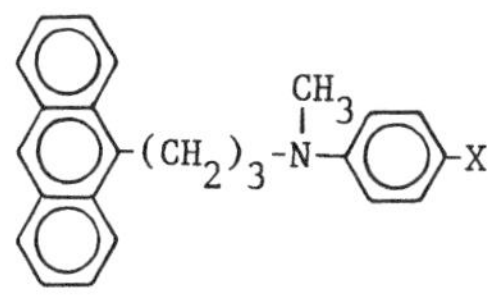

2a, X = H; 2b, X = OCH_3

EXPERIMENTAL PART

Compounds 1 and 2 were synthesized by conventional methods [6,7]. The steady state fluorescence measurements were made with a Perkin-Elmer MPF-66 Spectrofluorimeter using spectrograde solvents. The time-resolved fluorescence measurements were made with systems previously described in the

Published 1989 by Elsevier Science Publishing Co., Inc.
Photochemical Energy Conversion
James R. Norris, Jr. and Dan Meisel, Editors

literature [8,9].

RESULTS

The steady state fluorescence data of 1a and 2a are presented in TABLE I and those of 1b and 2b in TABLE II. Each emission is characterized by the 0→0 band of monomeric anthryl emission (±1nm), λ_M, the quantum yield of the monomeric anthryl emission (±10%), Φ_M, the maximum of the exciplex emission (±2nm), λ_E, the quantum yield of the exciplex emission (±10%), Φ_E.

TABLE I. Fluorescence Characteristics of 1a and 2a.

	1a				2a			
Solvent	λ_M(nm)	Φ_M	λ_E(nm)	Φ_E	λ_M(nm)	Φ_M	λ_E(nm)	Φ_E
n-pentane	389	0.28	a	<0.02	389	0.048	487	0.45
n-hexane	389	0.35	a	<0.02	389	b	487	b
n-heptane	b	b	b	b	389	0.061	487	0.43
Me-cyclo-hexane	391	0.45	a	<0.02	390	0.081	487	0.41
n-hexade-cane	391	0.45	a	<0.02	390	0.105	488	0.38
ether	390	0.34	490	0.057	390	0.015	506	0.49
ethyl acetate	391	0.16	520	0.095	c	-	528	0.16
diethyl succinate	391	0.16	520	0.046	b	b	b	b
dichloro-methane	393	0.08	525	0.082	c	-	533	0.11
2-propanol	391	0.01	550	0.018	b	b	b	b

a)due to the weakness of the emission, it is not possible to determine the wavelength accurately. b)not measured. c)not detected.

TABLE II. Fluorescence Characteristics of 1b and 2b.

	1b				2b			
Solvent	λ_M(nm)	Φ_M	λ_E(nm)	Φ_E	λ_M(nm)	Φ_M	λ_E(nm)	Φ_E
n-pentane	389	0.070	502	0.14	390	0.020	525	0.43
n-hexane	390	0.078	502	0.14	a	a	a	a
Me-cyclo-hexane	391	0.094	504	0.070	392	0.022	525	0.27
n-hexa-decane	391	0.10	505	0.074	393	0.023	525	0.22
ether	a	a	a	a	391	<0.01	565	0.03
ethyl acetate	392	0.073	b	b	b	-	604	<0.004
diethyl succinate	392	0.080	b	b	a	a	a	a
dichloro-methane	395	0.027	b	b	a	a	a	a

a)not measured. b)not detected.

Although we were not successful in detecting the anthracene radical anion absorption at 700nm [10], we monitored the decay of the anthryl group fluorescence as well as the rise of exciplex fluorescence of both 1b and 2a by kinetic spectroscopy. The results are tabulated in TABLE III.

TABLE III. Fluorescence Decay and Rise of 1b and 2a.

Compound	Solvent	τ(decay,420nm)	τ(rise,530nm)
1b	n-pentane	0.60±0.04ns(a)	1.00±0.10ns(a)
	Me-cyclohexane	1.09±0.01ns(b)	1.63±0.07ns(b)
	n-hexadecane	1.67±0.02ns(b)	2.30±0.08ns(b)
2a	n-pentane	0.80±0.02ns(c,d)	0.84±0.08ns(c,d)
	Me-cyclohexane	1.10±0.15ns(d)	1.10±0.15ns(d)
	n-hexadecane	1.90±0.20ns(d)	1.90±0.20ns(d)
	ether	0.51±0.05ns(c)	0.65±0.06ns(c)

a)The value represents the average and range of 5 determinations. b)The value represents the average and range of 3 determinations. c)The value represents the average of 2 determinations. d) from Yang et al. [6].

Effect of Chain Length

In comparing 1a and 2a, we note that there is not only an appreciable increase in the quantum efficiency of the anthryl monomeric emission, Φ_M, but also an appreciable decrease in quantum efficiency in the exciplex emission, Φ_E. The results clearly indicate that the increase in chain length reduces the rate of exciplex formation. It is also interesting to note that due to the weakness of the intramolecular exciplex emission from 1a in hydrocarbons, it is not possible to determine the position of this emission accurately while it is a predominant process from 2a, and the exciplex from 1a is generally blue shifted from that of 2a. Such a blue shift has been noted previously in the intramolecular excimer formation [4]. The latter observation suggests that the increase in chain length from 3 to 7 may introduce some strain into the system similar to the medium-sized rings in alicyclic chemistry. A similar trend is also noted in comparing 1b with 2b.

Effect of an Electron-Releasing Group on Anilino Donor

Since no exciplex formation is detectable from 1a in n-hexane, while it is an important mode of action of excited 1b, the p-methoxy group promotes the intramolecular exciplex formation in hydrocarbon solvents. By comparing compounds in the b series with those in the a series, we found that the Φ_M is quenched to a larger extent in the b series. Since no exciplex formation is noted from the b series in solvents more polar than ether, the results indicate that an efficient competitive process has occured. Since this process is more efficient in a more polar solvent and the anisidino group in b is a better electron donor than the anilino group in a, a rational explanation is that this process may be the electron transfer from the donor to the excited anthryl group.

Viscosity Effect

It is known that solvent viscosity has a marked effect on the intramolecular exciplex formation from 2a in hydrocarbons [6]. In this case, going from n-pentane to n-hexadecane, Φ_M increases by a factor of approximately 2. On the other hand, viscosity seems to have very little effect on 2b, Φ_M does not vary from the range of experimental error. Although the solvent viscosity seems to play a role in Φ_M in 1a in hydrocarbons, the effect seems to level out in more viscous hydrocarbons, e.g., methylcyclohexane vs. n-hexadecane. Similarly, in more polar solvents, such as esters, the viscosity seems to have a minimal effect on Φ_M on both 1a and 1b. These results clearly indicate that there is a mode of decay in the excited state of these compounds which is independent of the solvent viscosity.

DISCUSSIONS

We chose these bichromophoric molecules containing a 9-anthryl and a tertiary anilino group linked by a polymethylene chain for our investigation because of (a) their well known photophysical and photochemical behaviors [1,11] and the facility and versatility of their synthesis [6,7]. In the current investigation, we studied the effect of chain length, from three methylene units to seven methylene units, and the effect of an electron-releasing substituent, p-methoxy group, on the anilino donor on the intramolecular exciplex formation in these bichromophoric molecules.

Since the fluorescence behavior of 9-alkylanthracenes is not appreciably modified by the alkyl substituent [12], we may assume that the changes in fluorescence properties are the results of the intramolecular interaction of the excited group with the anilino group, which result in either exciplex formation and/or electron transfer. For the sake of simplicity in our discussion, we assume that the intramolecular exciplex formation is irreversible [6]. The rate expressions are given below:

$$\text{An—N(Me)Ar} \xrightarrow{h\nu} \text{An*—N(Me)Ar} \qquad (1)$$

$$\text{An*—N(Me)Ar} \xrightarrow{1/\tau} \text{An—N(Me)Ar} + h\nu' + \text{heat} \qquad (2)$$

$$\text{An*—N(Me)Ar} \xrightarrow{k_r} \text{exciplex and/or ion pair} \qquad (3)$$

Therefore, the quantum yield of the monomeric anthryl emission, Φ_M, may be given the following expression, where Φ is the fluorescence quantum yield of the reference compound,

$$\Phi/\Phi_M = 1 + k_r\tau \qquad (4)$$

and the decay rate of the monomeric emission, k_d, is,

$$k_d = 1/\tau + k_r \qquad (5)$$

Although the acyclic bichromophoric molecules used in our investigation are easier to synthesize than their rigid analogs, a complexity in analyzing their excited state dynamics is the participation of the chain dynamics, i.e., the molecule may undergo conformational changes during the lifetime of the excited state. Since the rotational barrier of the 2,3-bond of n-butane is 4.5kcal·mole^{-1} in the gas phase, the rate of rotation of an alkyl chain in solution is in the order of 1x10^{9}s^{-1} (the rate of a first order reaction with an activation energy of 5 kcal·mole^{-1} is 1.33x10^{9}s^{-1}) which is higher than the decay rate of an excited 9-alkylanthracene [12].

Possible Role of Electron-Transfer in Exciplex Formation

In summary, we have noted that the electron releasing p-methoxy group exerts a marked influence on the excited state behaviors of these compounds. In particular, we noted that k_r of 2b in hydrocarbons, 3x10^{9}s^{-1}, is not only higher than those of 2a, 3.9-9.9x$10^8 s^{-1}$, but also is independent of the solvent viscosity. On the other hand, the Φ_E of 2b decreases markedly from 0.43 in n-pentane to 0.22 in n-hexadecane while of that of 2a does not. Since exciplex formation from 2a follows the previously suggested mechanism which is restricted by chain dynamics (equation 3), and we may assume that 2a and 2b will have similar chain dynamics in their excited state, the results clearly indicate that there is a different mechanism for the formation of exciplex from excited 2b. The results strongly suggest that electron transfer may have taken place between the anisidino donor and the photoexcited anthryl group. When such a process takes place in a hydrocarbon solvent in which the resulting radical ion pair is at a higher energy state than the corresponding exciplex, the radical ion pair 3 may be the intermediate in the exciplex formation (equation 6). The quenching of monomer fluorescence via electron transfer to intermediate 3 as the first step will be relatively independent of the solvent viscosity, while the collapse of 3 to give the exciplex will then be dependent on solvent viscosity.

$$An^{*}\!-\!\!-\!\!-\!N(Me)Ar \xrightarrow{k_r} \underset{\textstyle 3}{An^{\dot{-}}\!-\!\!-\!\!-\!N^{\dot{+}}(Me)Ar} \xrightarrow{k_{ex}} \text{exciplex} \qquad (6)$$

Such a suggestion would also account for the abnormally high quenching rates, which exceed the diffusion-controlled limit, of anthracene fluorescence by aniline derivatives in viscous hydrocarbons previously noted in our laboratory. We wish to emphasize, however, radical ion pairs as intermediates in intramolecular exciplex formation from a photoexcited aryl group and an amino donor has been suggested independently by Verhoeven and his coworkers. They named such a process, the "Harpooning Mechanism" in exciplex formation [13]. Their proposal was based on the wavelength dependence in the intramolecular exciplex formation of bichromophoric naphthyl systems in a supersonic jet and its substituent effect.

In an effort to detect the possible presence of the anthryl radical anion from the decay of these bichromophoric molecules in their excited state by its unique absorption around 700nm [10], we applied kinetic spectroscopy to study the behaviors of their excited state. Since 2b forms an exciplex highly efficiently and since we were concerned about the possible interference of both exciplex absorption and exciplex fluorescence from 2b on the detection of anthryl radical anion, the excited state of 1b was investigated first. We not only attempted to detect the anthryl radical anion directly by its absorption, but also analyzed the rate of decay of the monomeric anthryl fluorescence as well as the rate of rise of exciplex fluorescence. The data were compared with those from 2a as a reference. The results are given in TABLE III.

In comparing the decay rate of the monomeric anthryl fluorescence with the rate of exciplex formation in 2a, there was no measurable time lag between these two events. This suggests that the exciplex is derived directly from the decay of photoexcited 2a without an intermediate. Since there is a continuous decrease in k_d for excited 2a with a corresponding increase in the viscosity of the hydrocarbon solvent, this observation indicates that the exciplex formation from 2a is dependent on the bond rotation of the methylene chain in order to orient the components of the exciplex into an appropriate conformation for its formation. On the other hand, there is consistently a small but measurable time lag between the decay of excited 1b and the rise of its exciplex fluorescence, the result is consistent with our suggestion that there is an intermediate in the formation of exciplex from photo-excited 1b, and such an intermediate may be the radical ion pair 3 resulted from the electron transfer in excited 1b (equation 6) However, three trends are also noticeable in the kinetic data from 1b as the viscosity of the solvent increases from n-pentane (η = 0.225 cp at 20°) to methylcyclohexane (η = 0.72cp at 20°) to n-hexadecane (η = 3.44cp at 20°) [14]. These are: a) the decay time of monomer fluorescence increases from 0.60ns to 1.09ns to 1.67ns, b) the rise time of exciplex fluorescence increases from 1.00ns to 1.63ns to 2.30ns, and c) the lag time increases from 0.40ns to 0.54ns to 0.63ns respectively. These results suggest that bond-rotations or conformational changes may also play a role in these events. These results are consistent with an explanation that excited 1b must undergo a conformational change in order to bring the excited anthryl group and the anisidino group closer together for the formation of radical ion 3 which undergoes a further change for the formation of the exciplex.

According to equation 6, the steady state concentration of the radical ion 3, if it is formed, is $k_r \cdot I/k_{ex}$ where I is the intensity of exciting light (assuming the exciplex formation is not reversible). Judging from the small time lag between the monomer decay and exciplex rise, the steady state concentration is not appreciably higher than the exciting light intensity. Therefore, it is not surprising that we have failed to detect the anthryl radical anion absorption in our current investigation. Our experience will enable us to modify our current approach to increase the transient concentration of 3 in order to improve the probability of its detection.

ACKNOWLEDGEMENT

The work at the University of Chicago was supported by a grant from the National Science Foundation, and the work at the Argonne National Laboratory was supported by the Division of Chemical Sciences, Office of Basic Energy Sciences, U.S. Department of Energy, under Contract W-31-109-Eng-38. The authors also wish to thank Professor S. Tazuke and Dr. K. A. Zachariasse for some valuable discussions.

REFERENCES

1. H. Beens and A. Weller in: Organic Molecular Photophysics, Volume 2, J. B. Birks, ed. (Wiley, London 1975), pp. 159-215.
2. N. C. Yang, R. Gerald, and M. R. Wasielewski, J. Am. Chem. Soc., 107, 5331-5332 (1985).
3. N. C. Yang and Z-H. Lu, Tetrahedron Letters, 25, 475-468 (1984).
4. K. Zachariasse and W. Kuehnle, Z. Phys. Chem. N.F., 101, 267-276 (1976).
5. K. Zachariasse, unpublished results.
6. N. C. Yang, S. B. Neoh, T. Naito, L-K. Ng, D. A. Chernoff, and D. B. McDonald, J. Am. Chem. Soc., 102, 2806-2810 (1980).
7. D. W. Minsek, D. G. Johnson, and N. C. Yang, unpublished results.
8. M. R. Wasielewski, J. R. Norris, and M. K. Bowman, Faraday Discuss. Chem. Soc., 78, 279-288 (1984).
9. M. R. Wasielewski, J. M. Fenton, and Govindjee, Photosynth. Res., 12, 181-190 (1987).
10. R. Potashnik, C. R. Goldschmidt, M. Ottolenghi, and A. Weller, J. Chem. Phys., 55, 5344-5348 (1971).
11. N. C. Yang, D. M. Shold, and B. Kim, J. Am. Chem. Soc., 98, 8567-8596 (1976).
12. J. Rice, D. B. McDonald, L-K. Ng, and N. C. Yang, J. Chem. Phys., 73, 4144-4146 (1980).
13. R. M. Hermant, B. Wegewijs, J. W. Verhoeven, A. G. M. Kunst, and R. P. H. Rettschnick, Rec. Trav. Chim., 107, 349-350 (1988).
14. Landolt-Boernstein, Zehlenwerte und Funktionen, Band II, Teil 5, (Springer-Verlag, New York 1969) pp. 148-195.

PANEL DISCUSSIONS ON ELECTRON TRANSFER

Communicated by: J. R. Bolton[1] (Panel Chairman)

Panel Members: R. A. Friesner[2], S. F. Fischer[3], R. A. Marcus[4], M. Michel-Beyerle[5] and A. Scherz[6]

The topic of Electron Transfer Reactions has been of considerable interest since the Second World War when radioactive tracer techniques made it possible to study electron self exchange reactions. Modern theories of electron transfer (ET) are based on the pioneering work of Marcus, Levich and Dogonadze, Hopfield and Jortner which, along with other subsequent treatments, has established a solid theoretical base for understanding ET reactions. This theory has been widely applied with considerable success to organic, inorganic and more recently to biological systems. Nevertheless, there remain some important questions and processes which need answers and understanding. This is particularly true in regard to the ET reactions which occur in the reaction center of bacterial photosynthesis, since we now have detailed crystal structures of this remarkable solar energy converter. The availability of this structure has stimulated a feverish level of activity, both experimental and theoretical, aimed at elucidating the details of the *in vivo* ET reactions and how they relate to the structure of the reaction center. There has also been a flurry of activity in the design and construction of covalently-linked donor-acceptor molecules which mimic, to a greater or lesser extent, ET reactions in the *in vivo* structure. This work has led to a much better understanding of factors, such as energetics, distance, orientation, solvent, temperature and the nature of

[1]Photochemistry Unit, Department of Chemistry, University of Western Ontario, London, Ontario, Canada N6A 5B7.
[2]Department of Chemistry, The University of Texas at Austin, Austin, Texas 78712.
[3]Physik Department, Technische Universitat Munchen, D-8046 Garching, West Germany.
[4]Noyes Laboratory of Chemical Physics, California Institute of Technology, Pasadena, California 91125.
[5]Technical University Munich, Inst. für Physics und Theor. Chem., Lichtenbergstrasse 4, 8046 Garching, West Germany.
[6]Weizmann Institute of Science, Rehovot 76100, Israel.

Published 1989 by Elsevier Science Publishing Co., Inc.
Photochemical Energy Conversion
James R. Norris, Jr. and Dan Meisel, Editors

the bridge, which control electron transfer from a donor to an acceptor within a defined molecular structure.

The Panel was asked to address their remarks to one or more of three questions. The following summary Report is a condensation of the answers and discussion in relation to these questions.

1. In intramolecular photochemical electron transfer reactions what design criteria can be identified to maximize the ratio of the forward electron transfer rate constant from the excited state vs. the reverse electron transfer rate constant to the ground state?

Within the conventional theoretical framework for non-adiabatic ET reactions in which the ET rate constant is given by

$$k_{et} = (2\pi/\hbar)|V|^2 \text{FCWD}$$

where V is the electronic coupling coefficient and FCWD is the Franck-Condon Weighted Density of states, the maximum forward ET rate occurs when the reorganization energy λ is equal to $-\Delta G^o$, where ΔG^o is the standard Gibbs energy change in the ET process. Thus for a given singlet excitation energy U(S), the maximum ratio k^f_{et}/k^r_{et} of the forward to reverse rate constants will prevail when λ is made as small as possible. This result pertains because then the reverse ET reaction will have a very large negative ΔG^o and hence be strongly in the inverted region. Other factors which may affect this ratio favorably are:

(i) A larger V for the forward vs. reverse reaction - this may be achieved by either a superexchange mechanism in which there is a low-lying state accessible to the excited state which is not accessible to the radical-ion-pair (RIP) product state, or by a symmetry difference between the coupling of the excited state with the RIP state vs. the RIP state with the ground state.

(ii) A rapid relaxation of the RIP state which moves the donor and acceptor into a position which strongly reduces |V| for the reverse ET.

There was general agreement that the monomeric bacteriochlorophyll B in the photosynthetic reaction center is required to bridge the large distance (~17 Å) between the centers of the primary donor P (a special pair of bacteriochlorophyll molecules) and the bacteriopheophytin intermediate acceptor I (Fischer, Friesner, Michel-Beyerle, Jortner, Marcus). One possibility is that B plays a role as a superexchange mediator, either in the form of the virtual state $P^{+}B_{L}^{-}$ or as $B^{+}I_{L}^{-}$.

A series of five aromatic amino acids in the protein chain may also contribute to the superexchange coupling between P and I (Scherz and Norris). This chain was also shown to be in close contact with the porphyrin chain possibly leading to a strong coupling of states as well as stabilizing charge-transfer (CT) states. Such states were suggested to be of two kinds: an inner CT of P and a CT state between P and B (virtual in the superexchange theory). There appears to be little evidence for a separate CT state within the P* absorption band since P* carries the full oscillator strength and exhibits a change in dipole moment on excitation of only 8 Debye (Fisher). Another possible mechanism to be considered for predictions is a nondiabatic/adiabatic mechanism. The key unknown is the energy position of B^{-}. On the other hand Friesner argued that P* should be looked on as a mixture of two states, one of which is an internal CT state, but Fischer pointed out that such a model would give the "wrong" Stark spectrum (first derivative instead of second).

Still, one cannot rule out the possibility of having relatively fast (2-3 ps) ET from P* to B_L or from P* to $B^{+}I_{L}^{-}$ followed by hole transfer from $B^{+}I_{L}^{-}$, which may also go via superexchange. This would then be followed by ultrafast (<1 ps) ET from B to I (Fischer).

A suggestion was made that the first step (from P^{*} to $P^{+}I_{L}^{-}$) is so fast because this ET reaction may be *adiabatic* rather than *non-adiabatic* but with a relatively slow (~10 Å/ps) velocity for passage through the crossing region (Fischer). This would require an effective V ~20 cm^{-1} and protein relaxation lifetime of ~1 ps. This adiabatic pass-over could account for the "universality" of the rate of charge separation found in bacterial and green-plant systems (Photosystem II) (Fischer).

New data was presented on the ratio of the ET rate constant between the "active" L side and the "inactive" M side in the reaction center which indicates that this ratio is ≥20 at 80°K (Michel-Beyerle,Tiede). It was suggested that this differentiation may be due to the aromatic amino acids which are conserved on the L side but which are absent on the M side (Scherz, Norris, Schiffer), or due to different electronic couplings on the M-branch and the L-branch (Fischer, Michel-Beyerle).

A question was posed as to why a large energy loss (0.7 - 0.8) eV is incurred in forming stabilized charge separation in the reaction centers of both bacterial and green plant systems. One answer is that the lifetime of the final charge-separated state must be long enough (~10 - 100 ms) to couple effectively with the relatively slow biochemical electron-transport processes which occur in the membrane. A simple transition-state argument for a unimolecular reaction indicates that the activation barrier for the reverse reaction would have to be ~0.8 eV to achieve such a lifetime (Bolton).

2. *What is the role and importance of solvent dynamics in photochemical electron transfer reactions?*

Several recent theoretical treatments have shown that solvent relaxation can become a rate-limiting process when ET reactions become very fast. It appears that the solvent longitudinal relaxation time τ_L is the important relaxation time. There was some debate as to how important this effect is in the *in vivo* reaction center. Recent molecular dynamics simulations of Schulten, Karplus and coworkers indicate that the protein dielectric relaxation time is much faster than the first electron transfer step (Michel-Beyerle). If experimental evidence will show that the protein relaxation time is relatively slow, then the fast ET may be explained by a breakdown of the continuum theory (Jortner) or possibly a critical protein vibration may serve to overcome the need for "solvent" relaxation (Marcus, Fischer and others). It was pointed out that better experimental information on the thermodynamic properties of the reaction-center processes is needed, particularly estimates of ΔS. More work on temperature dependence is required (Gunner, Marcus).

3. *Is it possible to model quantitatively the role of the bridge in non-adiabatic electron transfer reactions?*

From studies of donor-acceptor model compounds covalently linked by various bridges, it is now clear that the bridge plays a very important role in mediating intra-molecular ET, perhaps by a superexchange mechanism or via some other route. Some theoretical treatments have been carried out to address this point using semiempirical methods such as extended Hückel and INDO-CI methods.

A question was posed as to the effect of the "gaps" in the pathway for ET in proteins (Closs, Scherz). It would be useful to have model compounds where a donor and acceptor are covalently linked by arching bridges which would allow the insertion of various molecules by Van der Waals contact between the donor and acceptor (Closs, Scherz).

It is clear that there is a need to develop more sophisticated models which will allow quantitative predictions of the effect of various molecular groups in the bridge. Perhaps a "conductivity coefficient" could be established for each group such that the effect of any given bridge could be estimated by adding (or multiplying?) the conductivity coefficients. The ultimate goal would be the design of efficient "molecular wires". Here it may be useful to focus on models for the coupling factor V, rather than on FCWD. One such system where the kind of spacer between the same donor and acceptor is changed was shown (Bolton) to result in a dramatic change of $\sim 10^3$ in the rate constant from a totally aliphatic bridge to an aromatic bridge.

How can the various theories for ET be tested *in vivo*? A major difficulty in the various theories is the *in vivo* verification. It must be shown that the various phenomena are invariant among species. It has been noted that *R. sphaeroides* has a somewhat different location for the primary components, yet similar rates to other reaction centers. Future work should focus on mutations (Scherz), the perturbation of states by electric fields (Boxer) and the use of Raman spectroscopy (Friesner).

GUEST-HOST INCLUSION COMPOUNDS AS SPECTROSCOPIC MEDIA: A NEW ORGANIZED ENVIRONMENT

DAVID F. EATON*, JON V CASPAR and WILSON TAM, E. I. du Pont de Nemours and Company, Central Research and Development Department, Experimental Station, Wilmington, Delaware 19898 USA

Contribution No. 4714

INTRODUCTION

A unifying characteristic of many systems which are investigated for conversion of incident solar irradiation into useful chemical energy is the high degree of organization exhibited by the entity. Arrays of organized chromophores often act to gather the light; energy acceptors disposed at specific distances and orientations with respect to the antenna chromophores then act to channel the solar energy to a specific site where redox or other chemical changes occur. Nature, and man, have devised highly organized systems to effect solar energy conversion.

Many types of organized systems have been studied to assess the influence of chromophore ordering on the photophysical properties of molecules that are of interest in solar conversion (or in the use of photons to do other useful work, such as photoelectron generation for xerography, or nonlinear optical conversion of one frequency of light to another). The degree of order imposed on the contents of the system can range from very low to extremely high. Examination of energy transfer among chromophores in solution or frozen glasses where the arrangement is isotropic represents the total lack of any system order. However, if the chromophores are dispersed in a polymeric glass such as poly(vinyl alcohol) and the film stretched to elongate it in one direction, then a small degree of anisotropy can be introduced into the system. At higher levels of imposed order are micelles, bilayers, liquid crystal media, monomolecular Langmuir-Blodgett films, mixed crystals and other single crystal configurations. Incorporation of chromophores into these environments has been reported by many workers, and valuable information has been gathered about energy transport and localization in these systems. [1]

We have used inclusion chemistry to prepare ordered complexes of organic and organometallic chromophores. In previous studies we have reported the use of cyclodextrin complexation as a technique for capturing select molecular conformations within the inclusion host so that the photophysics of the included multichrompohoric species can be examined at leisure, without care that conformational equilibria which interconvert various configurations can complicate the study. [2] In this

Photochemical Energy Conversion
James R. Norris, Jr. and Dan Meisel, Editors

way we were able to unravel the complicated photochemistry and photophysics which ocurs on excitation of a trichromophoric species, bis(α-naphthylmethyl)-1,3-dithiane.[3] In another series of studies we used the paradigm of guest-host inclusion complexation to prepare polar, crystalline materials in which the guest was ordered such that the bulk materials were capable of frequency doubling incident laser light, a useful nonlinear optical property of hyperpolarizable, acentric solids. [4] In the present paper, we report on the use of guest-host inclusion complexes as a medium in which to study the spectroscopy, especially fluorescence, of organic solids. Unlike other solid media which can be used as spectroscopic media, such as Sh'polski matrices, or doped molecular crystals, where the concentration of emissive guest is usually extremely low, in crystalline, stoichiometric, inclusion complexes the emitting material is present at high levels, up to 30-40% by weight of the solid. Also, since in principle the molecular three dimensional structure of the inclusion compounds can be solved by x-ray diffraction techniques, the spatial relation among chromophores, and between the chromophore guest and the matrix host can be known with precision. Thus we anticipate being able to study chromophore-chromophore (excimer, exciton) interactions as well as chromophore-matrix (exciplex, charge transfer) interactions in these solids.

In the following sections, we first review in a general way the subject of guest-host inclusion complexes. Then we present data on the preparation, characterization and emission of a series of simple aromatic chromophores included in host matrices of thiourea, tris-*ortho*-thymotide, or hexakis-(*p*-*t*-buylbenzylthio)benzene. We compare and contrast the data in the inclusion complexes with the emissive properties of the compounds themselves as pure solids or in solution.

INCLUSION COMPLEXES

A wide variety of molecules are capable of including other molecules within their crystalline network on co-crystallization from a suitable solvent. The new materials are true molecular solids, not merely examples of solid solutions of guest in host. That is, the inclusion complexes possess three dimensional order, with the guest situated periodically on specific site(s) within the three dimensional lattice. One feature of these solids is ability to construct situations in which the environment surrounding the guest molecule is controlled in some specific manner. By taking advantage of this aspect of inclusion complexation, chemists are able to devise models for complex biological and physical phenomena and study them under controlled conditions.[5]

Many species are capable of acting as hosts for molecular guests. Crowns, cryptands and cyclodextrins can function as molecular wastebaskets into which a chemist may toss a favorite molecule. The restricted space within these rings and cavities can act to select specific conformations of a molecule from many that may be accessible in solution. The specific complexation and selection of one of many conformers can be used advantageously by the chemist to study the properties of that conformer in relative ease. Breslow [6] has used this technique to organize reagents for reaction, effectively catalyzing the solvolysis of an ester to mimic enzymic processes. We have employed the same method to organize photochemically active molecules within a CD cavity.[2,3] In this way we have shown that the radiative and radiationless paths available for decay of an electronically excited molecule can be manipulated and directed to follow complexation dependent pathways. This represents a powerful method in principle to direct photochemistry along desired pathways. Another medium which can be used to direct such photochemical reactivity is the cage structure within zeolites, the aluminosilicate, three dimensional analog of the single ring cyclodextrin. Turro [7] has pioneered in the use of zeolites to control the pathway and the timing of photochemical processes. We have recently provided another example of size restriction directing reactivity within a zeolite, and in addition shown that the presence of metal ions within the zeolite cage can be used to further influence photoreactivity in subtle, but predictable ways.[8]

A final series of molecules that have been discovered by chemists to act as hosts for molecular guests are the clathrate formers, such as urea, thiourea, tris-*ortho*-thymotide, deoxycholic acid, and many other organics, as well as inorganics such as Werner complexes (MX_2A_4, where M is a divalent cation, X is an anionic ligand, and A is an amine) and Hofmann type clathrates, and organometallics such as cyclophospazenes. These materials have been extensively studied and reported on in reviews. [5] They form van der Waals solids with a variety of guests. The molecular solids possess void spaces (channels, cavities, cages, lattices) in three space that can be occupied by guests which conform to the spatial dimensions of the cavity present in the structure. Our present work employs this class of inclusion complexes.

Clathrate inclusion complexes can be formed with a wide variety of guests, and the final three dimensional structure of the complex is established in a cooperative way between the host and guest, unless structural constraints present in the host dictate structure on the complex. Such a situation arises in the urea and thiourea clathrates, where hydrogen bonds among the urea units imposes limits on the size of the channel structure available for complexation of a guest. No such limitations are present in clathrates prepared from many other compounds, such as tris-*ortho*-thymotide (TOT). In these systems, the structure can expand or contract to accommodate guests of many different sizes. A similar flexibility obtains with the so-called hexahosts, hexa-substituted benzene derivatives originally prepared by MacNicol and coworkers.[9] In these materials, the substituents act as arms on the "hexapus" to enfold the guest(s) in spatial positions above and below the plane of the aromatic nucleus, forming cavities of variable size which encapsulate the guest using only van der Waals forces to bind the complex.

We describe complexes of aromatic chromophores with TOT, thiourea and with the aromatic hexahost hexa(*p-t*-butylphenylthiomethyl)benzene.

EXPERIMENTAL

Materials.

TOT was prepared according to Green and coworkers.[10] The hexahost, hexa(*p-t*-butylphenylthiomethyl)benzene, was prepared following MacNicol.[9] All aromatic chromophores were commercial materials, used without special prior purification.

Inclusion complexes were prepared as described below. All complexes were analyzed by elemental and spectroscopic methods, especially NMR, to determine stoichiometry. In some cases powder x-ray methods were used to prove that the products were not simply admixtures of the ingredients.

TOT and Anthracene. 150 Mg of TOT was added to 50 mL of methanol and the mixture heated to solution. The solution was filtered hot and cooled to room temperature. To this solution was added 100 mg of anthracene and the mixture heated for solution and again filtered hot. The solution was cooled to room temperature and then allowed to evaporate. Thus 73 mg of the inclusion complex were obtained and determined to be a 3 TOT : 2 anthracene complex by elemental analysis. Found: C: 78.72; H: 6.44; calculated for 3 TOT: 2 anthracene: $C_{127}H_{128}O_{18}$: C: 78.53; H: 6.64.

TOT and 9-Methylanthracene. 9-Methylanthracene (150 mg) was added to 40 mL of methanol and heated to dissolve. To this warm solution was added 150 mg of TOT. The solution was filtered and allowed to evaporate at room temperature. After several days, crystals of the inclusion compound were filtered. Elemental Analysis: Found: C: 77.48; H: 6.82; calculate for 2 TOT:1 9-methylanthracene: $C_{81}H_{84}O_{12}$: C: 77.86; H: 6.78. Obtained 46 mg of the inclusion complex.

TOT and 2-Methylanthracene. 150 Mg TOT and 200 mg of 2-methylanthracene were placed in 50 mL methanol and heated to solution and filtered hot. Upon cooling 101 mg of a solid was isolated which was a mixture of complex and the starting guest. The filtrate was allowed to evaporate and 64 mg of the complex was isolated. ^1H NMR indicated the complex was 1:1 guest to host.

TOT and 2-Methylnaphthalene. 150 Mg of TOT was dissolved in 50 mL of hot MeOH and the solution filtered hot and then cooled to room temperature. 400 Mg of 2-methylnaphthalene was added. Slow evaporation of the solution yielded 100 mg of 1:1 TOT to 2-methylnaphthalene as determined by ^1H NMR in CD_2Cl_2.

TOT and 2,3-Dimethylnaphthalene. The procedure was the same as for 2-methylnaphthalene except 300 mg of 2,3-dimethylnaphthalene was used. Thus 89 mg of 2 TOT : 1 2,3-dimethylnaphthalene was obtained. The molar ratio of host: guest was determined by integration of the ^1H NMR spectrum in CD_2Cl_2.

TOT and 2,6-Dimethylnaphthalene. The procedure was the same as above but 300 mg of 2,6-dimethylnaphthalene was used to yield 82 mg of the 2 TOT : 1 2,6-dimethylnaphthalene inclusion complex. Molar ratio was determined by 1H NMR spectrum in CD_2Cl_2.

Hexahost and 1-Methylnaphthalene. This complex was prepared according to the literature.[9] 300 Mg of the host was added to about 3 mL of 1-methylnaphthalene and the mixture heated until dissolved. The solution was filtered hot and allowed to cool to room temperature. 327 Mg of the 2:1 guest: host inclusion complex were obtained. Elemental analysis : Found: C: 79.17, 78.94; H: 7.61, 7.87; calculated for 2 guest: 1 host: $C_{94}H_{110}S_6$: C: 78.83; H: 7.74.

Hexahost and 2-Methylnaphthalene. This complex was also prepared according to the literature.[9] 150 Mg of the host and 150 mg of 2-methylnaphthalene were heated until dissolved and filtered hot. After cooling to room temperature, the inclusion complex eventually precipitated from solution. Thus obtained was 88 mg of the 2:1 guest : host complex.

TOT and Pyrene. The procedure was the same as above but 150 mg of pyrene was used. Thus obtained was 243 mg of 1:1 TOT/pyrene inclusion complex. Elemental analysis: Found: C: 79.96; H: 6.48; calculated for 1 TOT: 1 pyrene: $C_{48}H_{46}O_6$:C: 80.20; H: 6.45. In another preparation, excess molar amounts of pyrene were used. Admixtures of 1:1 complex and pyrene were presumed to be formed in this case.

Hexahost and Pyrene. 150 Mg of the host and 150 mg of pyrene was added to about 3 mL of mesitylene; the mixture was heated until a solution was obtained. It was then filtered hot and allowed to cool to room temperature. The solution was allowed to evaporate. 49 Mg of the inclusion complex were obtained. A second preparation gave 40 mg of the inclusion complex which was determined to be a 1:1 host: guest complex by ^1H NMR.

Spectroscopic Methods. Emission spectra were determined using a SPEX Industries Fluorolog 222 Fluorimeter with photon counting detection and were corrected for detector sensitivity. Excitation spectra were measured on the same instrument using a quantum counter (rhodamine B in glycerol) to correct for intensity variations in the lamp (450 W xenon arc). Survey spectra were taken on a SPEX Fluorolog 212 single beam instrument and were also corrected.

Samples used in measurements of low temperature spectra were mounted in sapphire cells sealed with indium gaskets. Temperature control was achieved using either an Oxford Instruments model CF1204 Continuous Flow Cryostat (T<77K) or an Oxford Instruments model DN1704 Cryostat (T> 77K). In either case temperature stability was maintained to better than +/- 0.3 K using an Oxford Instruments model 3120 Temperature controller.

Emission lifetimes were determined using single photon counting techniques which have been described.[11]

RESULTS AND DISCUSSION

In the discussion which follows, we note that none of the complexes prepared by us has been characterized structurally. The discussion is therefore necessarily empirical, though comparisons will be made to molecular and polycrystalline spectra of the guests when available. Further work will allow us to make more definitive statements about the spectra we have measured to date.

Hosts. Solid TOT fluoresces in the region 330 nm when excited at 260 nm. The lifetime of the emission is very short (250-500 ps measured by single photon counting). The excitation spectrum maximized at 230 nm. Similarly, solid hexahost emits at 330 nm with short lifetime (225-380 ps).

Anthracene Complexes. Complexes with a variety of anthracene derivatives were prepared for examination because so much prior work had been done on the emission spectroscopy of this class of materials in a wide range of media. Figure 1 shows fluorescence spectra of anthracene, anthracene mixed physically with one mole of TOT and the 1:1 stoichiometric complex of anthracene and TOT. Anthracene is a "type A" solid in the Stevens nomenclature.[12,13] That is, the fluorescence is structured, nearly molecular in bandshape, and nearly mirrors the absorption spectrum. This is because in the crystal structure, neighboring molecules are either not parallel (in anthracene neighbors are edge-to-face oriented with a 55° cant to one another), or are far apart (low overlap). In anthracene the so-called "55° dimer" dominates the emission at long wavelengths. Further, anthracene is notorious for exhibiting self-absorption effects in its fluorescence spectra. As a solid, anthracene emission is skewed by self-absorption, and only a weak, "solid state shifted" 0-0 band is seen. The solid is in fact excitonic as proved by the lifetime of the emission, but short wavelength emission originates from defect sites that are molecular in nature rather than delocalized. The emission spectrum of an intimate physical mixture of anthracene and TOT exhibits only anthracene emission. The complex, however, gives emission that is considerably different in band shape, if not position, from solid anthracene. The high energy bands are enhanced relative to the low energy bands, but the overall emission is quite similar to anthracene microcrystals themselves. We conclude that the effect of complexation is to situate molecules far enough apart that they behave more like isolated anthracene than " 55° dimer" anthracene. However, at some sites (possibly at surface defects), anthracene pairs exist which are similar to those formed in microcrystals, so a portion of the emission still persists at long wavelength.

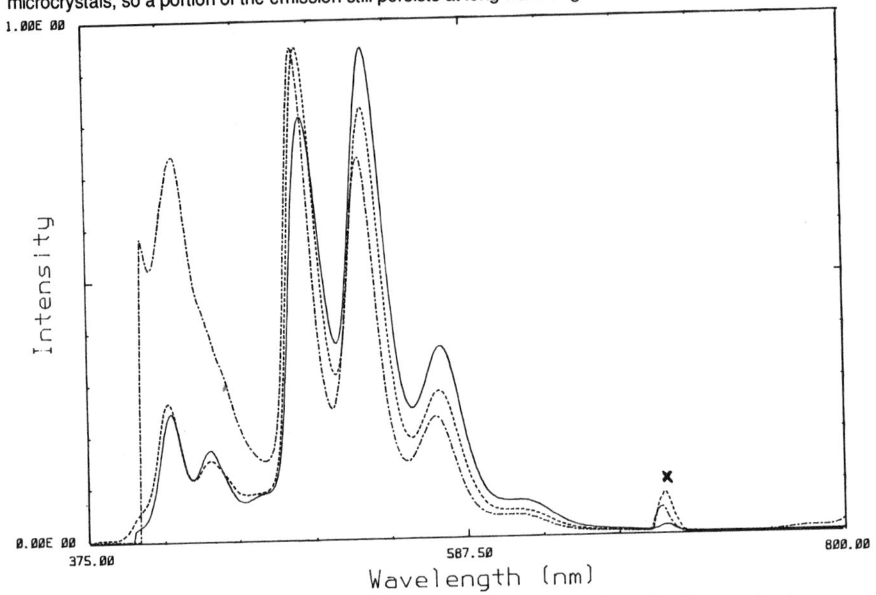

FIG. 1. Emission of solid anthracene (----), an intimate mixture of TOT and anthracene (- - -) and the 1:1 TOT:anthracene complex (-· - · -). Excitation 350 nm. Features marked X are second order scatter. The anomaly near 400 nm in the TOT:anthracene spectrum is an instrument artifact; a peak at 407 nm is present in that region (see Table I).

Complexation of TOT and 9-methylanthracene (9-MA) produces quite a different result than the simple anthracene system. Figure 2 shows spectra of 9-MA as microcrystals, and as the crystalline 1:1 complex with TOT. Crystalline 9-MA, which is a Stevens "type B" solid [12], is one in which strong overlap obtains between neighbors, leading to excimer emission observed as a broad, structureless feature at long wavelength. The TOT complex, on the other hand, clearly exhibits emission which is molecular in appearance. The arrangement of molecules within the complex must not be face to face. Nearest neighbor interaction can occur, however, as the lifetime of the complex emission is longer (6 ns) than that of isolated 9-MA in solution (4.2 ns in cyclohexane), though shorter than excimeric 9-MA (26 ns in the cyrstal; 200 ns in the Chandross sandwhich photodimer excimer.[14] This lifetime suggests weak excitonic interaction may occur among 9-MA units in the complex which leads to extension of the lifetime beyond the solution molecular value via multisite hopping before trapping at a defect site.

The situation in 2-methylanthracene (Figure 3) may be intermediate between anthracene and 9-MA. A broad emission is seen as well as some structure emission. It is possible that both molecular and dimeric sites exist within the lattice structure.

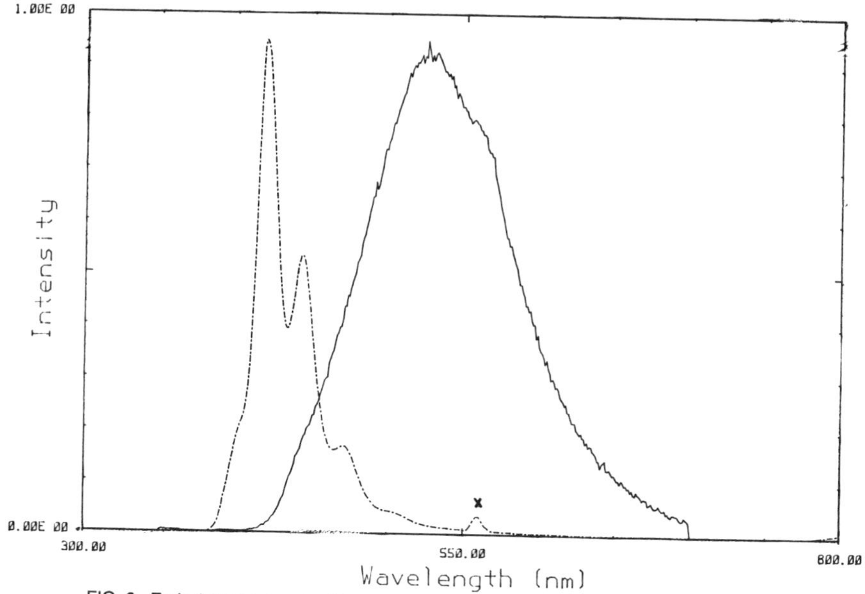

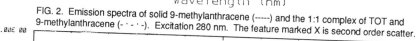

FIG. 2. Emission spectra of solid 9-methylanthracene (-----) and the 1:1 complex of TOT and 9-methylanthracene (- · - · -). Excitation 280 nm. The feature marked X is second order scatter.

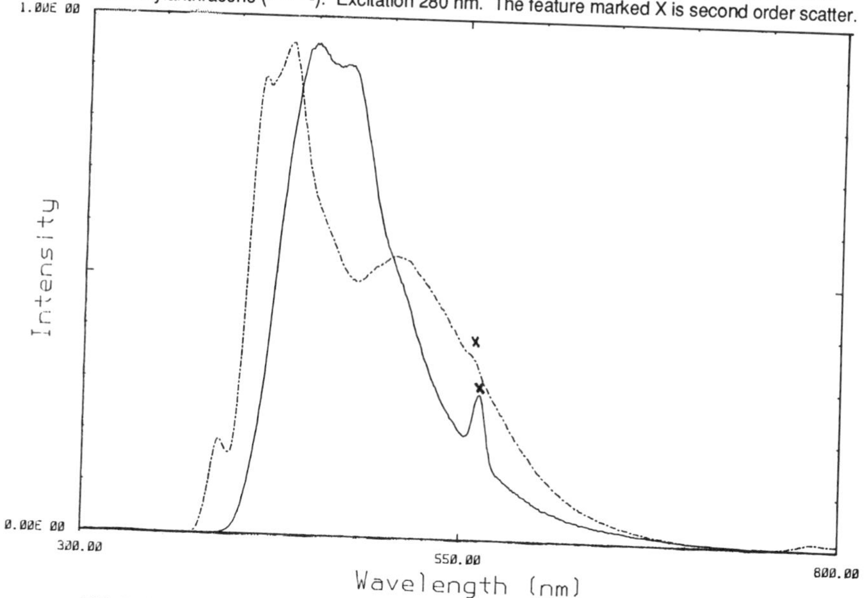

FIG. 3. Emission spectra of solid 2-methylanthracene (-----) and the 1:1 complex of TOT and 2-methylanthracene (- · - · -). Excitation 280 nm. The features marked X are second order scatter.

Table I lists the anthracene complexes examined and their spectral data.

TABLE I. Emission Maxima of Solid Anthracene Complexes.

Complex	Wavelength of Maxima (nm)	Lifetime(s)[a]
Anthracene·TOT	407, 425, 493, 529, 571	
9-Methylanthracene·TOT	405 (sh), 419, 444, 480	6 (1.30)[b]
2-Methylanthracene·TOT	391(sh), 417, 434, 505 (brd)	

a. In ns; see text for discussion b. Chi squared value for fit

Naphthalenes. Several naphthalene derivatives were complexed with various hosts. Table II lists complexes and spectral maxima. 2-Methylnaphthalene was complexed with TOT. Three separate preparations of this 1:2 complex were made. All showed identical spectral behavior. The emission is monomer like, with moderate structure in the 320-370 nm range. The lifetime of the emission was cleanly two component (680 ps, 98% and 13.1 ns, 2%) when excited at 260 nm, where both 2-MN and TOT absorb. Since the lifetime of 2-MN in hexane is long (59 ns) we conclude that either some neighbor-neighbor interaction exists in the complex or that the emission is "quenched" by interaction with the host matrix.

2,3-Dimethylnaphthalene was complexed with both TOT and thiourea. Emission spectra were monomer like (maxima near 329 and 339 nm in TOT; 331 and 338 for thiourea), but also exhibited a strong phosphorescence component at longer wavelength (structured emisison at 380, 403, 453 and 480 nm). The phosphorescence was stronger in TOT than in thiourea. The differences may only reflect changes in oxygen permeability through the matrix crystals, or they may reflect matrix induced intersystem crossing effects. 2,6-Dimethylnaphthalene included in TOT also showed phosphorescence in addition to monomer like fluorescence, but the phosphosescence was minor. Finally, 1,4-dimethylnaphthalene included in hexahost exhibited only monomer emission and showed no phosphorescence component. (Table II)

A markedly different emission behavior was observed for the hexahost complex of 1-methylnaphthalene, a complex that contains 2 moles of 1-MN per host molecule. This material exhibited dual emission (Figure 4). Structured, monomer-like emission was observed in the range 320-370 nm, and a broad emission, centered at 420 nm was also present. These emissions each produced the same excitation spectrum: weak excitation near 320 nm (1-MN excitation locus) and strong excitation peaking at 235 nm (the TOT maximum). The lifetime of the emissions each was two component, with approximately the same value independent of wavelength of observation, but differing weights: at 330 nm, τ_1 = 1.6 ns (88%), τ_2 = 5.6 ns (12%); at 420 nm, τ_1 = 1.4 ns (64%), τ_2 = 6.8.ns

(36%). The lifetimes were not kinetically coupled, but appear to be independent and both arise within the pulse width of the excitation source. We tentatively conclude that two sites must exist in this complex, one of which is occupied by dimer like pairs of molecules, which give rise to the long wavelength emission; and a singly occupied site (or one in which the 1-MN molecules are not near

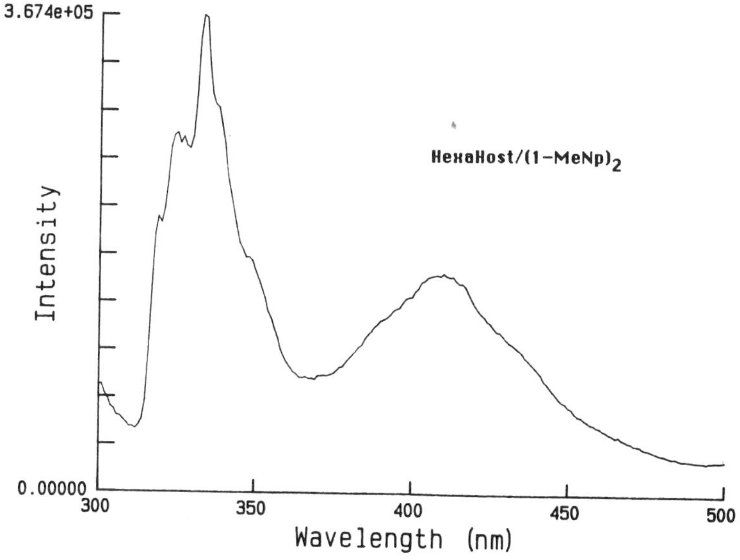

FIG. 4. Emission spectrum of the 1:2 complex of hexahost and 1-methylnaphthalene. Excitation is 280 nm.

enough or oriented correctly for excimeric interaction) which gives the short wavelength emission. The lifetimes, however, suggest that the sites communicate, on a time scale short with respect to their emission lifetimes. Only in that case can the same lifetimes be observed at each monitoring wavelength. The communication may occur through the host matrix. Similar dual banded emission was observed for the hexahost:$(2\text{-MN})_2$ complex (Table II), though the emission appears at longer wavelengths. Experiments currently in progress suggest that the spectral shape depends on excitation wavelength, so that the dual emission may result from excitation of two distinct absorbing species. Energy transfer from host to guest may occur on excitation at short wavelengths. Further data is required to firmly establish these possibilities.

TABLE II. Emission Maxima of Solid Naphthalene Complexes.

Complex	Wavelength of Maxima (nm)	Lifetime(s)[a]
2-Methylnaphthalene·TOT	328,335, ~348 (sh)	0.68, 98 % 13.1, 2 % (1.01)[b]
2,3-Dimethylnaphthalene·TOT	329,339, ~355 (sh) 381, 403, 453, 480[c]	
2,6-Dimethylnaphthalene·TOT	~330 (sh), 341, ~353 (sh) 392, 416, 448[c]	
1-Methylnaphthalene·HH[d]	317, 325, 331, 410 (brd)	1.6, 88 % 5.6 12 % (1.86)[b,e] 1.4 64 % 6.8, 34 % (1.32)[b,e]
2-Methylnaphthalene·HH[d]	388, 402, 508 (brd)	

a. In ns; see text for discussion b. Chi squared value for fit c. Phosphorescence
d. HH = Hexahost e. Top numbers at 330 nm; bottom at 420 nm

Pyrene. The prototypical fluorore is pyrene. We have prepared complexes of TOT and pyrene. In one preparation, excess pyrene (on a molar basis) was used, while in the other, excess TOT was used. The spectra of solid pyrene and the two complexes are shown in Figure 5. Microcrystalline

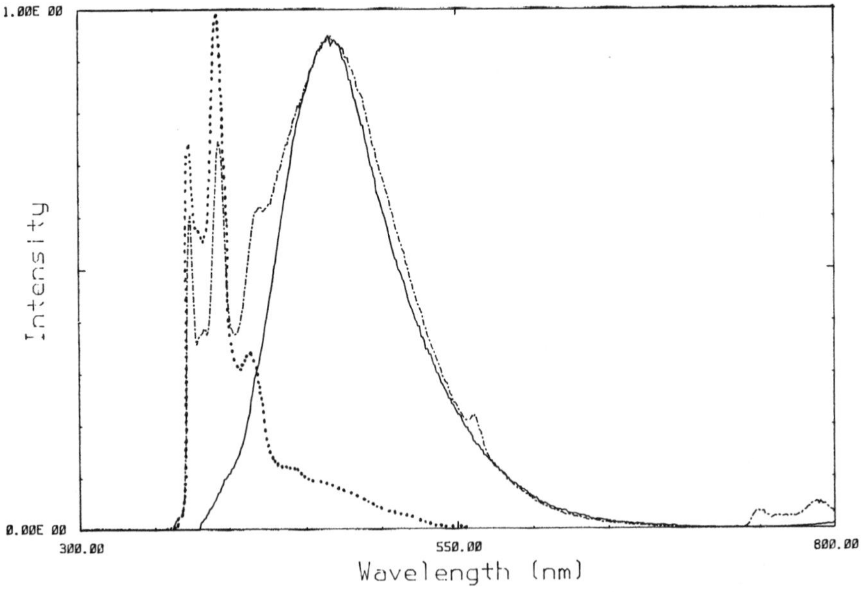

FIG. 5. Emisison spectra of solid pyrene (----), an off stoichiometry preparation of the 1:1 TOT pyrene complex (- · - · -) containing excess pyrene, and the 1:1 TOT:pyrene complex (- - - -). Excitation 302 nm.

pyrene is a "type B" solid [12], with many excimer sites. No monomer emission is seen in this solid. Our complexes appear to include largely only monomeric pyrene emission. When excess pyrene is used to load the complex, the emission is consistent with an admixture of a stoichiometric 1:1 complex and excess pyrene. We conclude that most of the included pyrene is in an isolated environment, incapable of attaining a face to face geometry typical of the pyrene excimer. However, some excimer sites may exist as evidenced by lifetime data as a function of wavelength. At short wavelength (390 nm), the emission lifetime of the nominal 1:1 pyrene:TOT complex is two-component (τ_1 = 4.1 ns, 15%; τ_2 = 16.4 ns, 85%). The long wavelength emission also deconvolves to nearly the same two lifetimes, but with different weighting factors: τ_1 = 7.4 ns, 60%; τ_2 = 16.3 ns, 40%. This behavior is unlike crystalline pyrene, which has an (excimer) lifetime of 113 ns [15]. Similar to the case of Hexahost:(1-MN)$_2$ described previously, we tentatively conclude that there are at least two pyrene occupancy sites which emit in different regions; one site may be excimer like, emitting at long wavelengths, though it need not be a face to face dimer-like species. Both emitters are clearly influenced by the TOT matrix, since the lifetimes observed are much shorter than either pyrene "monomer" in solution (450 ns in cyclohexane; 330 ns in acetone, which might be a good model for TOT as a "solvent"), excimer in solution (65 ns in cyclohexane, 43 ns in acetone), or, *vide supra,* crystalline excimeric pyrene. The hexahost:pyrene complex was substantially the same in steady state emission behavior to the TOT complex. Lifetimes were not determined.

The examples just described provide a wide variety of empirical possibilities for the manner in which inclusion can alter the emission behavior of the guest. We do not have a clear understanding of the parameters which govern the effects, nor even an understanding of the full nature of the interactions which are possible within these kinds of structures. At this stage of the investigation we conclude that electronic interactions between the host and guest occur in nearly all the systems studied, as evidenced by drastic changes in emission lifetimes for the complexes compared to either isolated chromophores or native crystals. We also believe that guest-guest interactions occur in at least some systems, but that multi-site occupancy can complicate the interpretation of the nature of the interactions.

These systems promise to provide interesting information about chromophore-chromophore interactions. Future work will address these questions. We plan to prepare "doped" inclusion complexes in which energy transfer traps can be introduced in such a way as to probe hopping or exciton migration lengths. We also hope to crystallographically analyze some of the reported materials to determine directly the orientational relations among guests. Finally, temperature and excitation wavelength dependence of the emission spectra and lifetimes will help to unravel multisite emission behavior from energy transport phenomena.

REFERENCES

1. A useful review is K. Kalyanasundaram, Photochemistry in Microheterogeneous Systems (Academic Press, N. Y., 1987).
2. R. Arad-Yellin and D. F. Eaton, J. Phys. Chem., 87, 5051 (1983).
3. R. Arad-Yellin and D. F. Eaton, J. Am. Chem. Soc., 104, 6147 (1982).
4. D. F. Eaton, A. G. Anderson, W. Tam and Y. Wang, J. Am. Chem. Soc., 109, 1886 (1987).
   D. F. Eaton, A. G. Anderson, W. Tam and Y. Wang in: ACS Symposium Series No. 346, Polymers for High Technology: Electronics and Photonics, M. J. Bowden and S. R. Turner, eds. (American Chemical Society, Washington, D. C., 1987), chapter 32.
   A. G. Anderson, J. C. Calabrese, W. Tam and I. D. Williams, CHem. Phys. Letts., 134, 392 (9187).
5. A series of monographs presents details on inclusion complexation: Inclusion Compounds, J. L. Attwood, J. E. Davies, D. D. MacNicol, eds. (Academic Press, N. Y., 1984), volumes 1,2,3.
6. R. Breslow, Science, 218, 532 (1982).
7. N. J. Turro and B. Kraeutler, Accts. Chem. Res., 13, 369 (1980).
   N. J. Turro, Pure Appl. Chem., 58, 1219 (9186).
8. D. R. Corbin, D. F. Eaton and V. Ramamurthy, submitted to J. Am. Chem. Soc.
9. A. D. U. Hardy, D. D. MacNicol and D. R. Wilson, J. Chem. Soc. Perkin II, 1011 (1979). See also reference 5, Volume 2, Chapter 5.
10. R. Arad-Yellin, S. Brunie, B. S. Green, M. Knossow and G. Tsoucaris, J. Am. Chem. Soc., 101, 7529 (1973).
11. J. J. McCullough, W. K. MacInnis and D. F. Eaton, Chem. Phys. Letts., 125, 155 (1986).
12. B. Stevens, Spectrochimica Acta, 18, 439 (1962)
13. J. B. Birks, Photophysics of Aromatic Molecules (John Wiley and Sons, N.Y., 1970), pps. 317-319, 528-537.
14. Ref 13, p. 323, and references therein.
15. Ref 13, p. 353 (Table 7.3), and references therein.

LONG-DISTANCE PHOTOINITIATED ELECTRON TRANSFER THROUGH POLYENE MOLECULAR WIRES

Michael R. Wasielewski, Douglas G. Johnson, Walter A. Svec,
Kristen M. Kersey, Denae E. Cragg, and David W. Minsek
Chemistry Division, Argonne National Laboratory, Argonne, IL 60439 USA

ABSTRACT

Long-chain polyenes can be used as molecular wires to facilitate electron transfer between a photo-excited donor and an acceptor in an artificial photosynthetic system. We present data here on two Zn-porphyrin-polyene-anthraquinone molecules possessing either 5 or 9 all trans double bonds between the donor and acceptor, 1 and 2. The center-to-center distances between the porphyrin and the quinone in these relatively rigid molecules are 25 Å for 1 and 35 Å for 2. Selective picosecond laser excitation of the Zn-porphyrin in 1 and 2 results in the very rapid transfer of an electron to the anthraquinone in < 2 ps and 10 ps, respectively. The resultant radical ion pairs recombine with τ = 10 ps for 1 and τ = 25 ps for 2. The electron transfer rates remain remarkably rapid over these long distances. The involvement of polyene radical cations in the mechanism of the radical ion pair recombination reaction is clear from the transient absorption spectra of 1 and 2, which show strong absorbances in the near-infrared. The strong electronic coupling between the Zn-porphyrin and the anthraquinone provided by low-lying states of the polyene make it possible to transfer an electron rapidly over very long distances.

INTRODUCTION

In biological electron transport proteins, such as the photosynthetic reaction center, a series of electron transfer steps is used to transport charge across membranes.[1] Electron transfer reactions have been shown to occur through the participation of the intervening molecular framework linking two redox centers.[2] For example, the rate of electron transfer through saturated C-C bonds is attenuated by about a factor of ten for each C-C bond linking the redox centers. In order to design molecular devices that separate and store charge, and can be used for solar energy conversion or information storage, it is desirable to provide a means of connecting redox centers with "molecular wires" that can facilitate rapid electron transfer.[3-5] Long chain polyenes, most notably the carotenoids, were originally proposed as transmembrane electron transfer intermediates in photosynthesis.[6] Recent experiments have shown that carotenoids normally function only as energy transfer molecules in photosynthesis,[7] although electron transfer from carotenoids can occur under unusual conditions.[8] Nevertheless, we can make use of the low energy electronic states of polyenes to design artificial photosynthetic systems which employ polyenes as molecular wires to facilitate electron transfer between a photo-excited donor and an acceptor. In this paper we present data on two Zn-porphyrin-polyene-anthraquinone, ZnP-C-AQ, molecules possessing either 5 or 9 all trans double bonds between the donor and acceptor, 1 and 2. The center-to-center distances between the porphyrin and the quinone in these relatively rigid molecules are 25 Å for 1 and 35 Å for 2. The strong electronic coupling between the ZnP and the AQ, provided by the low-lying states of the polyene, C, should make it possible to transfer an electron rapidly over very long distances.

MATERIALS AND METHODS

The synthesis of compounds **1** and **2** along with reference compounds **3-7** will be discussed in a future publication. The polyenes are the all-E-isomers as shown in the structures. Compounds **1-7** were purified before use by HPLC or preparative TLC. Measurements of one-electron redox potentials vs SCE were carried out at a Pt disc electrode in butyronitrile/1 % pyridine containing 0.1 M tetra-n-butylammonium perchlorate using AC voltammetry as described previously.[9] The data are given in Table 1.

Published 1989 by Elsevier Science Publishing Co., Inc.
Photochemical Energy Conversion
James R. Norris, Jr. and Dan Meisel, Editors

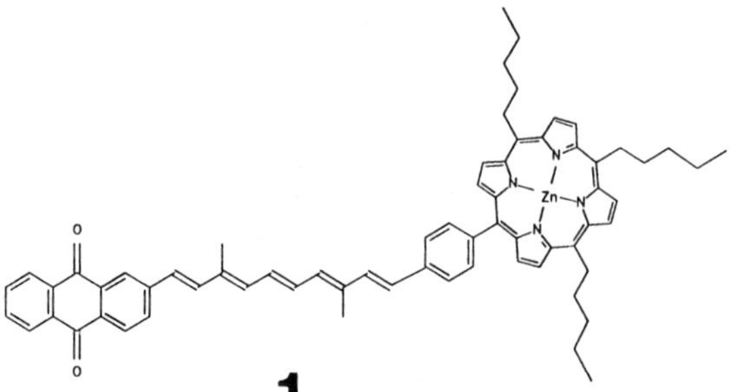

1

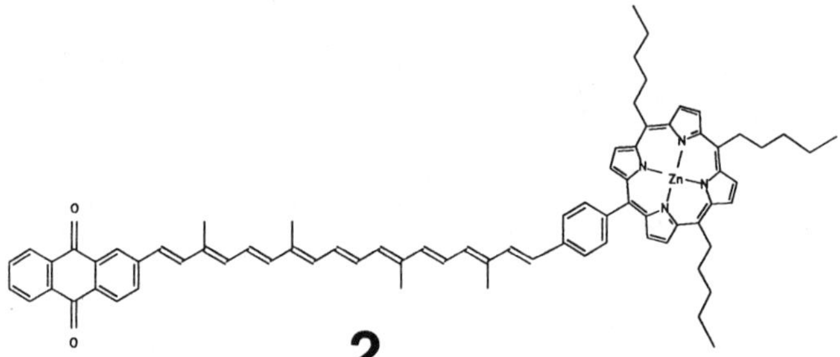

2

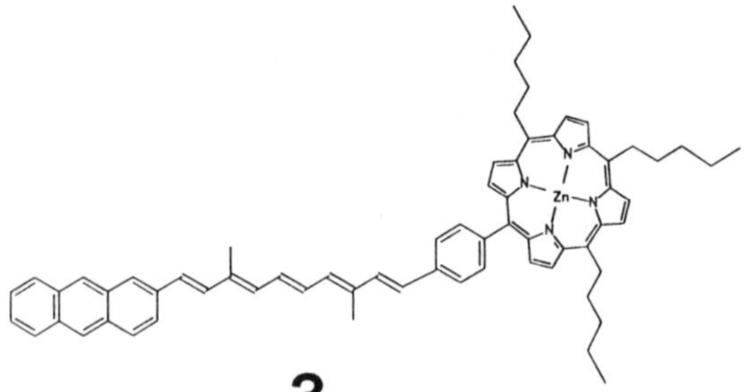

3

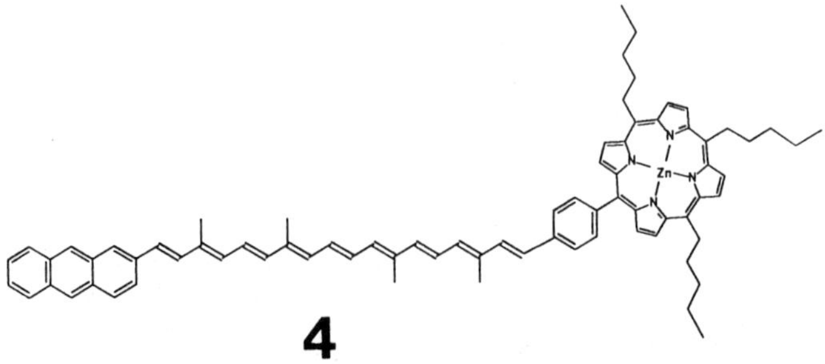

4

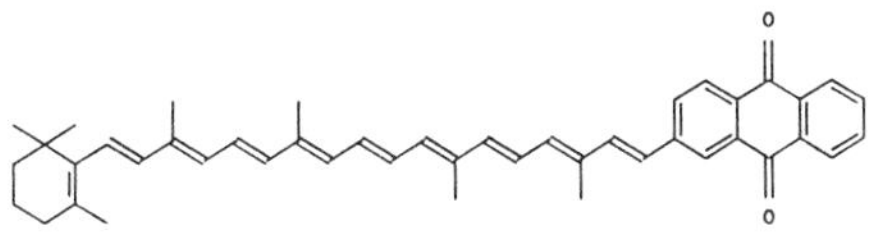

6

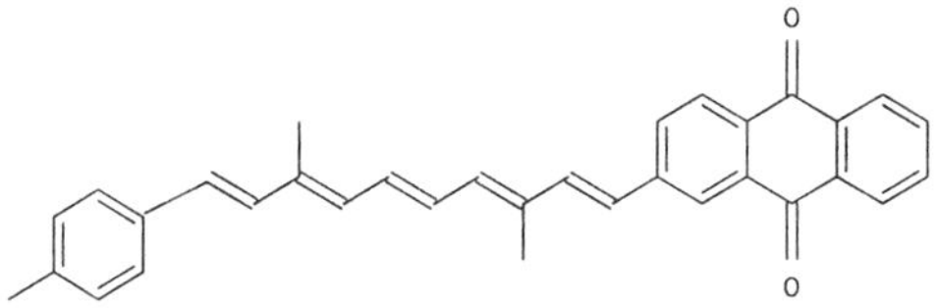

7

Fluorescence quantum yields for **1-5** in butyronitrile/1% pyridine were determined using 10^{-7} M solutions in 1 cm cuvettes at the 90° geometry. The samples were excited at 430 nm and the resultant fluorescence emission spectra were digitized. The emission spectra of **1-5** differed only in relative intensity. A direct comparison between the integrated and normalized fluorescence spectra of **1-5** was made with that of ZnTPP.[10] The data are listed in Table 2. Compounds **6** and **7** are non-fluorescent.

Picosecond transient absorption measurements were obtained using an apparatus described in detail previously.[11] Briefly, the absorbance at the 610 nm excitation wavelength was about 0.3. A 2 mm diameter spot on the sample cell was illuminated with the pump and probe beams derived from the transient absorption apparatus. The 514 nm output of a mode-locked Ar^+ laser operating at an 82 MHz repetition rate was used to synchronously pump a rhodamine-6G/DQOCI dye laser, which resulted in 610 nm, 0.4 ps, 0.5 nJ pulses from the dye laser. These pulses were amplified to 1.5 mJ using a 4-stage rhodamine-640 dye amplifier pumped by a frequency-doubled Nd-YAG laser operating at 10 Hz. The amplified pulses, which broadened slightly to 0.5 ps, were split with a dichroic beam splitter. A 610 nm, 0.5 ps, 400 μJ pulse was used to excite the samples. The remaining 610 nm, 0.5 ps, 1.1 mJ pulse was used to generate a 0.5 ps white light continuum probe pulse. Pulse lengths were determined by autocorrelation techniques. The total instrument response function was 0.5 ps. Samples of each compound in butyronitrile/1% pyridine were prepared in 2mm pathlength cells. Typically, 64 laser shots were averaged at each time point to obtain the data presented here. Transient absorbance measurements were made with a double beam optical configuration which employed optical multichannel detection. Time delays between pump and probe pulses were accomplished with an optical delay line. Time constants for kinetic data were determined by the method of Provencher.[12]

TABLE 1. REDOX POTENTIALS[a] AT 294K (V vs SCE)

Compound	ZnP/ZnP^+	ZnP/ZnP^-	C/C^+	C/C^-	AQ/AQ^-
1	0.63	-1.63	0.86	-1.85	-0.87
2	0.63	-1.65	0.69	-1.58	-0.88
Zn-Tripentyl-Monophenyl-Porphine, **5**	0.61	-1.65	--	--	--
β-apo-8'-carotene-AQ, **6**	--	--	0.69	-1.60	-0.86
tolyl-C_{10}-AQ, **7**	--	--	0.86	-1.85	-0.87

[a]Electrochemical half-wave potentials were determined in butyronitrile containing 0.1 M $n\text{-}Bu_4N^+BF^-_4$; Pt disc working electrode; SCE reference electrode.

TABLE 2. FLUORESCENCE QUANTUM YIELDS

Compound	1	2	3	4	5
ϕ_F	$<10^{-4}$	$<10^{-4}$	0.025	$<10^{-4}$	0.027

RESULTS AND DISCUSSION

The ground state optical spectra for **1** and **2** are shown in Figure 1, while the spectra for **3** and **4** are shown in Figure 2. Attachment of the porphyrin to the polyene does not significantly perturb either the Soret band or the Q bands of the ZnP. The polyene-5 absorption in **3** overlaps the Soret band resulting in its broadened appearance, while the longer polyene-9 in **4** possesses a distinct band at 460 nm very similar to typical carotenoids. The presence of the anthracene, AN, in **3** and **4** does not grossly perturb the optical spectra of the attached polyene. However, attachment of AQ to the polyene does result in some broadening and a red shift of the polyene absorption band. This broadening and shifting is somewhat more pronounced for **1** than it is for **2**. The absorbance of the ZnP band at 610 nm is about 9 times larger than the residual polyene absorbance at this wavelength so that ZnP can be excited selectively. The absorption spectra of reference compounds **6** and **7** are shown in Figure 3. The presence of the AQ results in broadening and red shifting of the polyene absorbance. The broadening of the AQ compounds relative to the AN compounds may be due to charge transfer state contributions to the optical spectra.

The fluorescence quantum yields for compounds **1-5** are given in Table 2. Attachment of polyene-5 to the porphyrin with anthracene at the other terminus results in little of no decrease in fluorescence yield. However, attachment of polyene-9 to the porphyrin results in a sharp decrease in fluorescence yield. Substitution of AQ for AN results in a very high degree of quenching both for **1** and **2**. The extent of the π systems in **3** and **4** are similar to those of **1** and **2**, respectively.

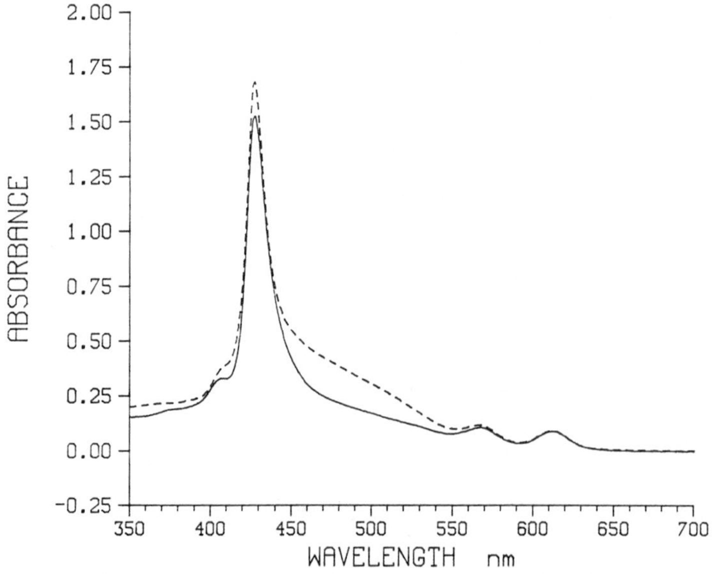

Fig. 1. Ground state absorption spectra of **1**—— and **2** ------.

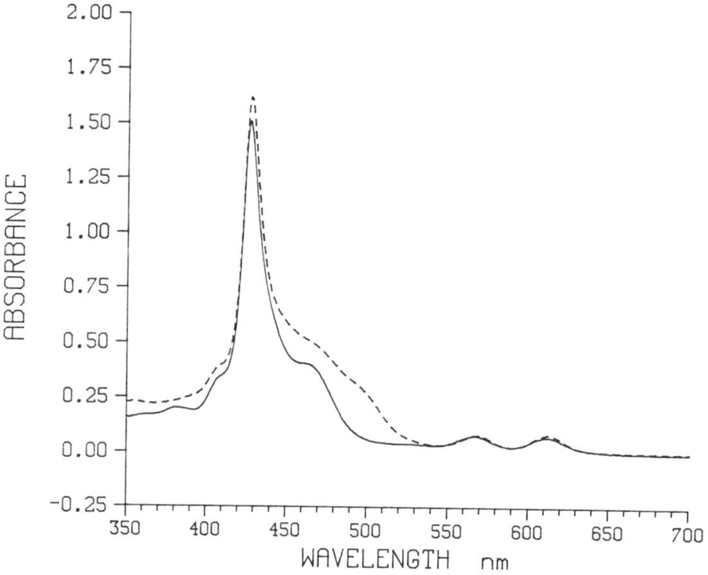

Fig. 2. Ground state absorption spectra of **3** —— and **4** ------.

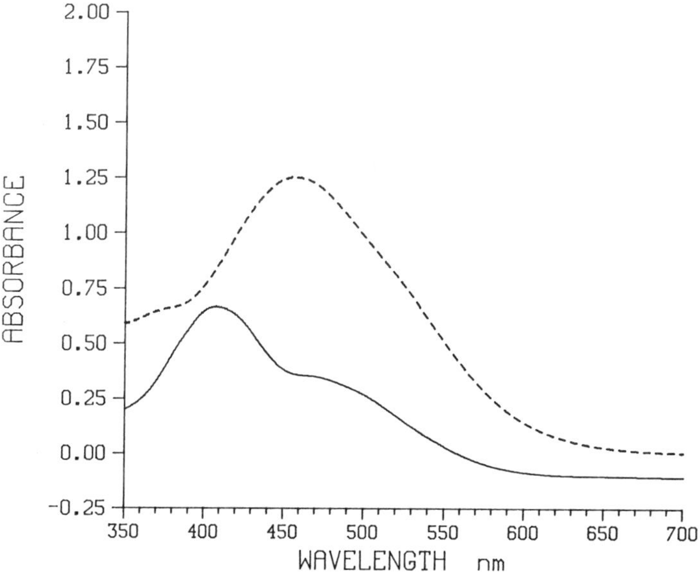

Fig. 3. Ground state absorption spectra of **6** ------ and **7** —— .

The fluorescence quenching in compound **4** may be ascribed to energy transfer from ZnP to the polyene. The ground state absorption band centered at 460 nm in **4** is due to a fully allowed dipole transition to the $^{1*}B_u$ state of the polyene. However, the lowest excited singlet state of polyene-9 is an $^{1*}A_g$ state, which is much lower in energy than its $^{1*}B_u$ state, probably about 2 eV.[13] The energy of the $^{1*}A_g$ state is probably below that of ^{1*}ZnP. Electric dipole transitions between the polyene-9 ground state and its $^{1*}A_g$ state are symmetry forbidden. On the other hand, the energy of the $^{1*}A_g$ state of polyene-5 is probably above that of ^{1*}ZnP. These conclusions are supported by two facts: First, the $^{1*}B_u$ state of polyene-5 is about 2650 cm^{-1} higher in energy than that of polyene-9. Second, the $^{1*}B_u$ - $^{1*}A_g$ energy gap becomes smaller as the length of a polyene decreases.[14] Picosecond transient absorption spectra of **3** and **4** show that ionic states of the polyene are not involved in the quenching mechanism. The cation and anion radicals of the polyenes, which have distinct, intense absorptions in the near-infrared, were not observed. The intramolecular energy transfer properties of **4** will be discussed fully in a subsequent paper.

Transient absorption spectra of **1** obtained 2 ps and 7 ps following a 0.5 ps, 610 nm laser flash are shown in Figure **4**. The spectrum at 2 ps is characteristic of the formation of ZnP^+. The bands at 470 nm and especially between 650 nm and 750 nm are characteristic of this ion.[15] At 7 ps a large broad absorption band centered at 780 nm appears, along with another absorbance increase near 540 nm. The absorption increase at 780 nm is diagnostic the formation of the polyene cation in **1**.[16] The sequence of electron transfer events in **1** can be followed from the kinetics of the transient absorption changes at several wavelengths. Figure **5** shows the transient absorption changes at 470 nm. An instrument limited rise is followed by a period of roughly constant ΔA and is in turn followed by a single exponential decay with τ = 10 ps. Similar behavior is seen at 650 nm. These data indicate that electron transfer from ^{1*}ZnP to AQ occurs in < 2 ps and is followed by decay of ZnP^+-C-AQ^- back to ground state in 10 ps. The details of the involvement of the polyene become apparent when kinetic data at 800 nm is examined, Figure **6**. The absorption at 800 nm displays a 3 ps risetime followed by a 10 ps decay. The same kinetics are also observed at 540 nm.

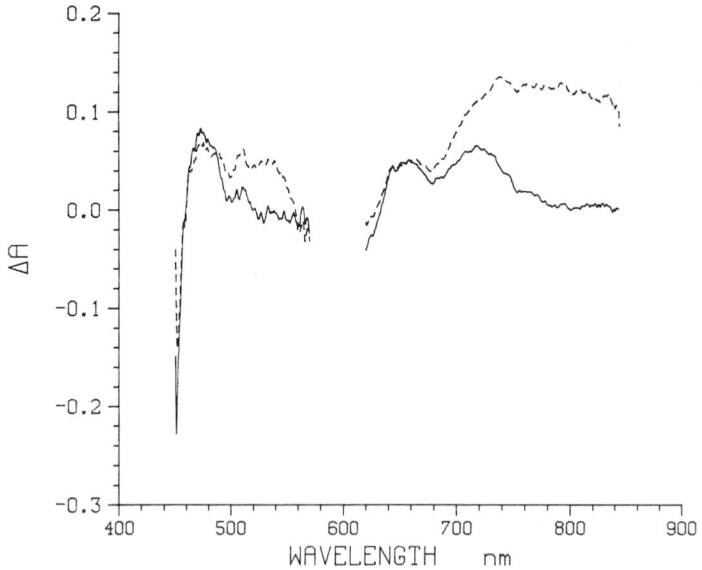

Fig. 4. Transient absorption spectra of **1** at 2 ps —— and 7 ps ------ following a 0.5 ps, 610 nm laser flash.

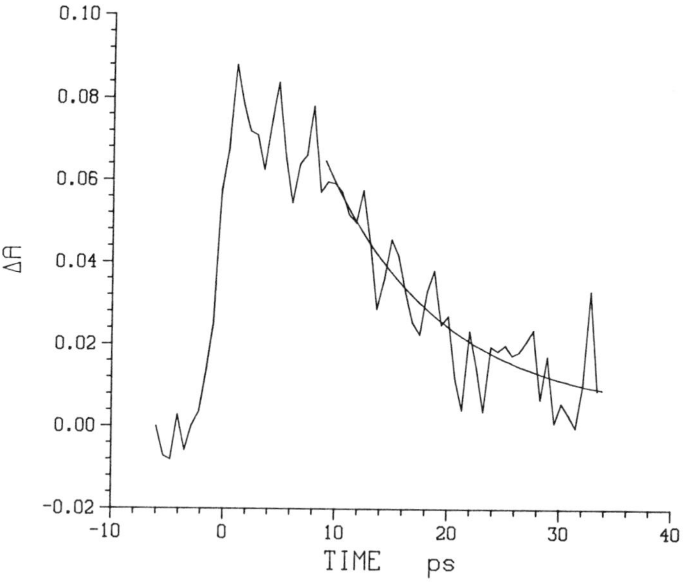

Fig. 5. Transient absorption changes of 1 at 470 nm following a 0.5 ps, 610 nm laser flash.

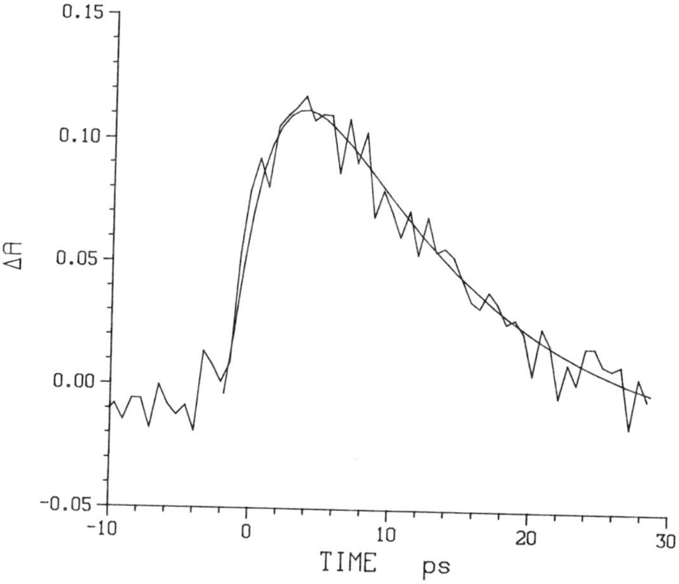

Fig. 6. Transient absorption changes of 1 at 800 nm following a 0.5 ps, 610 nm laser flash.

We propose the mechanism given in Figure 7, which is consistent with the data, for both the forward and back electron transfer reactions in **1**. The fluorescence quenching data suggests that energy transfer from $^{1*}ZnP$ to polyene-5 does not occur. Excitation of ZnP results in formation of $^{1*}ZnP$ which transfers an electron to the anthraquinone in < 2 ps. The strong electronic coupling provided by the conjugated polyene linkage between the donor and acceptor in **1** facilitates the charge separation. We find no direct evidence for the presence of the radical anion of the polyene as a distinct chemical intermediate. However, the state ZnP^+-C^--AQ is 0.46 eV above $^{1*}ZnP$ and may enhance the charge separation rate via a superexchange mechanism. There are two possible routes back to ground state for ZnP^+-C-AQ^- in **1**. The radical ion pair could recombine directly, or the low-lying cation radical state of polyene-5 could be an intermediate in this process. Our data show that both processes occur. The decay of ZnP^+ absorbance at 470 nm and 650 nm shows that ZnP^+-C-AQ^- decays in about 10 ps. Our electrochemical data show that ZnP-C^+-AQ^- is no more than about 0.2 eV above ZnP^+-C-AQ^-. Population of the polyene cation state occurs in 3 ps and is followed by collapse to ground state in 10 ps. The rapid rates of electron transfer observed for **1** are reasonable within the context of strong electronic coupling between the donor and acceptor via the polyene. Sequential radical ion pair recombination mechanisms have been observed previously in triad donor-acceptor molecules.[17]

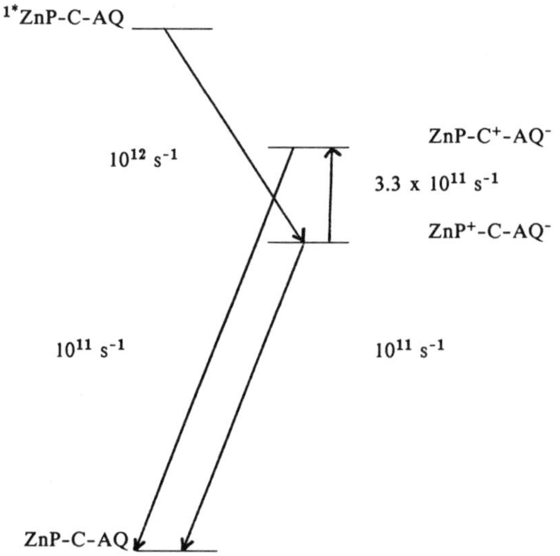

Fig. 7. Mechanism of electron transfer in **1**.

A somewhat different situation prevails for compound **2**. The transient absorption spectra of **2** at 5 ps and 15 ps following 0.5 ps, 610 nm laser excitation are shown in Figure **8**. At 5 ps the polyene absorbance is bleached near 450 nm and there is a residual band due to $^{1*}ZnP$ at 650 nm. As was mentioned previously, it is likely that the energy of the $^{1*}A_g$ state of polyene-9 is below that of $^{1*}ZnP$ and thus energy transfer from $^{1*}ZnP$ to C occurs. At 15 ps a distinct band at 540 nm appears and the band at 650 nm is diminished somewhat more than the 720 nm band. The kinetics of the polyene bleach are shown in Figure **9**. The polyene absorption band at 450 nm bleaches in < 1.5 ps. This indicates that an excited state of the polyene is produced very rapidly. The bleach recovers with distinctly biphasic kinetics. The fast recovery occurs with τ = 7 ps, while the slow phase occurs with τ = 25 ps.

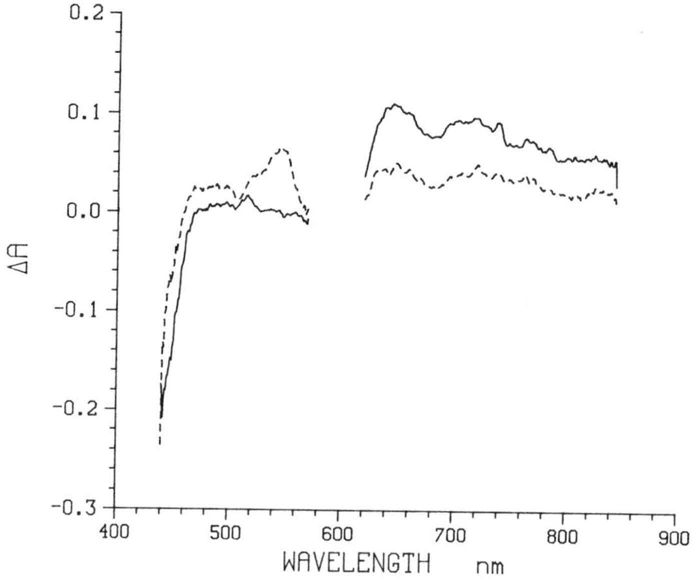

Fig. 8. Transient absorption spectra of 2 at 5 ps —— and 15 ps ------ following a 0.5 ps, 610 nm laser flash.

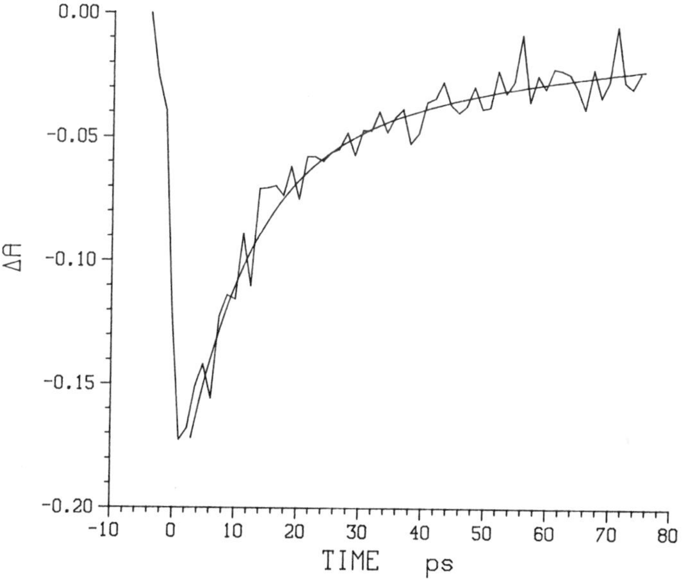

Fig. 9. Transient absorption changes of 2 at 450 nm following a 0.5 ps, 610 nm laser flash.

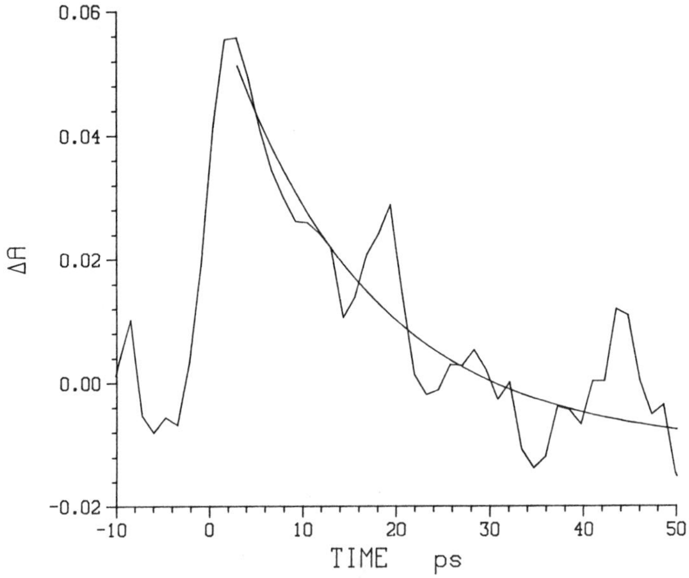

Fig. 10. Transient absorption changes of 2 at 950 nm following a 0.5 ps, 610 nm laser flash.

The appearance of ionic states involving polyene-9 can be ascertained from an examination of the near-infrared region of the spectrum. We find that 2 exhibits a positive absorption change at 950 nm, a wavelength that is diagnostic for the presence of the polyene-9 cation radical.[16] The polyene cation appears in < 3 ps and decays away completely with the same biphasic kinetics as exhibited by the recovery of the bleach at 450 nm, Figure **10**.

Figure **11** shows the recovery of the absorbance at 720 nm to ground state. This absorbance has contributions from both the polyene cation and ZnP^+. This makes the initial kinetics complicated. Yet, the longest component in the decay of this absorbance is 25 ps. This is consistent with the kinetics of radical pair recombination time of ZnP^+-C-AQ^- determined from absorbance changes involving the polyene.

Figure **12** shows an electron transfer mechanism for compound 2 that is supported by the data. ZnP is excited to ^{1*}ZnP, which transfers its singlet excitation to C, and is followed by transfer of an electron from ^{1*}C to AQ to yield ZnP-C^+-AQ^-. The entire process occurs in < 3 ps. The population of ZnP-C^+-AQ^- then partitions between recombination to ground state and a rapid secondary electron transfer from ZnP resulting in ZnP^+-C-AQ^- in $\tau = 7$ ps. Return of ZnP^+-C-AQ^- to ground state by the direct route occurs in $\tau = 25$ ps. However, since ZnP-C^+-AQ^- is only about 0.06 eV above ZnP^+-C-AQ^-in energy, radical ion pair recombination may also occur via ZnP-C^+-AQ^-. The relative magnitudes of the two components of the biphasic decay of ZnP-C^+-AQ^- at 450 nm suggest that the back reaction rate from ZnP^+-C-AQ^- to ZnP-C^+-AQ^- should be comparable to the forward 7 ps reaction.

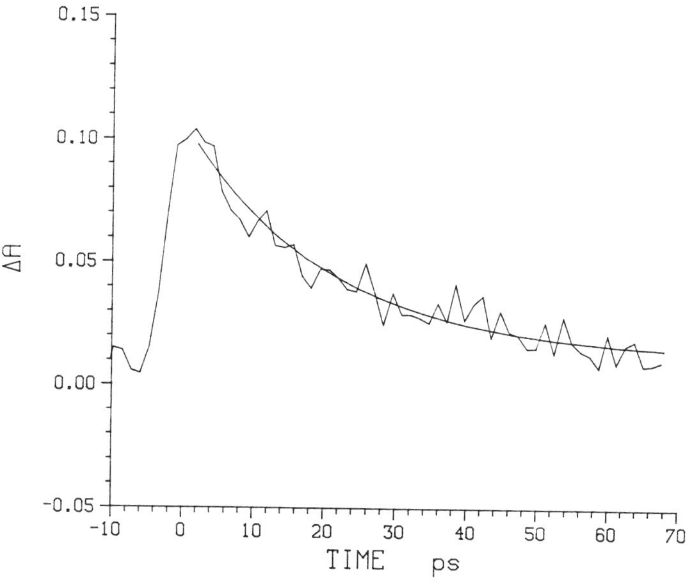

Fig. 11. Transient absorption changes of 2 at 720 nm following a 0.5 ps, 610 nm laser flash.

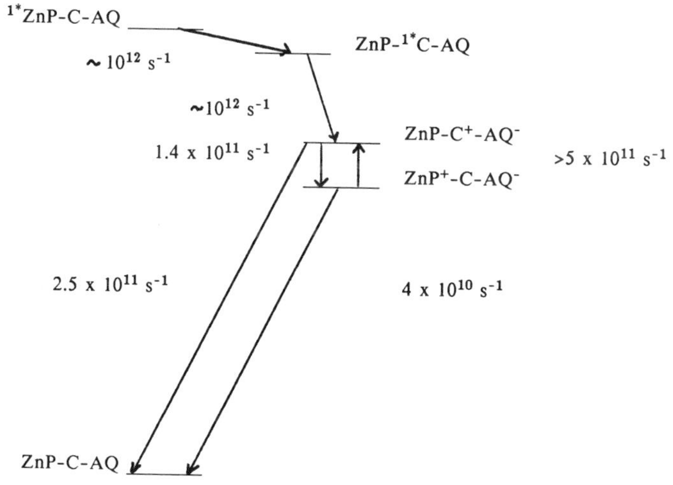

Fig. 12. Mechanism of electron transfer in 2.

In order to determine independently the rate of radical ion pair recombination of ZnP-C^{+}-AQ^{-}, we examined the electron transfer properties of β-apo-8'-carotene-AQ, reference compound **6**. The transient absorption spectrum of **6** at 2 ps following a 0.5 ps, 610 nm laser flash is seen in Figure **13**. The polyene exhibits a bleach at 450 nm, and positive absorption changes at 700 nm and at 950 nm. The 950 nm absorption is diagnostic for the formation of the polyene cation radical. The formation and decay times for C^{+}-AQ^{-} are τ = < 1.5 ps and τ = 4 ps as shown in Figure **14**. This result is consistent with the lifetime of ZnP-C^{+}-AQ^{-} obtained for **2**.

These results show that electron transfer from the lowest excited singlet state of a donor to an acceptor can occur in a few ps across long distances in molecules in which the donor and acceptor are tightly coupled electronically via a polyene molecular wire. The electrons traverse the wire very rapidly in both directions. Since most applications of molecular devices will require longer lifetimes of the stored charge separation, we are currently developing molecules in which the degree of electronic coupling of both the donor and the acceptor to the polyene molecular wire can be adjusted to prolong the lifetime of the charge separation.

ACKNOWLEDGEMENT

This work was supported by the Division of Chemical Sciences, Office of Basic Energy Sciences, of the U. S. Department of Energy under contract W-31-109-Eng-38. DWM is a graduate student at the University of Chicago. KMK and DEC are Undergraduate Student Research Participants supported in part by the Division of Educational Programs at ANL. The authors wish to thank Prof. L. Kispert of the University of Alabama and Hoffmann-La Roche AG, Basel for generous gifts of crocetin dialdehyde.

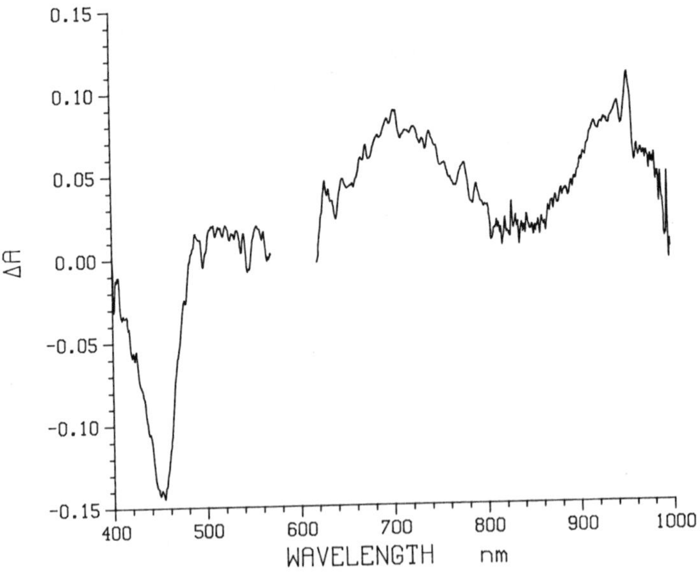

Fig. 13. Transient absorption spectrum of **6** at 2 ps following a 0.5 ps, 610 nm laser flash.

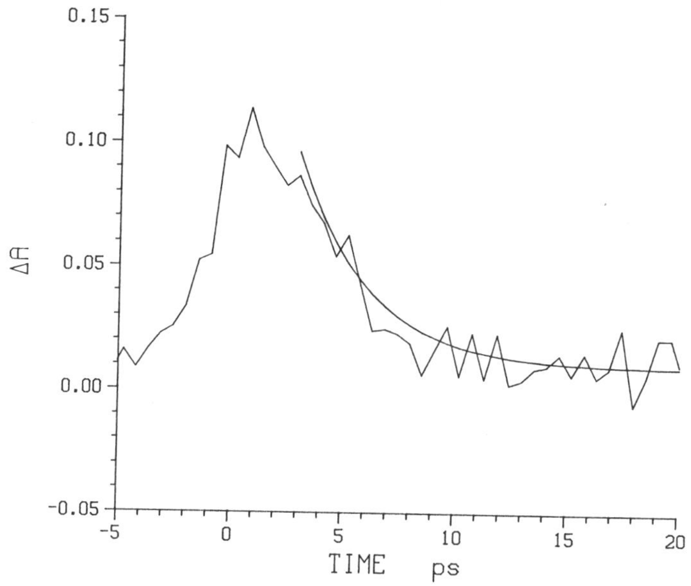

Fig. 14. Transient absorption changes of **6** at 950 nm following a 0.5 ps, 610 nm laser flash.

REFERENCES

1. D. Holten and C. Kirmaier, Photosynth. Res. 13, 225-260 (1987).
2. J. R. Miller, Nouv. J. Chim. 11, 83-89 (1987).
3. T. S. Arrhenius, M. Blanchard-Desce, M. Dvolaitzky, J.-M. Lehn, and J. Malthete, Proc. Natl. Acad. Sci. USA 83, 5355-5359 (1986).
4. J.-M. Lehn, Angew. Chem. Int. Ed. Engl. 27, 89 (1988).
5. F. Effenberger, H. Schlosser, P. Bauerle, S. Maier, H. Port, and H. C. Wolf, Angew. Chem. Int. Ed. Engl. 27, 281-284 (1988).
6. J. R. Platt, Science 129, 372-374 (1959).
7. D. Siefermann-Harms, Biochim. Biophys. Acta 811, 325-355 (1985).
8. C. C. Schenck, B. Diner, P. Mathis, and K. Satoh, Biochim. Biophys. Acta 680, 216-227 (1982).
9. M. R. Wasielewski, R. L. Smith, and A. G. Kostka, J. Am. Chem. Soc. 102, 6923-6928 (1980).
10. P. G. Seybold, and M. Goutermann, J. Molec. Spectroscopy 31, 1-13 (1969).
11. M. R. Wasielewski, J. M. Fenton, and Govindjee, Photosynth. Res., 12, 181-190 (1987).
12. S. W. Provencher, J. Chem. Phys. 64, 2772-2777 (1976).
13. M. R. Wasielewski and L. D. Kispert, Chem. Phys. Lett. 128, 238-243 (1986).
14. B. S. Hudson, B. E. Kohler, and K. Schulten in: Excited States, Vol. 6, E. C. Lim, ed. (Academic, New York 1982) pp. 1-95.
15. J. Fajer, D. C. Borg, A. Forman, D. Dolphin, R. H. Felton, J. Am. Chem. Soc., 92, 3451-3459 (1970).
16. R. V. Bensasson, E. J. Land, and T. G. Truscott, Flash Photolysis and Pulse Radiolysis (Pergamon, Oxford 1983) pp. 73-78.
17. D. Gust, T. A. Moore, L. R. Makings, P. A. Liddell, G. A. Menuth and A. L. Moore J. Am. Chem. Soc. 108, 8028-8031 (1986).

SEMICONDUCTOR PARTICLE MEDIATED PHOTOELECTRON TRANSFERS IN BILAYER LIPID MEMBRANES

Janos H. Fendler and Subhash Baral

Syracuse University
Department of Chemistry
Syracuse, NY 13244-1200

ABSTRACT

Semiconductor particles *in situ* generated on the *cis* surface of glyceryl monooleate (GMO) bilayer lipid membranes (BLMs), have been used to mediate photoelectric effects. The presence of semiconductors on the BLM surface has been established by voltage-dependent capacitance measurements, absorption spectroscopy, and optical microscopy. Subsequent to the injection of H_2S, the first observable change was the appearance of fairly uniform white dots on the black film. These dots rapidly moved around and grew in size, forming islands which then merged with themselves and with a second generation of dots, which ultimately led to a continuous film which continued to grow in thickness. Steady-state illumination of a CdS-containing BLM resulted in the prompt development of -150 mV to -200 mV (*cis* side negative) potential difference in an open circuit across the GMO BLM. This initial photovoltage, V_I, quickly decayed to a steady value, V_S (-100 mV to -150 mV). When the illumination was turned off, the potential difference across the GMO BLM decreased to its dark value in 3-4 minutes. Excitation of the CdS-containing GMO BLM by 20 ns, 354 nm laser pulses resulted in a transient photovoltage signal whose build-up fitted a single exponential ($t_{1/2}$ ~235±15 ns) and whose decay could be resolved into a faster (~40 μs) and a slower (1-2 s) component. The observed photoelectric effects have been rationalized in terms of equivalent circuits and chemical reactions.

Published 1989 by Elsevier Science Publishing Co., Inc.
Photochemical Energy Conversion
James R. Norris, Jr. and Dan Meisel, Editors

Bilayer (black) lipid membranes (BLMs) are formed by brushing an organic solution of a surfactant (or lipid) across a pinhole (2-4-mm diameter) separating two aqueous phases. Alternatively, BLMs can be formed from monolayers by the Montal-Mueller method.[1] In this method, the surfactant, dissolved in an apolar solvent, is spread on the water surface to form a monolayer below the Teflon partitioning, which contains the pinhole (0.1-0.5-mm diameter). Careful injection of an appropriate electrolyte solution below the surface raises the water level above the pinhole and brings the monolayer into apposition to form the BLM. An advantage of the Montal-Mueller method is that it permits the formation of dissymmetrical BLMs. The initially formed film is rather thick and reflects white light with a gray color. Within a few minutes the film thins and the reflected light exhibits interference colors that ultimately turn black. At that point, the film is considered to be bimolecular (40-60-Å thickness).

Separation of two aqueous solutions by the BLM allows electrical measurements by macroscopic electrodes. Precise capacitance, conductance, and impedance measurements, both in the absence and in the presence of ionophores, have contributed much to our understanding of impulse and ion transport mechanisms.[2] Particularly significant has been the development of voltage clamping (*i.e.*, holding the bilayer membrane at a predetermined potential and measuring the current flow) and single-channel recording.[3,4]

Investigations of BLMs suffer from two major drawbacks. First, BLMs are notoriously unstable. Very rarely do they survive longer than a couple of hours. Second, voltage clamping provides information only on the transition from an open state to a closed state in ion channels and not to events of the closed states. Current research in our laboratories is directed to overcoming these disadvantages by stabilizing BLMs by polymerization or by polymer coating and by developing simultaneous in situ spectroscopic and electrical techniques[5-7] for monitoring functioning BLMs.

Concurrent with our developing simultaneous electrical and spectroscopic techniques, studies have been initiated for the incorporation of semiconductor particles into BLMs.[7] Microcrystalline semiconductors have been fruitfully utilized in many photoconversion processes. Dispersed microcrystalline semiconductors offer a number of advantages. They have broad absorption spectra, high surface areas, and high extinction coefficients at appropriate band energies. They are relatively inexpensive and can be sensitized by doping (by chemical or physical modifications). Unfortunately, microcrystalline semiconductors also suffer from a number of disadvantages. Until recently they could not be reproducibly prepared as small (less than 20 nm in diameter), monodispersed particles. Small and uniform particles are needed to diminish nonproductive electron-hole recombinations. The smaller the semiconductor particle, the greater the chance of the escape of the charge carriers to the particle surface where the electron transfer can occur. There is a minimum size, however, which the particles must reach before absorption occurs at the bulk bandgap (i.e., before the polymolecular cluster becomes a semiconductor). The onset of semiconducting properties for CdS has been estimated to occur for particles whose diameters reach 6 nm. It is difficult to maintain semiconductors in a dispersed state in solution for extended items in the absence of stabilizers. Stabilizers are bound to affect, of course, the photoelectrical behavior of semiconductors. Their modification and their coating by catalysts are at present, more of an art than a science. Furthermore, the lifetime of electron-hole pairs in semiconductors is orders of magnitude shorter than the excited-state lifetime of typical organic sensitizers. This is due to the much faster electron-hole recombinations in semiconductors than the diffusion-limited quenchings observed with organic sensitizers in homogeneous solutions. Quantum yields for charge separations in colloidal semiconductors are, therefore, disappointingly low. Some of these difficulties have been overcome by incorporating semiconductor particles into reversed micelles,[11,12] polymer films,[13-17] surfactant vesicles,[18-21] clays,[22,23] vycor glass,[24] and zeolites.[25] Studies have been initiated by using BLM as a matrix for colloidal particles. Fundamental mechanistic information has been obtained by the investigation of photoelectron transfers, mediated by BLM incorporated colloidal semiconductor particles. Prior to CdS deposition, the resistance and the capacitance of a typical GMO BLM was determined to be $(3\text{-}5)10^8$ ohms and 2.0-2.2 nF. These values are in good agreement with those expected for a BLM of 0.07-0.08 cm diameter. Subsequent to the completion of CdS formation on the GMO BLM (about two hours after the addition of reactants to the bathing solutions), the resistance and capacitance across the semiconductor-containing BLM were found to be reduced to $(1\text{-}5)10^8$ ohms and 1.5-1.8 nF, respectively. This lowering in resistance and capacitance has been attributed to partial penetration of the BLM by the semiconductor particles on one surface.

Steady-state irradiation of a CdS-containing GMO BLM resulted in a further decrease in resistance to about $(0.5\text{-}3.0)10^8$ ohms. Usually the observed resistance drop was in the range of 20-25% with any particular CdS GMO BLM. Absence of a remarkable resistance change across the semiconductor-containing BLM upon steady-state irradiation is accountable in terms of incorporation of CdS into, rather than a fully spanned penetration across, the GMO BLM.

Subsequent to the CdS formation, a small potential difference (-30 mV to +10 mV) was measured between the Ag/AgCl electrodes across the BLM even in the absence of illumination. This potential difference was due to the asymmetry of the composition of the solutions bathing the BLM. In addition to the 5 mM KCl supporting electrolyte surrounding the BLM, the cis side contained unreacted $CdCl_2$, while H_2S and HS^- were present in the trans side. When the light shutter was opened, the potential difference across the BLM instantly reached a value between -150 and -200 mV, with the Ag/AgCl electrode on the cis side (the side of the BLM which contained the CdS particles) becoming negative. This initial photovoltage, defined as the change in the potential difference across the BLM caused by irradiation and designated as V_I, then quickly relaxed to a steady value, V_S (the electrode on the cis side still remaining negative). Over a short period of illumination, eg. 5-10 minutes, the photovoltage magnitude remained constant at V_S. When the illumination was turned off, the observed potential difference across the BLM decayed to its dark value in approximately 3-4 minutes. Typical photovoltage signals from a CdS-containing GMO BLM are shown in Figure 1.

Steady-state photocurrents were also determined at different bias voltages applied across the CdS-containing BLM. Advantage was taken of the ability of the patch clamp to hold a potential difference across the BLM at a predetermined value in the +400 and -400 mV range. Good linear current-applied voltage behavior was observed for the CdS GMO BLM, both in the dark and under constant illumination (Figure 2). The photocurrent at any particular voltage is given by the vertical difference between these two lines at that voltage. In the dark, the slope of the current-voltage curve was found to be small (1.6 pA/mV) corresponding to a composite resistance of the CdS-containing BLM (6.3×10^8 ohms). Upon illumination, the current-voltage slope increased to 4.6 pA/mV and the resistance decreased to 2.1×10^8 ohms.

The decreased resistance of the BLM, caused by the light, can be accommodated in a model equivalent circuit which includes a photoconduction pathway through a resistor, R_p, in parallel with R_m (Figure 3). The value of R_p is, of course, infinity in the dark, and is a finite function of the intensity of the steady-

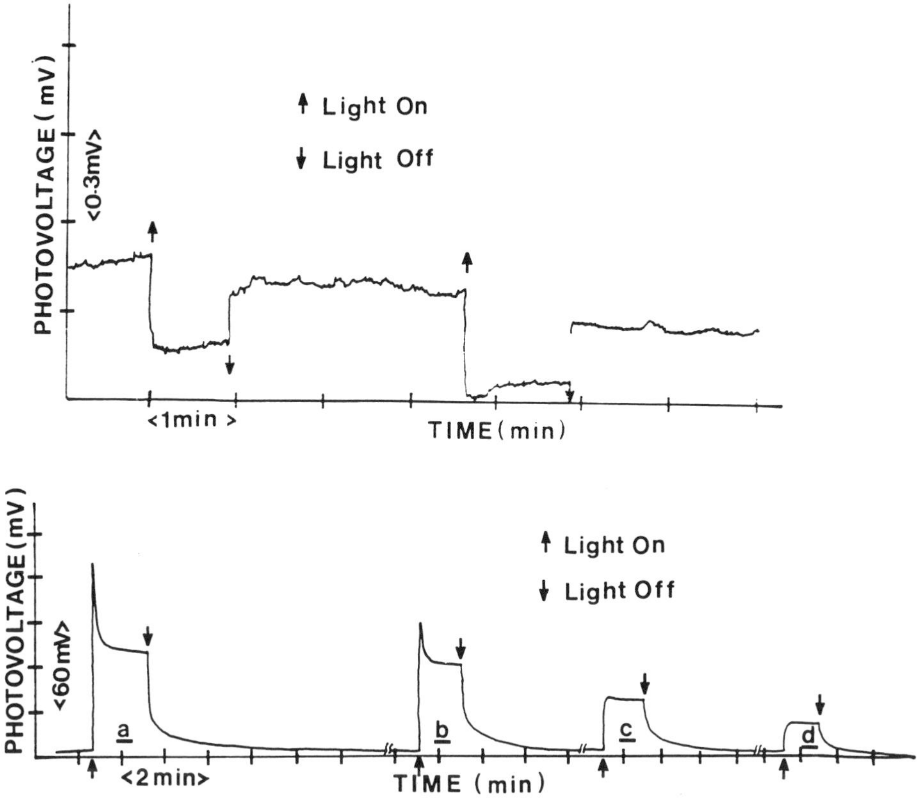

Photovoltage signals obtained from a CdS-containing GMO BLM when illuminated with a steady-state lamp with 4.08×10^{-6} watt (a), 0.89×10^{-6} watt (b), 0.042×10^{-6} watt (c), and 0.033×10^{-6} watt (d) energies incident upon the GMO BLM. The upper plot indicates results of irradiation of the electrode (with the teflon sleeve removed) in the absence of CdS.

FIGURE 1

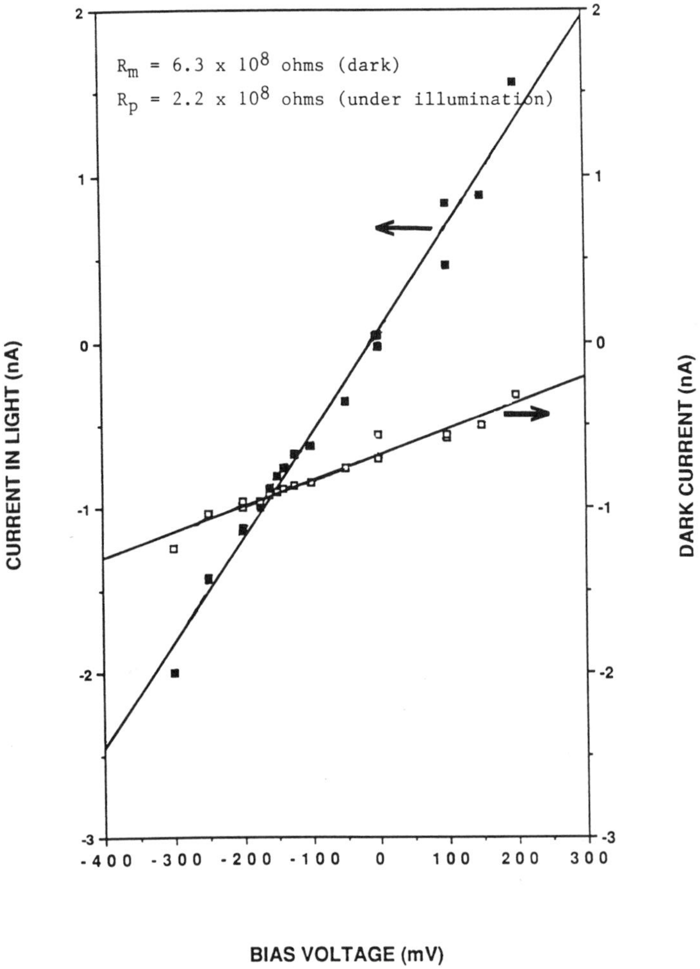

Current passing through a CdS-containing GMO BLM as a function of the potential difference held across the BLM with a patch clamp in the dark (open squares, □) and in the presence of light (4.0 x 10^{-6} watt of energy incident upon the BLM, full squares, ■). The magnitude of the photocurrent at any voltage is given by the difference in the ordinate of the two points with the same abscissa (*i.e.*, bias voltage).

FIGURE 2

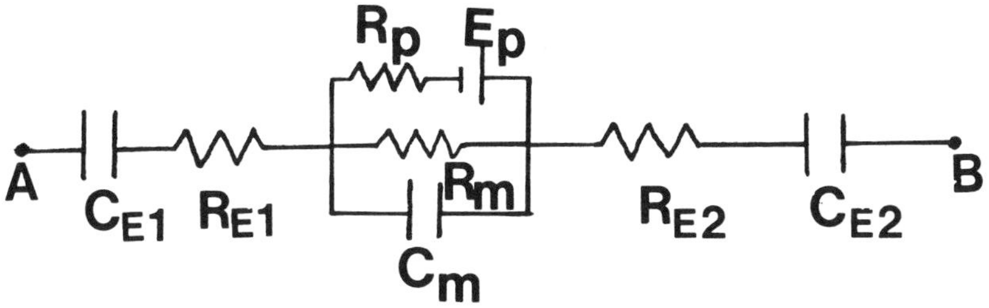

Equivalent circuit descriptions of a CdS-Containing BLM when irradiated.

FIGURE 3

state irradiation. Analysis of Figure 3 showed that, even under constant illumination, the observed photocurrent is an asymmetric, but linear, function of the potential difference held across the BLM by the patch clamp. This observation can be rationalized by introducing an emf source, E_p. In the dark, the magnitude of the emf produced by the source is zero, while in the presence of light its value is a function of the light intensity. Thus, when R_{E_2} and R_{E_1} are small, the photocurrent, I_p, is given by

$$I_p = \frac{(V_A - V_B) + E_p}{R_p}$$

When E_p is equal in magnitude, but opposite in sign, to V_A-V_B (*i.e.,* to the potential at which the BLM is held by the patch clamp between the two summing points, A and B) $I_p = 0$. Such point is given by the intersection of the dark and light current-voltage lines (see in Figure 2). From this intersection, E_p was found to be -155.4 mV for the CdS GMO BLM on which this experiment was performed. Slopes of these plots also yielded $R_p = 2.17 \times 10^8$ ohms and $R_m = 6.28 \times 10^8$ ohms. Similar measurements on another CdS-containing GMO BLM, prepared at a different time, resulted in $R_m = 3.8 \times 10^8$ ohms, $R_p = 1.2 \times 10^8$ ohms, and E_p = -184 mV. These results indicate satisfactory consistencies between the different preparations. The E_p values (-155.4 mV, -184 mV) also agree well with that determined in the open circuit photovoltage experiments under steady-state irradiations (120-150 mV). It is interesting to note that R_p values are several orders of magnitude larger than that expected had the BLM been shorted by complete penetration of the CdS particles. Assuming the resistance of the CdS to be negligible compared to that of the BLM, a ratio of $R_p/R_m \approx 1/3$ can be taken for the penetration depth of the semiconductor particles into the 50 Å thick membrane. This value (50 Å x 1/3 = 16.6 Å) is in good agreement with that previously assessed (17-18 Å) from simultaneous capacitance and reflectance measurements.[7]

The observed photoelectric effects are best accommodated in terms of light-induced vectorial transfer of charges across the BLM, carried by ions, in a direction opposite to the asymmetric BLM potential Φ_m (+426 mV). Absorption of light quanta with energy larger than 2.6 eV (λ < 530 nm), the bandgap of CdS,[25] results in electron transfer from the valence to the conduction band of the semiconductor. Most of the free carriers undergo quick radiative and non-radiative recombinations at impurity or defect sites. A small number of the electrons ($e^-_{CB(t)}$) and holes ($h^+_{VB(t)}$), however, escape recombination by being trapped both in the bulk and at the surface of the BLM-supported polycrystalline CdS particles. The trapped electrons are transferred, in turn, to oxygen molecules absorbed at the semiconductor-water interfaces. Under the influence of the asymmetric potential, Φ_m (+426 mV) in the CdS-containing GMO BLM, $e^-_{CB(t)}$ will move preferentially to the positive (*cis*) side and $h^+_{VB(t)}$ will migrate to the negative (*trans*) side of the membrane. The $O_2^{\cdot-}$ radicals formed at the semiconductor interface are subsequently replaced by oxygen molecules present in the solution, $O_{2(s)}$ either by direct absorption-desorption processes or by electron transfer process. At the same time, hydrogen sulfide, H_2S, or its dissociated form, SH^- (which are permeable across the BLM and, hence, can approach both of its surfaces), are oxidized by the trapped holes.

Generation of $O_2^{\cdot-}$ in the *cis* and H^+ in the *trans* side of the solution bathing the CdS-containing GMO BLM is the net chemical result of bandgap excitation. The observed photovoltage is then the consequence of vectorial transfer of charges,

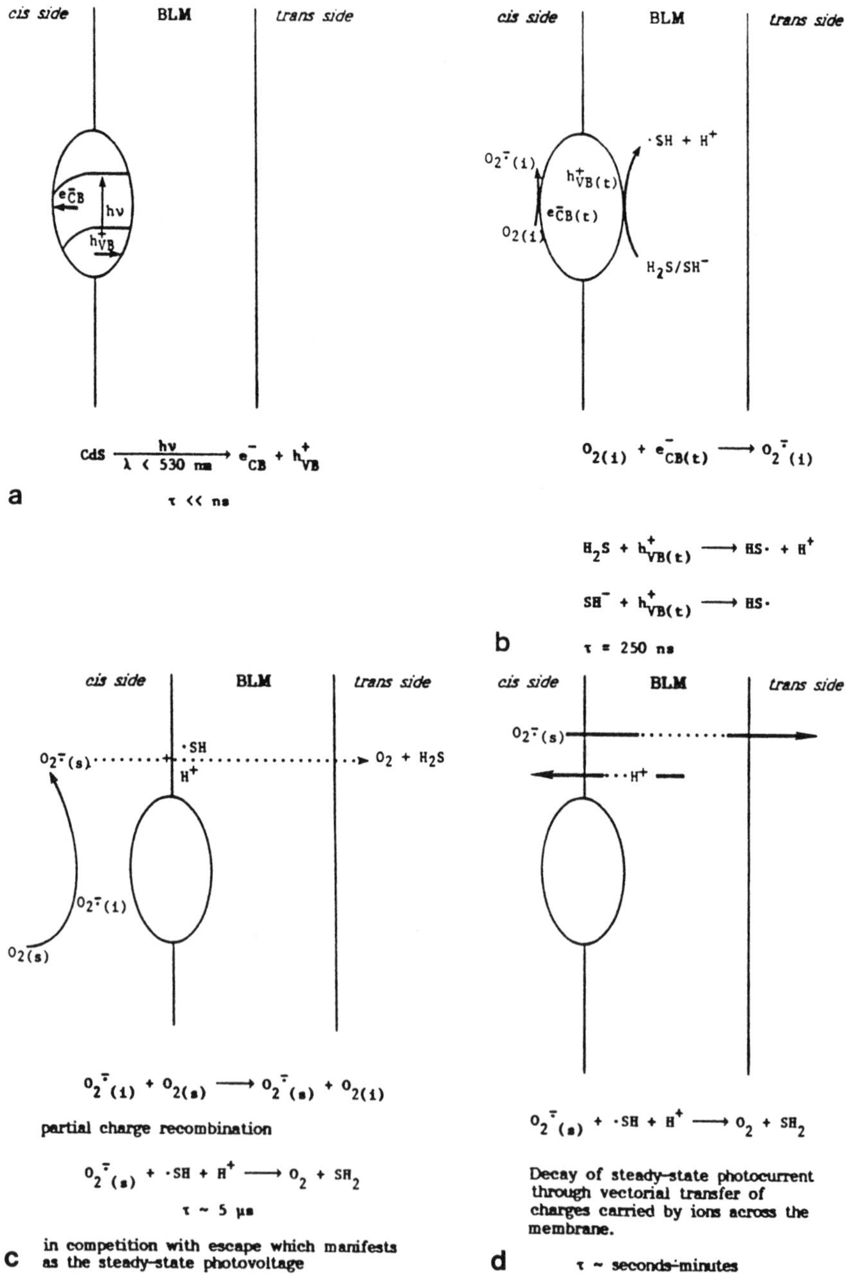

Proposed chemical model for photovoltage generation across the CdS-containing GMO BLM. Reaction sites are highly speculative.

FIGURE 4

carried by ions, in a direction opposite to the asymmetric membrane potential. Figure 4 illustrates the proposed chemical model compatible with the equivalent circuit, illustrated in Figure 3, for the generation of photovoltage across the CdS GMO BLM.

ACKNOWLEDGMENT:

Support of this research by the Department of Energy is gratefully acknowledged.

REFERENCES:

1) Fendler, J. H. *Membrane Mimetic Chemistry*; Wiley: New York, 1982.

2) Hoppe, W.; Lohmann, W.; Markl, H.; Ziegler, H. "Biophysics"; Springer Verlag: New York, 1983.

3) Sakmann, B.; Neher, E. "Single-Channel Recording"; Plenum Press: New York, 1983.

4) Hille, B. "Ionic Channels of Excitable Membranes"; Sinaver Associates, Inc.: Sunderland, Mass., 1984.

5) Rolandi, R.; Flom, S. R.; Dillon, I.; Fendler, J. H. *Prog. Colloid Polym. Sci.* **1987**, *73*, 0000.

6) Zhao, X. K.; Fendler, J. H. *J. Phys. Chem.* **1986**, *90*, 3886.

7) Zhao, X. K., Baral, S. Rolandi, R. and Fendler, J. H. *J. Am. Chem. Soc.* **1988**, *110*, 1012-1024.

8) Fox, M. A. *Acc. Chem. Res.* **1983**, *16*, 314.

9) Grätzel, M. "Energy Sources Through Photochemistry and Catalysis"; Academic Press: New York, 1983.

10) Fendler, J. H. *J. Phys. Chem.* **1985**, *89*, 2730.

11) Meyer, M.; Wallberg, C.; Kurihara, K.; Fendler, J. H. *J. Chem. Soc. Chem. Commun.* **1984**, 90.

12) Lianos, P.; Thomas, J. K. *Chem. Phys. Lett.* **1986**, *125*, 299.

13) Meissner, D.; Memming, R.; Kastening, B. *Chem. Phys. Lett.* **1983**, *96*, 34.

14) Krishnan, M.; White, J. R.; Fox, M. A.; Bard, A. J. *J. Am. Chem. Soc.* **1983**, *105*, 7002.

15) Mau, A. W. H.; Huang, C. B.; Kakuta, N.; Bard, A. J.; Campion, A.; Fox, M. A.; White, M. J.; Webber, S. E. *J. Am. Chem. Soc.* **1984**, *106*, 6537.

16) Tien, H. T.; Bi, Z. C.; Tripathi, A. K. *Photochem. Photobiol.* **1986**, *44*, 779.

17) Kuczynski, J. P.; Milosavljevic, B. M.; Thomas, J. K. *J. Phys. Chem.* **1984**, *88*, 980.

18) Tricot, Y.-M.; Fendler, J. H. *J. Am. Chem. Soc.* **1984**, *106*, 2475; Tricot, Y.-M.; Fendler, J. H. *J. Am. Chem. Soc.* **1984**, *06*, 7359.

19) Tricot, Y.-M.; Emeren, Å.; Fendler, J. H. *J. Am. Chem. Soc.* **1985**, *89*, 4721.

20) Rafaeloff, R.; Tricot, Y.-M.; Nome, F.; Tundo, P.; Fendler, J. H. *J. Phys. Chem.* **1985**, *89*, 1236.

21) Youn, H. C.; Tricot, Y.-M.; Fendler, J. H. *J. Phys. Chem.* **1987**, *91*, 581.

22) Enea, O.; Bard, A. J. *J. Phys. Chem.* **1986**, *90*, 301.

23) Stramel, R. D.; Nakamura, T.; Thomas, J. K. *Chem. Phys. Lett.* **1986**, *130*, 43.

24) Kuczynski, J.; Thomas, J. K. *J. Phys. Chem.* **1985**, 2720.

25) Wang, Y.; Herron, N. *J. Phys. Chem.* **1987**, *91*, 257.

COLLOIDAL SEMICONDUCTORS: SIZE QUANTIZATION, SANDWICH STRUCTURES, PHOTO-ELECTRON EMISSION, AND RELATED CHEMICAL EFFECTS

ARNIM HENGLEIN, HORST WELLER
Hahn-Meitner-Institut Berlin, Bereich Strahlenchemie,
1000 Berlin 39, Federal Republic of Germany

INTRODUCTION

During the past few years, colloidal semiconductors have been used to an increasing extent as sensitizers and catalysts of photochemical reactions. Besides the photocatalytic effects, the interaction of light with small semiconductor particles produces a number of physical effects, which were first considered to be interesting processes accompanying the chemical reactions but have recently attracted a lot of attention. A few of these effects, such as fluorescence and photo-electron emission, have already been described in one of the preceeding conferences [1]. Size quantization effects in extremely small particles was one of the main topics in the last conference [2]. In the meantime, non-linear optical effects, such as the shift in absorption experienced by small particles upon deposition of excess charge carriers [3] and conjugate reflection effects [4], have been found. Furthermore, electron transfer in colloidal sandwich structures [5] and electron emission via interaction of two excited states in small particles [6] have been reported.

SIZE QUANTIZATION EFFECTS

As the phenomena observed till 1986 were discussed in recent reviews [2,7], only a few new observations will be described in the present paper. In small particles a transition from semiconductor to molecular properties takes place with decreasing particle size. In this transition range the optical and photocatalytic properties change. For example, CdS becomes colorless [8] and Cd_3P_2, which usually is black, can be made in all colors of the visible [9]. It was proposed to use the prefix Q before the chemical formula to indicate that a material has unusual properties as compared to those of the macrocrystalline material [8].

Fig. 1 shows the most recent results on the dependence of the threshold of absorption of Q-CdS and Q-ZnO on the particle diameter [10,11]. In both

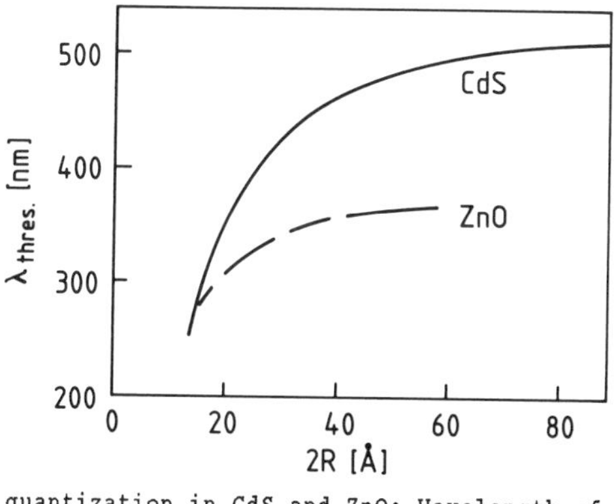

FIG. 1. Size quantization in CdS and ZnO: Wavelength of absorption threshold as function of particle diameter [10,11].

Published 1989 by Elsevier Science Publishing Co., Inc.
Photochemical Energy Conversion
James R. Norris, Jr. and Dan Meisel, Editors

cases, the quantum mechanical calculation was in good agreement with the experimental results. In this calculation, the electron-hole pair was treated as one body with the effective mass of the exciton in the macrocrystalline material, using a hydrogen atom-like wave function and a potential jump of 3.8 eV at the particle-solution interface [11]. The electron-hole-in-a-box model with effective mass approximation is useful to describe the absorption of the particles down to a certain size, which is about 10 A for CdS and ZnO but seems to be much larger for other materials with smaller band gap. In the case of PbS, small particles have been prepared recently in polymer films by Wang et al. [12]. The blue-shift in absorption with decreasing size eventually approached the transition energy of the first allowed excited state, X → A, of a PbS molecule. The authors also developed theoretical models, in which the nonparabolicity of the bands was included, that explained the observed size dependence down to about 25 A. They also showed that the effective mass approximation starts to break down for PbS particles smaller than 100 A. Fig. 2 shows the band gap of PbS as a function of particle size.

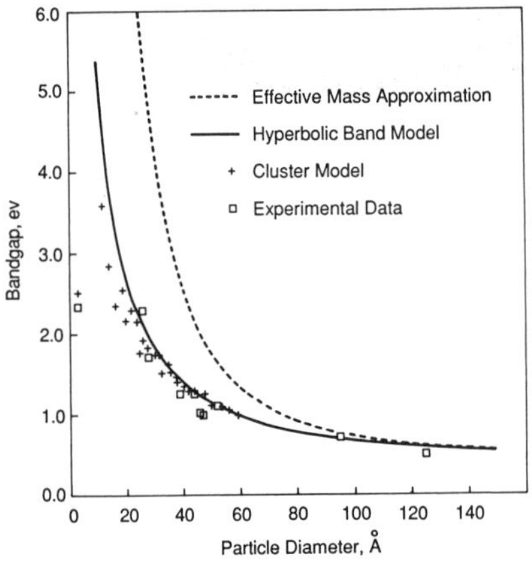

FIG. 2. Band gap as a function of particle size in PbS [12].

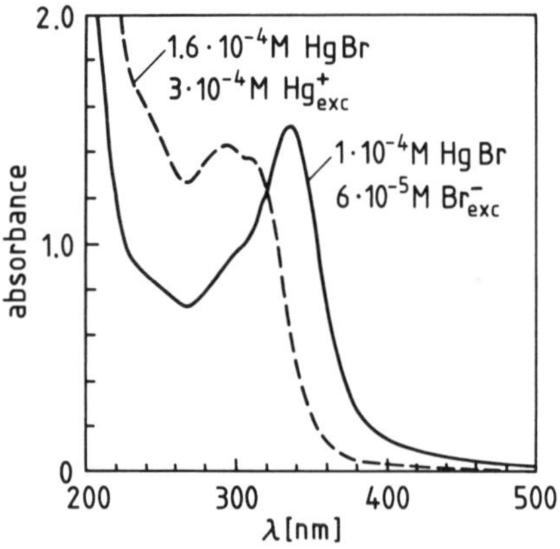

FIG.3. Size quantization in Hg_2Br_2: Absorption spectra of solutions containing particles of different size [13].

Size quantization effects in colloidal metal halides also have been observed. Most recently, the first example for a colloidal quantum rod was found [13]. Mercury-I-bromide, Hg_2Br_2, forms rod-shaped colloidal particles with a length of 500 → 1000 A and a diameter of about 80 A. In these structures, size quantization is operative only in two dimensions. Fig. 3 shows the absorption spectra of two samples of different size. The spectra contain the typical "exciton transition" peak which is blue-shifted with decreasing size. This shift is already seen upon a small change in diameter of the rods which can be detected by electron microscopy. A lot of attention has also been paid to small particles of layered semiconductors, such as PbI_2, HgI_2, and BiI_3 [14]. PbI_2, for example, is an anisotropic semiconductor whose structural repeat unit is a hexagonally closed-packed layer of lead sandwiched between two layers of iodide ions. The confinement effects in such particles should be reproduced by quantization in the two dimensions of the layer and quantization with another effective mass perpendicular to the layer. In fact, absorption spectra indicating the presence of Q-particles were obtained. The interpretation of the spectra, however, was rendered difficult as various metal-iodide complexes are often formed in these solutions which contribute to the absorption.

The question of "magic" agglomeration numbers plays an important role in the interpretation of the spectra of Q-particles [8]. These spectra often contain several maxima. A typical example is shown in Fig. 4. Two

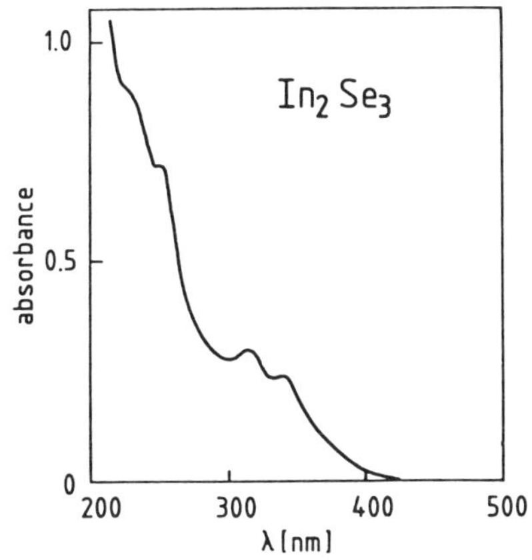

FIG. 4. Absorption spectrum of a solution of Q-In_2Se_3.

explanations are possible. 1. The maxima correspond to optical transitions to the first excited state in particles of different agglomeration numbers in a broad size distribution, certain agglomeration numbers being a little more abundant than adjacent ones. 2. The maxima correspond to optical transitions to various excited states in particles of almost equal size. A structured size distribution according to (1) may develop when the colloidal particles grow via agglomeration of the primary particles. The initially most abundant agglomeration number would then be preserved in the form of integer multiples. Ostwald ripening, on the other hand, does not lead to a structured size distribution. In fact, colloids with a structured absorption spectrum are often obtained under conditions of preparation where Ostwald ripening is slow, i.e. in solutions containing excess metal ions or a complexing agent for small particles such as polyphosphate.

In order to explain the maxima as transitions to higher excited states the energies of these states have often been calculated using the simple three-dimensional particle-in-a-box model with an infinitely high potential well. However, the well is only about 4 eV (difference in energies of a free electron in the vacuum and in the conduction band). The energies of the higher states lie too high when they are calculated assuming a well of inifinte height, and the correlation with the experimental data does not appear appropriate under these conditions.

SURFACE MODIFICATION AND SANDWICH STRUCTURES

After the first observations about fluorescence quenching in colloidal CdS solutions [1,15] a large number of reports have appeared including both continuous illumination and laser flash studies [7]. In most of the studies, no systematic preparative investigations were made to obtain samples having a high quantum yield of fluorescence. In fact, the quantum yield was often below 1 %, the reason being that the colloidal particles prepared had a lot of defect sites where radiationless recombination of the charge carriers occurred. If the defect sites are located at the surface of the colloidal particles, there appears to be a chance to influence these sites chemically.

In fact, it was recently reported in three studies that the fluorescence intensity may be drastically increased by certain procedures, such as exchanging the aqueous solvent by alcohol [10,16], covering the particles with cadmium hydroxide [10] or adsorbing triethylamine [16]. Fig. 5 shows the effect of covering with cadmium hydroxide. A CdS solution was

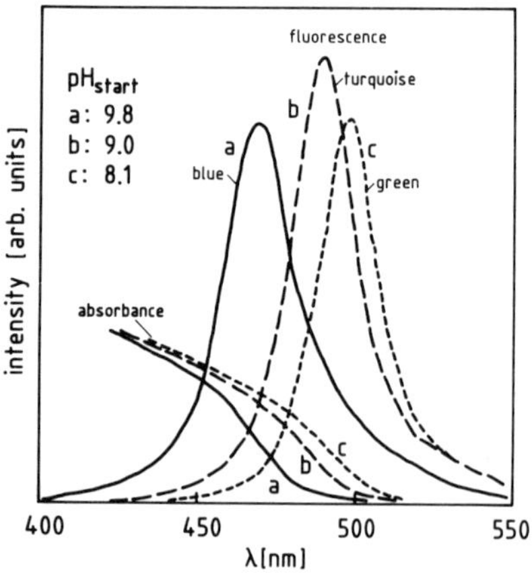

FIG. 5. Absorption and fluorescence spectra of activated CdS [10].

prepared by adding H_2S to a solution of $Cd(ClO_4)_2$ in the presence of sodium polyphosphate. The size of the colloidal particles depends on the pH of the solution before H_2S addition. With increasing pH, the particles become smaller as can be recognized from the blue-shift of their absorption begin. These solutions had a weak red luminescence, the quantum yield being less than 1 %. Activation of the solutions occurred by increasing the pH up to 11 and adding Cd-salt at a concentration three times higher than that of CdS. The solutions took on a bright green or blue fluorescence depending on particle size, the quantum yield lying close to 100 %. These new fluorescences were attributed to the band gap recombination of the charge carriers. It was proposed that SH^- ions at the surface of the colloidal particles are the centers that catalyse radiationless recombination of the charge carriers. In the presence of excess Cd^{2+} in alkaline solution these centers are substituted by S^{2-}--Cd^{2+}--OH^- structures which do not act as recombination sites [10].

The cadmium hydroxide layer of the strong fluorescing CdS samples also increases their photo-stability. Illumination of CdS colloids in the presence of air leads to a photo-anodic corrosion, $CdS + 2O_2 \rightarrow Cd^{2+} + SO_4^{2-}$, the quantum yield being 0.03 molecules CdS consumed per photon absorbed. In the strong fluorescing samples, this reaction occurs with a rate about 10,000 times lower, i.e. the photo-stability is comparable to that of the most stable organic dyes [10].

Fluorescence experiments also give information about the formation and properties of "sandwich structures", i.e. structures in which two particles of different semiconductors adhere to each other and mutually influence their electronic properties. Such structures may be made in two ways:

1. Metal ions are added to the solution of a colloid which substitute metal ions in the colloid. For example, when Cd^{2+} ions are added to a ZnS solution, a CdS phase adhering to the ZnS phase is built-up. Very small amounts of Cd^{2+} ions already quench the fluorescence of ZnS. A new fluorescence is produced which is typical for a phase of the composition $Zn_{0.5}Cd_{0.5}S$ [17].

2. The two colloids are prepared separately and then mixed. An equilibrium may exist between homo-particles and sandwiches, for example:

$$TiO_2 + CdS \rightleftharpoons TiO_2\text{-}CdS \qquad (1)$$

The equilibrium can be shifted to the right side by appropriately adjusting the pH of solution, the concentration of stabilizer (which in most of our experiments is sodium polyphosphate) and the concentration of excess metal ions which are chemisorbed by both kinds of colloids and act as "binding material". The formation of TiO_2-CdS structures is easily recognized as the fluorescence of CdS is quenched by TiO_2 [5]. Similarly, the fluorescence of Cd_3P_2 is quenched in sandwiches with TiO_2 and ZnO [18]. On the other hand, a strong luminescence is produced when a CdS and an Ag_2S phase are contacted [19] although the homo-colloids have little or no fluorescence.

The elementary process of quenching in sandwich structures is the rapid electron transfer from the CdS or Cd_3P_2 part to the TiO_2 or ZnO part. This transition is possible as the conduction band in the light absorbing colloids lies at slightly more negative potentials than in TiO_2 or ZnO. In the case of Cd_3P_2 the transfer is more efficient for smaller Cd_3P_2 particles. This is understood in terms of the shift of the electron potential to more negative values due to size quantization. The more negative the potential the greater the driving force for electron transfer. The positive hole produced in the CdS or Cd_3P_2 part of the sandwich cannot be transferred as the valence bands in TiO_2 and ZnO are on much more positive potentials. The net effect of electron transfer in sandwich structures is an efficient charge separation, the sandwich structure having in this respect a similar function to a p-n junction in macroscopic semiconductor techniques, although the principle of charge separation is different.

This primary charge separation should be very favorable for the initiation of chemical reactions. In fact, certain processes occur in sandwiches at a rate much higher than in the CdS homo-colloids. Fig. 6 shows

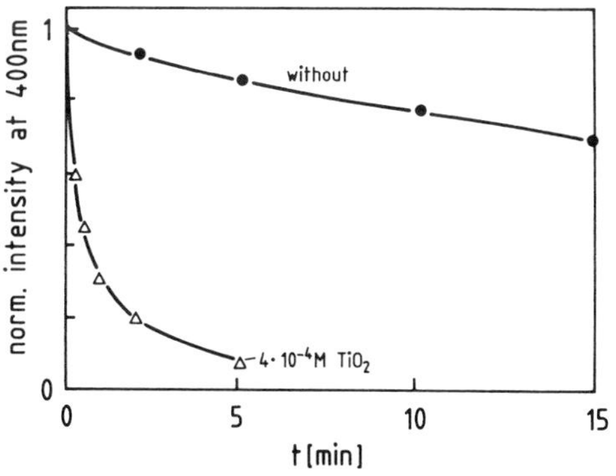

FIG. 6. Photo-anodic corrosion of Cd_3P_2 in aerated solution with and without TiO_2. The 400 nm absorption of the colloid was measured as a function of time [18].

that the photo-corrosion of Cd_3P_2 is accelerated by a factor of about 10. In connection with TiO_2, Cd_3P_2 corrodes with a quantum yield of 0.7 molecules consumed per photon absorbed. Similarly, the corrosion of CdS is enhanced. Methyl viologen is reduced on CdS-TiO_2 with a quantum yield of 0.9 molecules/photon absorbed (Fig. 7) [5,18].

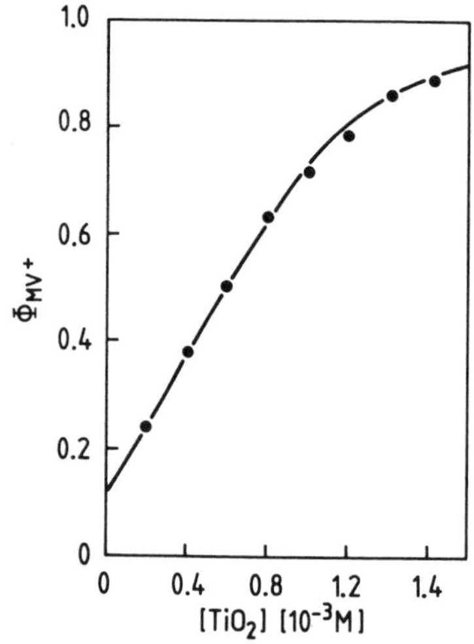

FIG. 7. Quantum yield of MV^+ formation as a function of TiO_2 content in $2\cdot10^{-4}$ M CdS solution containing $2\cdot10^{-4}$ M methylviologen (MV^{2+}).

EXCESS ELECTRONS AND HOLES

When excess electrons or holes are present on a colloidal particle its optical absorption often changes. Radiation and pulse radiolysis are useful tools to study such absorption changes. In these experiments, free radicals are produced in bulk solution, which subsequently diffuse to the colloidal particles to transfer an electron or inject a positive hole. For example, TiO_2 particles acquire a blue color upon deposition of an electron [20] and CdS or Cd_3P_2 get a broad absorption band in the visible when they carry a surface trapped hole [21].

However, deposition of an excess electron on a small particle may also be accompanied by a decrease in the optical absorption of the particle within a certain wavelength range below the threshold of absorption. The first observations were made by Albery et al. who carried out a 10 μs flash photolysis study on colloidal CdS [22]. They found a weak bleaching signal in the 460 to 510 nm range and attributed it to electrons which were trapped on the particles during the photoflash. A more quantitative study was made in our laboratory where the reaction of hydrated electrons with colloidal CdS was investigated [3]. Fig. 8 shows the spectrum of bleaching. The colloidal particles had a mean diameter of 30 A, they still were yellow and started to absorb slightly above 500 nm. It is most important to note that the negative absorption coefficient in the maximum of bleaching in Fig. 8 is greater by a factor of about 100 than the absorption coefficient of a CdS molecule. This means that the deposition of an electron on a colloidal particle does not influence the absorption of just one CdS molecule but of a great number of molecules in the particle. In other words, the excess electron influences an optical transition where the wavefunction extends over practically the whole particle. This effect was attributed to the influence of the excess electron on the energy of the lowest excitonic state formed by light absorption. In

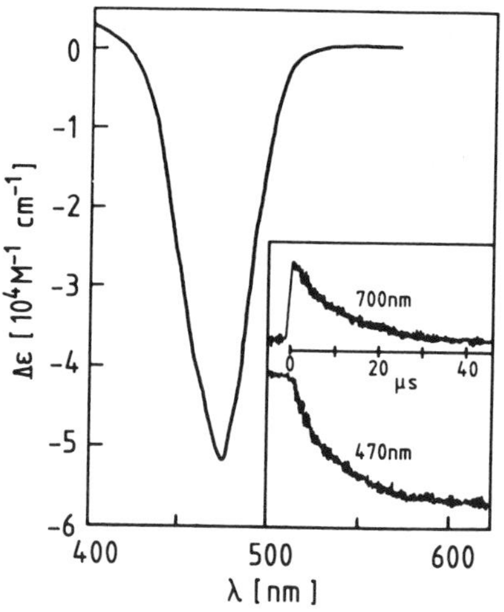

FIG. 8. Bleaching (expressed as negative absorption coefficient) as a function of wavelength in a $2\cdot10^{-3}$ M CdS solution in which $2\cdot10^{-7}$ M hydrated electrons were generated by pulse radiolysis. Particle concentration: $8.5\cdot10^{-6}$ M. Inset: Decay of the 700 nm absorption of e_{aq}^- and time profile for bleaching at 470 nm [3].

the electric field of the excess electron the excitonic state, consisting of a delocalized electron and a delocalized hole, is polarized.

Bleaching upon deposition of an electron was also observed in the case of colloidal ZnO [11,23,24]. Fig. 9 shows the absorption spectrum before and after electron storage. The electron was deposited by the radical CH_2OH produced by γ-irradiation.

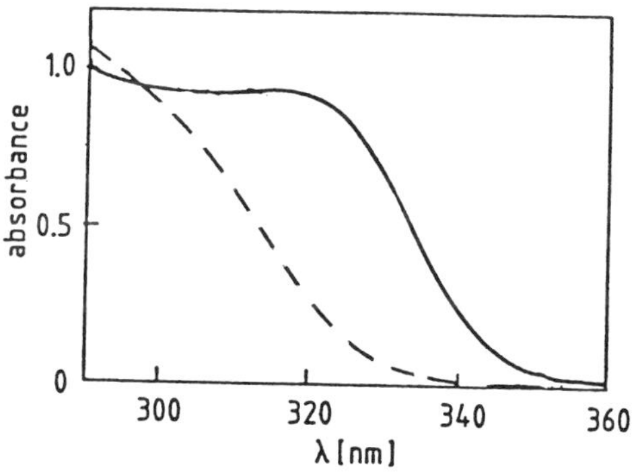

FIG. 9. Absorption spectrum of colloidal ZnO before (——) and after (----) deposition of one electron on each particle [11].

One can also see bleaching in laser flash experiments. In the case of CdS, the bleaching signal follows the laser profile while, in the case of ZnO, a long-lived signal is produced. Fig. 10 shows two absorption spectra of a CdS solution. One was measured using a commercial spectrophotometer, while the other was measured using a dye laser. It is recognized that the absorption threshold is blue-shifted in the latter case. In other words, the spectrum of the small particles depends on the intensity of the light used to

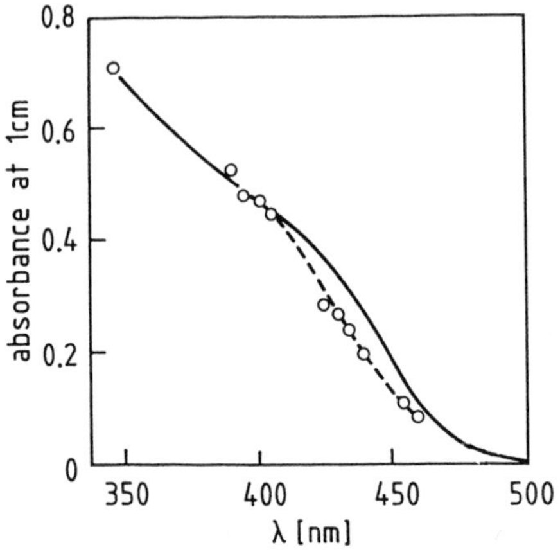

FIG. 10. Spectrum of colloidal CdS as measured in a commercial spectrophotometer (——) and using an intense dye laser (----) [6].

measure the spectrum. This non-linear absorption effect could perhaps be used to modulate one light beam by another one. The effect is weak in CdS, where the signals are fast, and much stronger in ZnO, where the signals are slow. It has to be seen how other semiconductor particles behave in this respect and whether a material can be found, where the signals are both strong and fast.

Two further experiments may be mentioned in which bleaching effects play an important role. When sandwich structures consisting of ZnO and CdS or Cd_3P_2 are illuminated with visible light, the absorption spectrum of the absorbing part of the sandwich does not change, while that of the ZnO part is blue-shifted. This was regarded as the most direct proof for electron transfer in illuminated sandwich structures [5,18]. The second study is concerned with colloidal titanium dioxide. As already mentioned above this colloid acquires a blue color upon deposition of electrons. Fig. 11 shows the

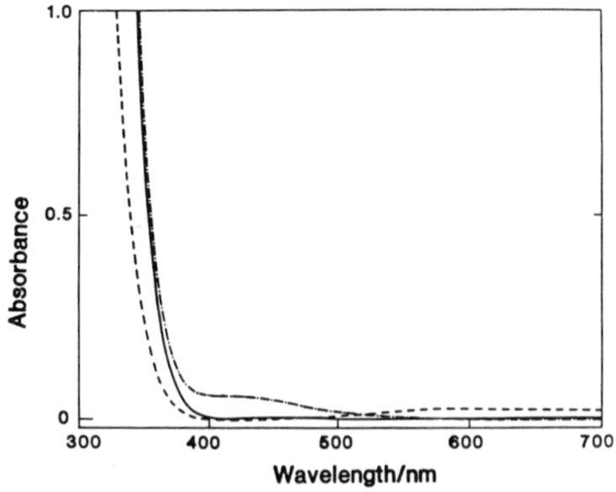

FIG. 11. Absorption spectrum of illuminated TiO_2 colloid before (——) and after illumination (----) in the absence of air, and after admission of air (-·-·-·) [25].

corresponding increase in the absorption spectrum at longer wavelengths. However, it can also be seen that the absorption of the semiconductor itself, which starts in the unirradiated sample at 380 nm, is blue-shifted by 8 nm. Both effects are reversed upon admission of air to the irradiated sample (in fact, a new absorption between 400 and 500 nm appears upon admission of air, which is attributed to Ti-peroxides resulting from the electron uptake by O_2) [25].

PHOTO-ELECTRON EMISSION

Laser illumination of small semiconductor particles in aqueous solution can lead to the formation of hydrated electrons. The effect was first detected in the cases of CdS and ZnS [26] and more recently for Cd_3P_2 [27]. No emission was observed for ZnO and TiO_2. A detailed study has recently been made for CdS solutions [28]. Fig. 12 shows the 700 nm absorption of the

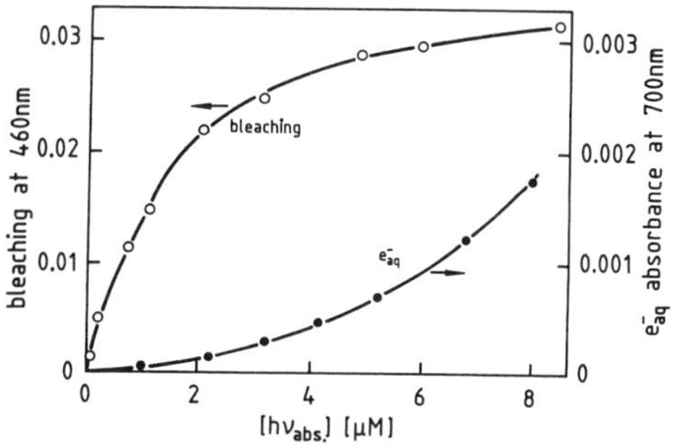

FIG. 12. Bleaching at 460 nm during the laser flash and e_{aq}^- absorption immediately after the flash as functions of the absorbed dose. The latter is expressed as concentration of absorbed photons.

hydrated electron immediately after the laser flash as a function of laser dose. It can be seen that a linear relation does not exist, the bending-upwards of the curve indicating that two photons are required. During the flash bleaching at 460 nm occurs which is ascribed to a certain stationary electron concentration on the particles. By measuring the bleaching signal one obtains information about this concentration. It can be also seen from Fig. 12 that the bleaching signal increases linearly with dose at low doses but increases less and less steeply at higher doses.

The emitted electrons can be scavenged by acetone in the aqueous solution. Under these circumstances one detects a long-lived absorption around 700 nm after the flash which is produced by surface trapped holes. Fig. 13 shows a comparison of the absorptions of the emitted electrons (absence of acetone) and the remaining holes (presence of acetone). At first sight, one would expect the two curves to parallel each other, since as many holes should remain as electrons are emitted. However, the h^+-curve starts to strive towards a limiting value at relatively low laser doses while the e^--curve still increases strongly. The reason lies in the fact that several electrons are emitted from a colloidal particle at higher doses, and the remaining holes, which chemically are S^- radical anions, deactivate each other: $2\ S^- \rightarrow S + S^{2-}$ (or S_2^{2-}).

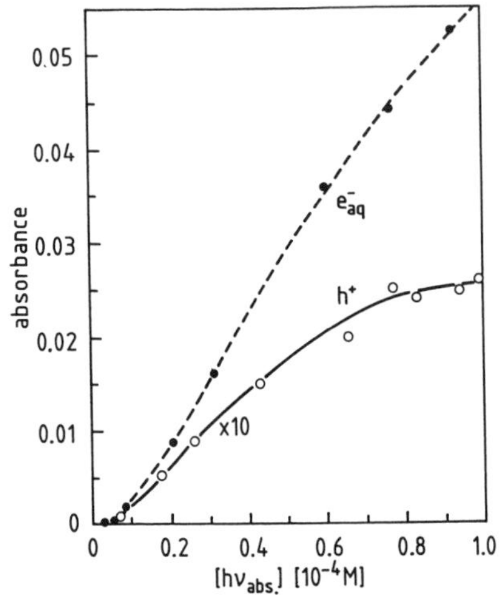

FIG. 13. Absorption at 700 nm of the emitted electrons and of the remaining holes as functions of the laser dose.

The following mechanisms of photo-electron emission was proposed [28]:

$$(CdS)_n \underset{k_1}{\overset{h\nu}{\rightleftarrows}} (CdS)_n(e^- - h^+) \underset{k_2}{\overset{h\nu}{\rightleftarrows}} (CdS)_n(e^- - h^+)_2 \xrightarrow[k_e]{} CdS(h^+) + e_{aq}^- \quad (2)$$

Absorption of the first photon yields a particle carrying one excited state (e^--h^+). This particle may either loose its excitation by recombination of the charge carriers with rate constant k_1 or absorb the second photon to yield a doubly excited particle. This particle in turn may loose its second excitation with rate constant k_2 or emit an electron into the aqueous phase with rate constant k_e. The dose dependencies of the e_{aq}^- yield and of the stationary electron concentration on the particles could be understood if the bimolecular rate constants k_1 and k_2 were not equal, but $k_2 > k_1$. Note that the photo-electron emission from small particles is a two-photon process which does not belong to either of the two processes which are generally found, i.e. either simultaneous photon absorption or absorption of the first photon to produce an excited state which absorbs the second photon. In the small particles two excitonic states are formed by the two photons, and it is the interaction between these two states that finally leads to the emission of an electron. This effect is particularly efficient in very small semiconductor particles, as the quantum yield for electron emission decreases with increasing particle size.

CONCLUDING REMARKS

The photo-chemistry of inorganic colloids has made rapid progress during the past few years. This kind of science is distinguished from the more conventional studies in colloid chemistry with respect to the preferential use of very small particles. The reason for this was originally to deal with solutions that scattered little and could be investigated using the powerful methods of flash photolysis and pulse radiolysis to elucidate reaction mechanisms. In preparing smaller and smaller particles it was soon recognized that they had unusual properties as their electronic structure changes with size. In fact, the investigation of very small particles constitutes a neglected dimension in colloid science. In ordinary colloid chemistry, quantum mechanical effects do not play a role as the particles consist of thousands and millions of molecules or atoms, the electronic structure being the same as in the compact materials. In going to extremely small particles, quantum mechanical effects occur as described above, and it can be expected that further interesting effects will be found in the future.

REFERENCES

1. A. Henglein in: Photochemical Conversion and Storage of Solar Energy, J. Rabani, ed. (The Weizmann Science Press of Israel, part A, 1982) p. 115.
2. L. Brus, Nouv. J. Chim. 11, 123 (1987).
3. A. Henglein, A. Kumar, E. Janata, and H. Weller, Chem. Phys. Lett. 123, 133 (1986).
4. Y. Wang and W. Mahler, Opt. Commun. 61, 233 (1987).
5. L. Spanhel, H. Weller, and A. Henglein, J. Amer. Chem. Soc. 109, 6632 (1987).
6. M. Haase, H. Weller, and A. Henglein, J. Phys. Chem., in press.
7. A. Henglein, Topics in Current Chemistry 143, 113 (1987).
8. A. Fojtik, H. Weller, U. Koch, and A. Henglein, Ber. Bunsenges. Phys. Chem. 88, 969 (1984).
9. H. Weller, A. Fojtik, and A. Henglein, Chem. Phys. Lett. 117, 485 (1985).
   A. Henglein, A. Fojtik, and H. Weller, Ber. Bunsenges. Phys. Chem. 91, 441 (1987).
10. L. Spanhel, M. Haase, H. Weller, and A. Henglein, J. Amer. Chem. Soc. 109, 5649 (1987).
11. M. Haase, H. Weller, and A. Henglein, J. Phys. Chem. 92, 482 (1988).
12. Y. Wang, A. Suna, W. Mahler, and R. Kasowski, J. Chem. Phys. 87, 7315 (1987).
13. K. Pohl, H. Weller, and A. Henglein, Chem. Phys. Lett. in press.
14. C.J. Sandroff, D.M. Hwang, and W.M. Chung, Phys. Rev. B33, 5953 (1986).
    O.I. Micic, M.T. Nenadovic, M.W. Peterson, and A.J. Nozik, J. Phys. Chem. 91, 1295 (1987).
    O.I. Micic, Li Zongguan, G. Mills, J.C. Sullivan, and D. Meisel, J. Phys. Chem. in press.
15. A. Henglein, Ber. Bunsenges. Phys. Chem. 86, 301 (1982).
16. T. Dannhauser, M.O'Neil, K. Johansson, D. Whitten, and G. McLendon, J. Phys. Chem. 90, 6074 (1986).
17. H. Weller, U. Koch, M. Gutiérrez, and A. Henglein, Ber. Bunsenges. Phys. Chem. 88, 649 (1984).
18. L. Spanhel, A. Henglein, and H. Weller, Ber. Bunsenges. Phys. Chem. 91, 1359 (1987).
19. L. Spanhel, H. Weller, A. Fojtik, and A. Henglein, Ber. Bunsenges. Phys. Chem. 91, 88 (1987).
20. A. Henglein, Ber. Bunsenges. Phys. Chem. 86, 241 (1982).
21. S. Baral, A. Fojtik, H. Weller, and A. Henglein, J. Amer. Chem. Soc. 108, 375 (1986).
22. W.J. Albery, G.T. Brown, J.R. Darwent, and E. Saiev-Iranizad, J. Chem. Soc. Faraday Trans. I. 81, 1999 (1985).
23. U. Koch, A. Fojtik, H. Weller, and A. Henglein, Chem. Phys. Lett. 122, 507 (1985).
24. D.W. Bahnemann, C. Kormann, and M.R. Hoffmann, J. Phys. Chem. 91, 3789 (1987).
25. C. Kormann, D.W. Bahnemann, and M.R. Hoffmann, J. Phys. Chem., in press.

26. Z. Alfassi, D. Bahnemann, and A. Henglein, J. Phys. Chem. 86, 4656 (1982).
27. M. Haase, H. Weller, and A. Henglein, Ber. Bunsenges. Phys. Chem., in press.
28. M. Haase, H. Weller, and A. Henglein, J. Phys. Chem., in press.

CATALYTIC AND PHOTOCATALYTIC OXYGEN EVOLUTION FROM WATER. CHEMICAL MODELS OF PHOTOSYSTEM II OF GREEN PLANT PHOTOSYNTHESIS

V.Ya. SHAFIROVICH and A.E. SHILOV
USSR Academy of Sciences Institute of Chemical Physics, Chernogolovka, 142 432 USSR

Oxygen evolution in green plant photosynthesis is one of the most important natural processes causing the existence of all fauna on our planet. The formation of dioxygen is also a necessary part of model catalytic systems of water photocleavage, which may be used in the future for accumulation of solar energy in the form of dihydrogen and dioxygen. The mechanism of dioxygen formation in photosynthesis is still not sufficiently clear. Therefore the creation of pure chemical model systems capable of evolving dioxygen from water photocatalytically under the solar light action, besides its own interest, may help to clarify the mechanism of the dioxygen formation in the photosynthetic process.

Thermodynamics of Water Oxidation and a Possible Mechanism of Catalytic O_2 Formation [1]

The necessity of a catalyst for water photodecomposition into dihydrogen and dioxygen under the action of solar light follows from the consideration of thermodynamics of possible stages in water oxidation:

$$H_2O - e \rightleftharpoons {}^{\bullet}OH + H^+ \qquad (+2.38\ V)$$

$$2H_2O - 2e \rightleftharpoons H_2O_2 + 2H^+ \qquad (+1.35\ V)$$

$$2H_2O - 3e \rightleftharpoons HO_2^{\bullet} + 3H^+ \qquad (+1.27\ V)$$

$$2H_2O - 4e \rightleftharpoons O_2 + 4H^+ \qquad (+0.82\ V)$$

The values in brackets are the redox potentials of the corresponding processes for pH 7. To have effective photocatalytic process using the most part of the solar spectrum the oxidant potential should not exceed 1.0-1.3 V. With such oxidants only four-electron oxidation of water will proceed with noticeable rate. A one-electron oxidant, $P_{680}^{+\bullet}$, is formed in the primary step of photosynthesis ($E^0 \cong +1.1$ V [2]). Therefore a catalyst is needed which would donate four electrons consecutively to the oxidant and then would form dioxygen from two water molecules in a four-electron process without intermediate formation of OH radicals or hydrogen peroxide. The catalyst redox potential should not change considerably in the four stages of the reaction with the oxidant, the redox potential of the whole process being close to that of each step. This condition is most easily met, when the catalyst is a cluster uniting n weakly bound metal ions, including at least one with two water molecules in its coordination sphere.

Published 1989 by Elsevier Science Publishing Co., Inc.
Photochemical Energy Conversion
James R. Norris, Jr. and Dan Meisel, Editors

$$M_{n-1}\cdots M(H_2O)_2 \xrightarrow{-4e,\,-4H^+} M_{n-1}\cdots M\langle{}^{O}_{O} \longrightarrow$$

$$M_{n-1}\cdots M\langle{}^{O}_{O}| \xrightarrow{+2H_2O,\,-O_2} M_{n-1}M(H_2O)_2 + O_2$$

There may be some variants of this general scheme. E.g., dioxygen may be formed from the O atoms bound to two neighbouring metal atoms rather than to a single one. The one-electron oxidant reacting with the cluster may itself participate in the catalytic complex promoting dioxygen formation. It may be expected that it is easier in a large cluster to create a suitable redox potential for the catalytic process. Indeed, if some mutual influence of the metal ions does exist, the redox potential of the fully oxidized state of the polynuclear complex should be higher and for the reduced state - lower than the potential for the same oxidation state of the mononuclear complex. Electrons and holes of the cluster should be at least partially collectivized and therefore the redox potential of the cluster should increase at the oxidation of the metal ions involved. Thus at the increase in n the probability increases for some intermediate oxidation state to be a suitable catalyst for dioxygen evolution from water.

However, a polynuclear catalyst may be difficult to realize in a biological system. Naturally even a mononuclear catalyst could be suitable, provided it could change its oxidation state by four units without substantial change in the redox potential at each one-electron step. For a binuclear catalyst, which is still more acceptable for biological systems than a polynuclear one, there should be a possibility of changing the oxidation state by two units for each of the ions in the catalytic complex.

Photosystem II of Plant Photosynthesis

According to the present-day knowledge [3,4] photosystem II (PSII) of the plant photosynthesis includes a dimer of chlorophyll P_{680}. Exitation of this dimer leads to the electron transfer to the pheophitine molecule, which is an intermediate electron acceptor, this process taking less than 1 ns and with the energy loss of 0.04-0.08 eV forming the ion-radical pair $[P_{680}^{+\bullet}Pheo^{-\bullet}]$. The recombination of charges in this pair (which would proceed for 2-4 ns) is prevented because of an even faster ($\leqslant$ 200 ps) electron transfer from $Pheo^{-\bullet}$ to the acceptor Q, which represents a complex of plastoquinone with Fe. The energy loss in this process is considerable ($\sim$0.5 eV) and the life-time of the $[P_{680}^{+\bullet}Pheo]Q^{-\bullet}$ state is $\simeq$150 mcs. There is no recombination of charges between $P_{680}^{+\bullet}$ and $Q^{-\bullet}$, since fast electron transfer ($\leqslant$ 1 mcs) takes place from the primary electron donor D to $P_{680}^{+\bullet}$. D^+ formed in the transfer oxidizes manganese-containing active center, which accumulates four oxidation equivalents consecutively and then oxidizes water forming dioxygen. According to the well-known Kok scheme [5] five oxidation states of the dioxygen-producing center are designated, S_o — S_4.

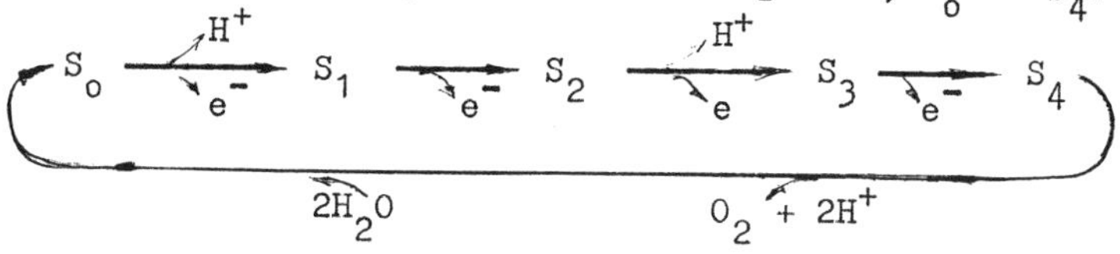

It is generally accepted now that O_2 evolution is produced as the result of the catalytic action of Mn ions, each S-state corresponding to the definite oxidation state of manganese. The number of Mn ions and their oxidation state remains, however, a matter of controversy. The minimum number of manganese ions present in the center is equal to four, however, there are some data (e.g., see [2]) indicating that only two Mn are necessary for O_2 evolution and two others play a different role and may be replaced by other bivalent metal ions, particularly Ca^{2+}.

Beginning with the work of Dismukes [6] Mn-containing center is investigated by EPR technique, however, no definite conclusion has been drawn, since EPR spectrum (belonging to S_2 state) may be interpreted as the result of the interaction of four but also of two magnanese ions. If to accept the number of Mn ions in PSII, which are directly responsible for O_2 evolution, as equal to two (this opinion seems to be shared by many investigators), the following Mn oxidation states may be attributed to S_o — S_4 states (A is an acceptor)

$$\underset{S_o}{Mn^{II}Mn^{III}} \longrightarrow \underset{S_1}{Mn^{III}Mn^{III}} \longrightarrow \underset{S_2}{Mn^{III}Mn^{IV}} \longrightarrow$$

$$\underset{S_3}{Mn^{IV}Mn^{IV}} \longrightarrow \underset{S_4}{Mn^{IV}Mn^{V}} \text{ (or } Mn^{IV}Mn^{IV}A)$$

Thus, manganese turned out to be the element, which was chosen by nature for the catalysis of dioxygen evolution from water in the process of photosynthesis. The conclusion that two Mn ions participate in the active center is consistent with the general scheme above. Two-electron oxidation of Mn^{II} to Mn^{IV} and catalytic O_2 formation in the presence of Mn^{IV} are very well known in manganese chemistry and it may be thought that these properties of Mn were used in PSII with two Mn ions participating in the active center. The structure of the active center and its surrounding in PSII suggested in [4] is presented in Fig. 1.

The important role of other metal ions such as Cl^- and Ca^{2+} was recognized for a long time [3,4]. The removal of Cl^- from thylakoids causes the loss of the catalytic activity, which may be restored at the addition of Cl^-. Some other ions (Br^-, NO_3^-, HCO_2^-, HCO_3^-) may replace Cl^- in restoring activity in O_2 evolution. At present the role of Cl^- as well as Ca^{2+} is not yet clear [3,4].

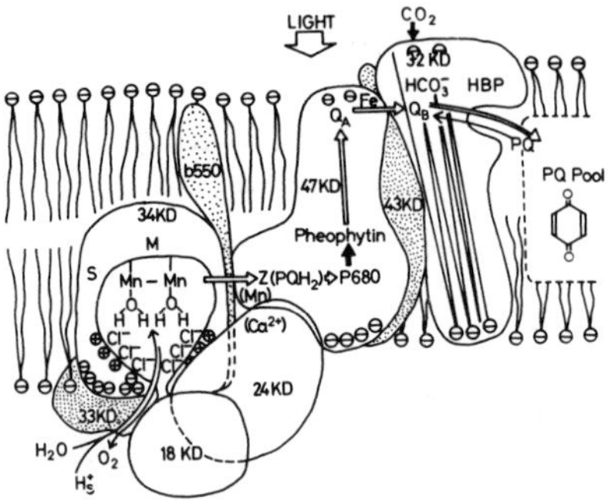

Fig. 1:
A working model for the organization of PSII components in the thylakoid membrane
(Reproduced with permission of Govindjee [4]).

Great attention was paid to the protein molecules surrounding the active center [7]. Three proteins designated as Y-33, Y-23, and Y-16 are particularly important in the catalytic activity (see Fig. 2). However, the attempts to isolate a protein containing manganese did not so far produce definite results. Presumably the role of Y-23 and Y-16 is regulatory and structural, rather than catalytic [7]. They may be removed under certain conditions without the loss of the catalytic activity of the manganese center. Y-33 is apparently more closely bound to manganese, however, it does not directly coordinate manganese to make it catalytically active.

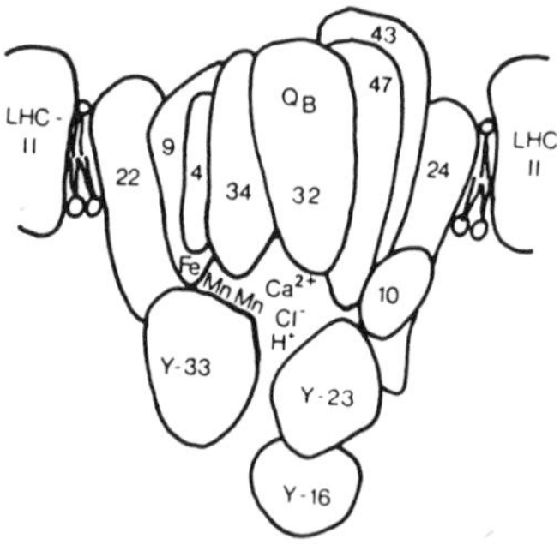

Fig. 2:
The architecture of the oxygen-evolving complex
(Reproduced with permission of B. Andersson [7]).

Thus O_2-evolving manganese-containing center is fixed in the phospholipid thylakoid membrane. The proteins and inorganic ions surrounding the manganese center help to stabilize the cluster in the form, which contains the optimum number of Mn ions (apparently two in the center directly connected with O_2 evolution, other two playing a regulatory and structural rôle).

In order to create a functional chemical model of PSII it is necessary to solve two problems:

(1) using solar light to form a one-electron oxidant capable of forming O_2 from water ($E \geqslant 0.8$ V at pH 7). It is es-

sential to prevent its recombination with simultaneously formed reduced forms;

(2) to perform coupled oxidation of water catalytically, first oxidizing the catalytic complex by the oxidant in four consecutive steps and then oxidizing water in four-electron process.

We shall start with the second problem, which has been solved better by now than the first one.

Chemical Catalysts of O_2 Formation in the Oxidation of Water

According to the scheme given above the formation of O_2 may proceed with the participation of the catalyst capable (at least in its highest oxidation state) of oxidizing water to form dioxygen. This requirement may be met in the case of transition elements situated in the Periodic Table beginning with the 7th group, i.e. for the following 3d elements: Mn, Fe, Co, Ni, Cu. It is convenient to use tris-bipyridyl complexes of Ru^{III}, Fe^{III}, and Os^{III} with the redox potentials of 1.26, 1.06, and 0.82 V, respectively. Indeed, it was shown [8,9], that hydroxocomplexes of all the elements mentioned may be used as catalysts of O_2 evolution from water in neutral and alkaline media (see Table I). Their activity is

TABLE I. The dioxygen yield as a function of the pH values in catalytic water oxidation by $Ru(bpy)_3^{3+}$ [8,9] ($[Ru(bpy)_3^{2+}] = 10^{-3}$ M, $[catalyst] = 10^{-4}$ M)

Catalyst \ pH	4	5	6	7	8	9	10	12	13
Mn(II)	2	6	7	14	18				
Fe(II)					78	86	84	51	6
Co(II)		22	68	87	90	88	94	76	56
Ni(II)					58	75	77	15	
Cu(II)					80	87	86	24	2
without catalyst					7	11	11	7	1

sometimes so high that considerable rates of dioxygen evolution have been observed even at concentrations $\sim 10^{-6}$ M. E.g., iron ions concentration in the distilled water is often sufficient to observe catalytic dioxygen evolution under the action of $Ru(bpy)_3^{3+}$. This was the reason of earlier erroneous conclusion [10], that non-catalytic O_2 evolution may be observed in water solution in the presence of $Ru(bpy)_3^{3+}$.

The yields of dioxygen evolved at different pH's catalyzed by cobalt hydroxocomplexes are presented in Fig. 3. It may be seen that the weaker is the oxidant, the more alkaline is the solution, which contains the hydroxocomplex capable of catalyzing dioxygen evolution. Presumably in alkaline media a less strong oxidant is needed to oxidize deprotonated complex to the state capable of forming dioxygen from two water molecules. At the first sight, it is surprising that hydroxocomplexes of all metals after manganese are capable of catalyzing O_2 evolution under the action of $Ru(bpy)_3^{3+}$. Accord-

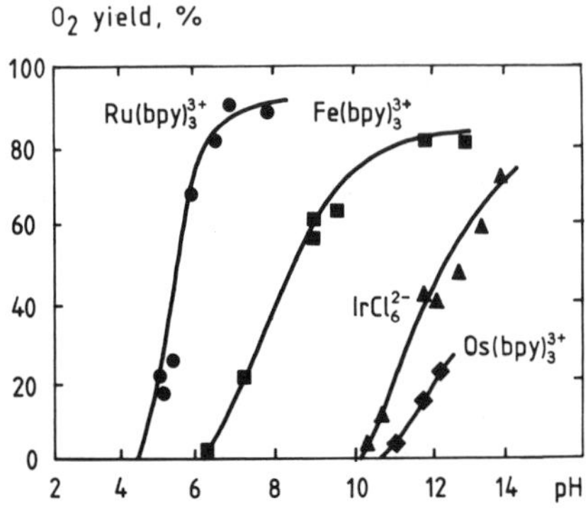

Fig. 3:
The dioxygen yield as a function of the pH values at water oxidation by one-electron oxidants in the presence of Co(II) [oxidant] = 10^{-3}M, [Co^{II}] = 10^{-4} M [8].

ing to the mechanism of catalysis the redox potential of the catalytic complex in its highest oxidation state must be higher than $E(O_2/H_2O)$ but at the same time it must not exceed significantly E for the oxidant. These two conditions seem to limit the number of complexes capable of catalyzing dioxygen evolution. Two reasons for comparative non-specificity may be suggested. First, while increasing pH we can make catalytic less and less active complexes. Secondly, as it was already stated the redox potential of the complexes increase with the number of ions oxidized. Therefore for polynuclear complexes the probability is increasing that some intermediate oxidation state will be catalytically active for O_2 evolution.

Manganese Compounds as Catalysts of O_2 Evolution

Freshly precipitated MnO_2 is catalytically active in O_2 evolution under the action of sufficiently strong oxidants. E.g., it is known since 1968, that MnO_2 catalyzes the formation of dioxygen under the action of Ce^{4+} [11]. O_2 is evolved in the presence of MnO_2 also when $Ru(bpy)_3^{3+}$ is an oxidizing agent [9]. As compared with cobalt hydrocomplexes O_2 formation is observed at lower pH (compare Figs. 4 and 3) as soon as E for four-electron oxidation, $H_2O = O_2 + 4e + 4H^+$,

Fig. 4:
The dioxygen yield as a function of the pH values at water oxidation by $Ru(bpy)_3^{3+}$ in the presence of MnO_2 [9].

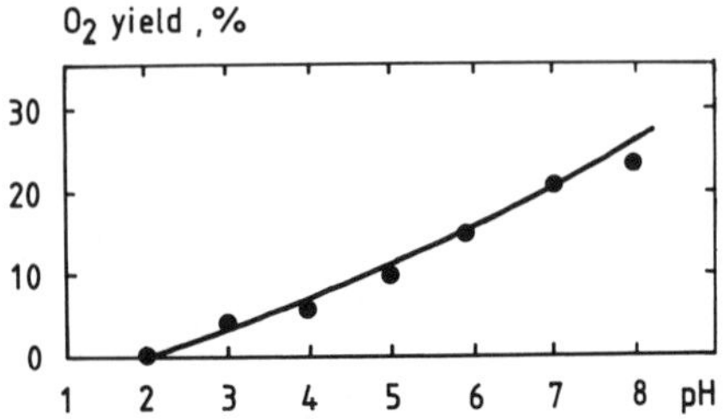

reaches the value of the oxidant redox potential. Hence, polynuclear MnO_2 ensures optimum correspondence in catalytic formation of O_2. With this respect RuO_2 has similar properties being an effective catalyst for dioxygen evolution [12].

Both MnO_2 and RuO_2 can be effective catalysts of water photooxidation under solar light [9,12]. In our work in 1981 we observed catalytic evolution of dioxygen in the system containing MnO_2 as a catalyst, Mn^{IV} pyrophosphate complex as an electron acceptor, and $Ru(bpy)_3^{2+}$ as a photosensitizer [8]. The following cyclic mechanism may be suggested:

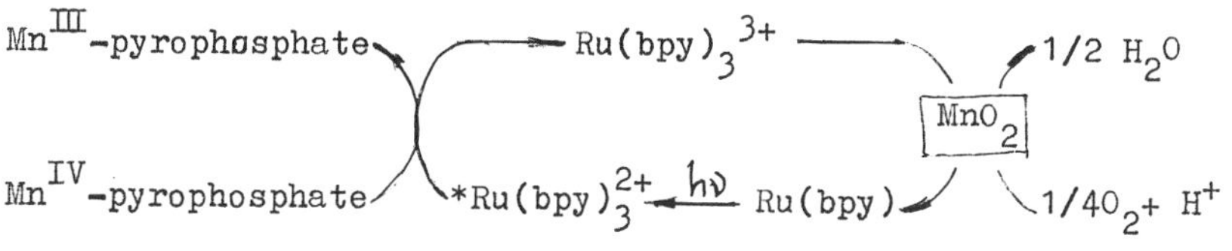

The system evolves dioxygen with a constant rate in the stationary process. The quantum yield for the formation of $Ru(bpy)_3^{3+}$ is 0.26. Catalytic activity of MnO_2 for O_2 evolution is, however, low and the latter process is the rate controlling step with all the Ru complexes being in the ruthenium(III) state.

Vesicular Catalytic and Photocatalytic Systems of O_2 Formation from Water

The results and considerations presented above allow to suggest a hypothesis, how and why manganese was chosen by nature to form a catalyst for dioxygen evolution in photosynthesis. It may be suggested that Mn hydroxocomplexes incorporated by biological membranes from water turned out to be optimum catalysts for O_2 evolution and became an active center of PSII [9]. If so, a simple Mn hydroxocomplex precipitated at the membrane has to be an active catalyst for dioxygen evolution and has to demonstrate definite advantages as compared with similar hydroxocmplexes of other metals. It should be noted that incorporating the catalyst in the vesicle wall is an important step in the creation of organized molecular systems capable of transformation of solar energy into the energy of chemical substances.

In our recent works [13,14] we confirmed the possibility of catalytic dioxygen formation in the presence of manganese hydroxocomplexes precipitated in the phospholipid membrane forming a wall of a vesicle and compared catalytic activity of hydroxocomplexes of various metals. Catalysts were prepared by ultrasonic dispergation (30 min) of dipalmitoyl-α-DL-phosphatidylcholine in a buffer solution containing salts of several metals. During this procedure metal hydroxides precipitated on the lipid membrane (in the case of Mn, Fe, Co in the oxidized forms). Vesicles were separated from the solution by gel-filtration and then were analyzed gel-chromatographically and also for metal content. Monolamellar vesicles containing tightly bound manganese were formed in the case of manganese. To determine the catalytic activity the oxidant was introduced in the solution and the amount of dioxygen formed measured. The activity of the manganese hydroxocomplex fixed on the vesicle was found to be considerably higher as compared with free MnO_2. It should be noted that preservation of the catalytic activity of the complex bound to the membrane is not a trivial fact. A priori it could be expected that hydroxocomplexes incorporated in the phospholipid membrane when oxidized by $Ru(bpy)_3^{3+}$ might oxidize organic surrounding rather than react with water to form dioxygen. Indeed, RuO_2 stopped catalyzing dioxygen evolution when fixed at the membrane. Instead CO_2 was formed presumably because of oxidation of organic molecules.

Preferential water oxidation with the dioxygen formation was an important precondition for participation of manganese complexes in photosynthesis. Comparison of the catalytic activity of hydroxocomplexes of different metals supported by the phospholipid membrane is given in Table II. Two oxidants were used: $Ru(bpy)_3^{3+}$ (E = 1.26 V) and $Fe(bpy)_3^{3+}$ (E = 1.05 V). In the case of the former, dioxygen is evolved in the presence of hydroxocomplexes of all 3d metals beginning with Mn. However, it can be seen, that at pH 7 and lower the activity of a manganese complex is higher than that of complexes of all other metals.

TABLE II. The pH-dependent O_2 yield per oxidant (%) for various catalysts fixed on the lipid membrane [14]($[Ru(bpy)_3^{3+}] = 5x10^{-4}$ M, $[Fe(bpy)_3^{3+}] = 5x10^{-4}$ M, $[catalyst] = 3x10^{-5}$ M).

Catalyst pH	5.7	7.0	7.6	8.4	9.0
		O x i d a n t	-	$Ru(bpy)_3^{3+}$	
Cr	-	0	-	-	0
Mn	21.5	51.0	43.0	43.3	43.5
Fe	0	0	52.1	71.5	81.6
Co	0	8.1	79.5	90.9	94.3
Ni	0	5.7	33.7	36.9	32.3
Cu	0	0	4.1	22.3	27.0
		O x i d a n t	-	$Fe(bpy)_3^{3+}$	
Cr		0	-	-	0
Mn		0	3.8	6.8	7.6
Fe		0	0	0	0
Co		0	4.4	16.6	29.8
Ni		0	3.1	17.3	31.1
Cu		0	0	0	0

In the case of $Fe(bpy)_3^{3+}$, which is even a weaker oxidant than $P_{680}^{+\bullet}$ in PSII, only Mn, Co, and Ni complexes remain catalytically active, dioxygen evolution being observed only at pH > 7. These results confirm and make more precise the suggestion about manganese hydroxocmplexes choice as the catalyst of dioxygen evolution in photosynthesis. It is seen, that the Mn complex turned out to be suitable not only because of its optimum thermodynamic and kinetic properties in O_2 evolution but also due to the ability to bind to a membrane without making it unstable both chemically and mechanically. The spectral characteristics of the Mn hydroxocomplex on the membrane show that it is first incorporated in the form of manganese(III) and then oxidized to manganese(IV) under the action of the oxidant.

Investigation of the reversible $Mn^{IV} \rightleftharpoons Mn^{III}$ - e transfer in the lipid membrane shows that redox potential dependence on pH may be expressed by the formula E = 1.54 - 0.14 pH It gives $E(Mn^{IV}/Mn^{III})$ = 0.5 V at pH 7. This value is not sufficient for dioxygen formation (E = 0.8 V). Manganese(IV), which remains after all oxidant used, does not evolve O_2 from water even at elevated temperatures. Therefore in its active form manganese must be oxidized to higher oxidation state (perhaps to Mn^{IV}) or the oxidant, e.g., $Ru(bpy)_3^{3+}$, must be involved in the catalytic complex increasing its redox potential. Apparently similar situation exists in PSII. The highest oxidation state S_4 may correspond either to a dimer $Mn^{IV}Mn^{V}$ or $Mn^{IV}Mn^{IV}A$ (A is an aceeptor).

Thus the Mn hydroxocomplexes bound to the phospholipid membrane may be considered as a functional model of PSII active center. Simple Mn hydroxocomplexes may be thought to have been used when PSII originated in living nature on the base of the bacterial photosynthesis. In the subsequent evolution PSII has changed to its present-day very complicated state (Fig. 1). The number of Mn atoms reached the minimum (perhaps two), which is important for a number of reasons. The proteins, which surround the complex, must keep it at the definite place and prevent the complex association to a polynuclear hydroxide. Possibly it is their most important function as well as the function of Ca^{2+} and Cl^- ions. At any rate no specific interaction of Mn with polar group of the proteins has been detected so far. However, the detailed structure of Mn-containing dimer (or tetramer) remains to be determined.

There are quite a few hypotheses about the nature of the Mn complex in PSII and many new and interesting complexes have been prepared recently (see [15]). So far the activity of the complexes towards water oxidation has not been reported and presumably has not been found as distinct from simple hydroxocomplexes. Is it worthwhile to search for new and sophisticated models of PSII or better to concentrate the attention at the structure of the hydroxocomplex and the possibility of preparing it in binuclear or tetranuclear state?

Photocatalytic Models of PSII with Participation of Vesicular Systems

Dioxygen can be a product of the photocatalytic reaction in a model system, when $Ru(bpy)_3^{2+}$ is used as a photosensitizer, Mn hydroxocomplex fixed on the vesicular membrane is a catalyst, and Mn(IV)-pyrophosphate is a sacrificial electron acceptor (Fig. 5). The quantum yield is close to that of $Ru(bpy)_3^{3+}$ formation in the absence of the catalyst and with respect to the rate of dioxygen evolution the system is about two orders of magnitude more active than that of the same composition but without vesicles [14]. Presumably Mn hydroxocomplex on the vesicle wall remains to be in finely divided state and this preserves its catalytic activity. Thus the problem of catalytic dioxygen formation under the action of oxidants similar to $P_{680}^{+\bullet}$ with respect to their redox potentials is being solved successfully.

In the absence of a sacrificial acceptor it is necessary

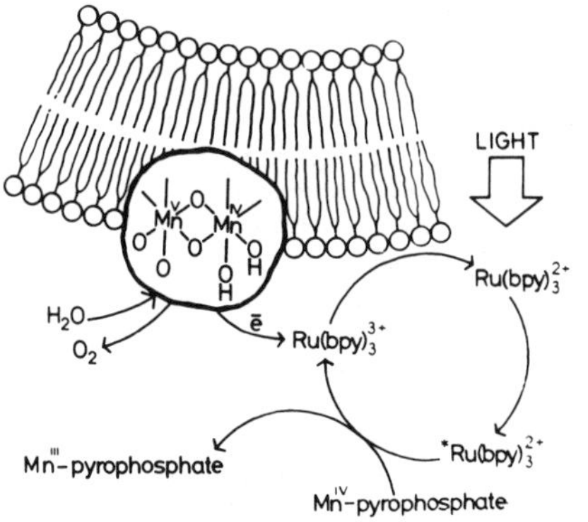

Fig. 5:

Photochemical oxygen evolution in the presence of Mn ions fixed on the phospholipid membrane.

to stabilize the divided charges from the primary photochemical act to ensure the subsequent catalytic reactions, e.g., the formation of dihydrogen and dioxygen from water. This is possible only if the oxidative and reductive parts of the system (formed photochemically) remain divided in space, since accumulation of positive and negative equivalents to form O_2 and H_2, respectively, is impossible in a homogeneous system.

Fig. 6:

Light-induced electron transfer in the vesicular system.

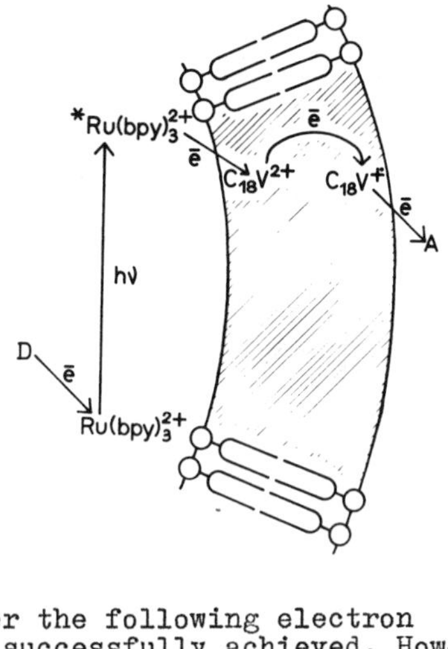

To realize this spatial division we can use general principles of PSII functioning described above. An organized molecular system containing a photosensitizer (S), primary and secondary acceptors (A_1 and A_2), must be designed and constructed. The systems most frequently used so far involved the electron transfer between the reagents colliding in solution in diffusion process. It should be noted that high yields of primary ion-radical products ($S^{+\bullet}$ and $A^{-\bullet}$) could be obtained on irradiation of such colliding partners. Moreover the following electron transfer to form $A_2^{-\bullet}$ can be also successfully achieved. However, it is difficult or impossible to prevent the further recombination (e.g., $S^{+\bullet}$ and $A_2^{-\bullet}$), if they are not fixed in a spacially organized system.

There are only few and so far rather imperfect systems specially constructed to divide spatially $S^{+\bullet}$ and $A_2^{-\bullet}$ with no sacrificial donor or acceptor used [16]. One of them is illustrated by Fig. 6. Electron transfer agent $A_1 = C_nV^{2+}$ (1,1'-

dialkyl-4,4'-bipyridine, n = 12-18) is fixed in the vesicle wall with $Ru(bpy)_3^{2+}$ (S) placed in the inner volume and the acceptor A_2 being in the outer solution.

The effectiveness of the spatial charge separation (ratio between the oxidant (A_2^-) reduced in the outer solution and the amount of the $C_nV^{+\bullet}$, which escaped from the ion-radical pair) reaches 0.1-0.2. Though the most of the divided charges formed in the primary event ($Ru(bpy)_3^{3+}$ and $C_nV^{+\bullet}$) recombine and lose the energy, the system allows to transfer noticeable part of electrons with the partial accumulation of light energy. The further development and improvement of this and similar systems will no doubt lead to the creation of effective models of PSII and the whole system of plant photosynthesis.

REFERENCES

1. V.Ya. Shafirovich and A.E. Shilov, Kinet.Katal. 23, 1311 (1982).
2. V.V. Klimov, S.I. Allakhverdiev, Sh. Demeter, and A.A. Krasnovskii, Dokl. Akad. Nauk SSSR 149, 227 (1979).
3. V.V. Klimov, Light Stages of the Electron Transfer in Photosystem II of Green Plants and Algae, Dissertation, (Pushchino, 1986).
4. Govindjee, Photochem. Photobiol. 42, 187 (1985).
5. B. Kok, B. Forbush, and M. McGloin, Photochem. Photobiol. 11, 457 (1970),
6. G.Ch. Dismukes, Y. Siderer, FEBS Lett. 121, 78 (1980); G.Ch. Dismukes, K. Ferris, and P. Watnick, Photobiochem. Photobiophys. 3, 243 (1982).
7. B. Andersson and H.-E. Åkerlund, Electron Transfer Mechanism and Oxygen Evolution, J. Barber, ed. (Elsevier, BV 1987) pp. 379-420.
8. V.Ya. Shafirovich, N.K. Khannanov, and V.V. Strelets, Nouv. J. Chim. 4, 81 (1980); N.K. Khannanov and V.Ya. Shafirovich, Kinet.Katal. 22, 248 (1981).
9. V.Ya. Shafirovich, N.K. Khannanov, and A.E. Shilov, J. Inorg. Biochem. 15, 113 (1981).
10. C.Creutz and N. Sutin, Proc. Natl. Acad. Sci. USA 72, 2858 (1975).
11. T.S. Glihkman and I.S. Shchegoleva, Kinet.Katal. 9, 461 (1968).
12. J.-M. Lehn, J.-P. Sauvage, and R. Ziessel, Nouv. J. Chim. 3, 423 (1979); J. Kiwi and M. Grätzel, Chimia 33, 289 (1979).
13. E.I. Knerelman, N.P. Luneva, V.Ya. Shafirovich, and A.E. Shilov, Dokl. Akad. Nauk SSSR 291, 632 (1986); N.P. Luneva, E.I. Knerelman, V.Ya. Shafirovich, and A.E. Shilov, J. Chem. Soc., Chem. Commun. 1504 (1987).
14. N.P. Luneva, E.I. Knerelman, V.Ya. Shafirovich, and A.E. Shilov, Dokl. Akad. Nauk SSSR (in press).
15. K. Wieghart, V. Bossek, D. Ventur, and J. Weiss, J. Chem. Soc., Chem. Commun. 347 (1985); Ch. Christmass, J.B. Vincent, J.C. Huffman, G. Christou, and H.-R. Chang, J. Chem. Soc. Chem. Commun. 1303 (1987).
16. E.E. Yablonskaya, and V.Ya. Shafirovich, Kinet.Katal. 24, 1022 (1983); N.K. Khannanov, V.A. Kuzmin, P.P. Levin, V.Ya. Shafirovich, and E.E. Yablonskaya, New J. Chem. 11, 688 (1987).

PHOTOSYNTHETIC WATER SPLITTING

ELIAS GREENBAUM
Chemical Technology Division, Oak Ridge National Laboratory, Bldg. 4500N, P. O. Box 2008, Oak Ridge, Tennessee 37831-6194 U.S.A.

ABSTRACT

This paper presents recent progress in the field of photosynthetic water splitting for both in vitro and in vivo systems. It is demonstrated that platinized chloroplasts are a novel photocatalytic material that is capable of the sustained simultaneous photoevolution of hydrogen and oxygen when irradiated with visible light. By using the technique of single-turnover saturating flashes of light an upper limit can be set on the number of platinum atoms that are necessary for the formation of a catalytically active cluster. The results of these experiments indicate that the onset of catalytic hydrogen evolution during the creation of platinum metal catalyst can be observed with 50 atoms or less, and that a stable metal catalyst is created at about 500 atoms or less.

Absolute thermodynamic efficiencies of conversion of light energy into chemical free-energy of molecular hydrogen by intact microalgae have been measured with an original physical measuring technique using a tin-oxide semiconducting gas sensor. Thin films of microalgae comprising 5–20 cellular monolayers have been entrapped on filter paper, thereby constraining them in a well-defined circular geometry. Based on absolute light absorption of visible polychromatic illumination in the low-intensity region of the light saturation curve, conversion efficiencies of 6 to 24% have been obtained. These values are the highest ever measured for hydrogen evolution by green algae.

INTRODUCTION

Chloroplasts and Platinum

Photosynthesis is the oldest and most reliable method of photochemical conversion and storage of solar energy. It is well known that photosynthesis can be separated into light and dark reactions [1–2]. The light reactions occur exclusively in the chloroplast membranes, where the overall process is energized. This ability of the chloroplast to serve as the photochemical "factory" of photosynthesis led to the consideration of systems that drive photoreactions other than carbon reduction.

In the results reviewed here, colloidal platinum was precipitated directly onto photosynthetic membranes in an aqueous suspension. The resulting chloroplast-colloidal platinum composition was then entrapped on filter paper. This moistened material was capable of sustained simultaneous photoevolution of hydrogen and oxygen when irradiated with visible light. Since no electron relay was added to the system and the overall reactions occurred in an immobilized matrix, it was concluded that the precipitated colloidal platinum directly contacts the reducing end of Photosystem I in such a way that electron flow occurs across the biological membrane-metal colloid interface with preservation of charge continuity and catalytic activity. In addition to the special photocatalytic properties of platinized chloroplasts, the specific ionic species used to prepare the material yield information on the physicochemical properties of the Photosystem I reduction site on the thylakoid membrane surface. Photoactive samples were obtained by precipitating platinum from the hexachloroplatinate(IV) ion, $[Pt(Cl)_6]^{2-}$. Platinized chloroplasts prepared by precipitating platinum from the tetraammine-platinum(II) ion, $[Pt(NH_3)_4]^{2+}$, resulted in no hydrogen activity and only a transient oxygen gush. The presence of insoluble platinum on the entrapped filter paper composition was determined by x-ray fluorescence analysis.

Flash Precipitation and Platinum Clusters

The first application of pulsed, flash-lamp techniques to elucidate the physicochemical aspects of photosynthesis began with the pioneering experiments of Emerson and Arnold which led to the

Photochemical Energy Conversion
James R. Norris, Jr. and Dan Meisel, Editors

modern concept of the photosynthetic unit [3–4]. The ability to drive the photochemical reactions of photosynthesis by pulsed illumination plus a knowledge of the orientation and vectorial nature of the photochemical reactions of photosynthesis suggest that reactions other than those associated with the normal function of photosynthesis can be driven synchronously. One well-known example is the classic Hill reaction of photosynthesis in which ferric ions are reduced to ferrous ions by isolated chloroplasts [5]. This technique has been used to photoprecipitate metallic platinum onto the external surface of chloroplast thylakoid membranes. The conversion of hexachloroplatinate ions to zero valent platinum requires a four-electron reduction:

$$PtCl_6^{2-}(aq) + 4e^- \rightarrow Pt(s) + 6Cl^-(aq) .$$

In the currently accepted molecular model of the structural organization of photosynthesis, each Photosystem I reaction center is oriented so that the external surfaces of the chloroplast thylakoid membranes contain the loci at which electrons emerge and provide reducing equivalents for the enzymatic reduction of NADP+ to NADPH [6]. It has been shown previously that the chemical precipitation of metallic platinum onto the surface of thylakoid membranes produces a photocatalytically active composite that is capable of simultaneous photoproduction of hydrogen and oxygen [7].

In the experiments reviewed here it is demonstrated that electrons from Photosystem I itself can serve as the source of reductant for platinum precipitation. By illuminating the chloroplast-hexachloroplatinate mixture with brief saturating flashes of light, it is possible to drive the reaction centers synchronously one step at a time. This in effect, allows for the photochemical titration of electrons in an experimentally countable way. As mentioned before, four reducing equivalents are required to convert hexachloroplatinate to metallic platinum. By counting the number of flashes and continuously monitoring for the onset of hydrogen, it is possible to set a limit on the number of platinum atoms that are necessary to form an active catalyst in a given experiment.

Hydrogen Evolution by Algal Water Splitting

Among the intriguing aspects of modern quantitative biological physics are the determination of the conversion efficiency of light energy into chemical energy by green plants and the analytical shape of the light-saturation curve of photosynthesis. Photosynthesis saturates with increasing light intensity because it is comprised of serially linked light-driven and thermally activated electron-transfer reactions. It is well known that measurement of conversion efficiencies under strictly light-limiting conditions assesses the inherent capabilities of the photophysical machinery of photosynthesis independently of subsequent non-light-dependent biochemistry [8].

Although obviously interrelated, maximum *quantum* conversion efficiency bears closely on molecular mechanism, whereas maximum *energy* conversion efficiency relates to net productivity. As reviewed by Pirt [9], there is, as yet, no clear consensus on values for maximum photosynthetic efficiency. However, measurements by Ley and Mauzerall [10] indicate a quantum efficiency for aerobic oxygen evolution by *Chlorella vulgaris* in the range of 9–11% for pulsed illumination at 596 nm. The limiting aspects of production of energy-rich compounds by photosynthetic and photochemical conversion and storage of light energy has been treated authoritatively by several authors [11–13]. In particular Parson [14] has presented a lucid analysis of the thermodynamics of the primary reactions of photosynthesis by deriving expressions for the change in free energy that occurs when a photochemical system is illuminated. In this, as well as the earlier work of Ross, et al. [15], the important distinction was made between midpoint redox potentials, which are molecular properties, and the actual redox potential of a system which is determined by the midpoint potential *and* the ratio of the concentrations of oxidized and reduced molecules.

Energy conversion efficiency measurements based on continuous light-induced water splitting by algae are presented in this review. Unlike aerobic terrestrial photosynthesis whose terminal electron acceptor is atmospheric carbon dioxide and whose energy-rich photoproduct is a carbon dioxide fixation compound, the energy-rich photoproduct in this work is molecular hydrogen. As first discovered by Gaffron and Rubin [16] and reviewed by Weaver et al. [17] and Bishop and Jones [18], certain classes of eukaryotic algae are capable of evolving molecular hydrogen under appropriate physiological conditions: when placed in an oxygen-free atmosphere, certain green algae are capable of synthesizing the enzyme hydrogenase. If, moreover, the atmosphere is devoid of carbon dioxide, the normal terminal electron acceptor of photosynthesis, hydrogen ions can serve as the acceptor by being reduced to molecular hydrogen in a reaction catalyzed by the enzyme hydrogenase. Since it has been previously shown that eukaryotic green algae are capable of sustained simultaneous photoevolution of hydrogen and oxygen [19], hydrogen production by algal water splitting is strictly analogous to normal photosynthesis in that water can serve as the source of reductant and, as mentioned above, hydrogen is the energy-rich product. However, from a quantitative point of view, the analogy breaks down because with hydrogen ions as the terminal electron acceptor, the light-saturated rate of oxygen evolution is at best a few percent of the light-saturated rate of oxygen evolution with carbon dioxide as the terminal electron acceptor.

EXPERIMENTAL SECTION

Chloroplasts and Platinum

Type-C chloroplasts were prepared by the procedure of Reeves and Hall [20]. In this preparation the chloroplast envelop is osmotically ruptured, exposing the thylakoid membranes to the external aqueous medium. A solution of chloroplatinic acid (5.34 mg/ml), neutralized to pH 7 with NaOH, was prepared separately. One milliliter of this solution was combined with 5 ml of chloroplast suspension (containing 3 mg of chlorophyll) in Walker's assay medium. The 6-ml volume was placed in a temperature-controlled, water-jacketed chamber fitted with O-ring connectors to provide a hermetic seal and inlet and outlet ports for hydrogen flow. The mixture was gently stirred and purged with molecular hydrogen in the headspace above the liquid. The temperature of the sample was held at 21°C.

In the platinum precipitation step, it was determined empirically that a hydrogen incubation time of ~30 minutes was needed to obtain photoactive material. Times of 60 to 90 minutes were typically used. After incubation, the reactor chamber was opened to air and the contents were filtered onto fiberglass filter paper (AP40, Millipore). The platinum precipitation reaction had a marked effect on filtration properties. Whereas control experiments without hydrogen incubation produced a chloroplast mixture that filtered immediately, the platinized chloroplasts required a considerably longer time typically 5 to 30 minutes. Also, the platinized chloroplasts were dark green, as opposed to the normal bright green of higher plant chloroplasts. The presence of insoluble platinum on the platinized-chloroplast filter paper composition was identified by x-ray fluorescence analysis after rinsing the filter paper in 2 liters of continuously stirred distilled water for 1 hour. Insoluble platainum was positively identified in this way for platinized chloroplasts prepared by precipitation from $[Pt(Cl)_6]^{2-}$ or $[Pt(NH_3)_4]^{2+}$. As mentioned above, however, only $[Pt(Cl)_6]^{2-}$ yielded photoactive material. The essential components of the measuring system are illustrated in Figure 1 [21].

Flash Precipitation and Platinum Clusters

For the flash precipitation experiments chloroplatinic acid was neutralized to pH 7 and added in the dark to a suspension of dark-adapted chloroplasts with a chlorophyll content of 3 mg. This value of chlorophyll was chosen for analytical and optical reasons. It was large enough to give an amount of hydrogen that was easily measurable with the tin-oxide hydrogen sensor, and it was small enough so that it could be light-saturated over the circular 40 cm^2 cross-sectional area of the photoreactor that was illuminated by the xenon flash lamps. The final concentration of hexachloroplatinate was 3.9mM which was obtained by dilution of a standard stock solution. The actual value is not crucial as long as there is a clear stoichiometric excess of hexachloroplatinate ions to Photosystem I reaction centers. Since it is well known that for higher plant chloroplasts such as spinach there are ~500 chlorophyll molecules for each Photosystem I reaction center, it follows that for these experimental conditions the stoichiometric ratio of hexachloroplatinate ions to Photosystem I reaction centers is ~4000:1. This ensures that hexachloroplatinate ions saturate Photosystem I binding sites. The requirement for saturation is discussed below in the context of hydrogen yield. The 8.0 ml liquid suspension was gently stirred so that equilibrium occurred between gases produced in the liquid and the carrier gas phase above the liquid.

When hydrogen is photoproduced in the reaction cell, a delay is observed in the hydrogen sensor output. This delay is composed of one or two parts, depending upon prior illumination. First, delay is associated with the response time of the flow apparatus and is always present. Second, an additional delay is associated with the photochemistry of zero-valent platinum formation for converting hexachloroplatinate ions to metallic platinum. This delay is associated only with the dark-adapted system that is being irradiated for the first time. The two sources of delay in hydrogen signal can be used to calculate an upper limit for the minimum number of atoms needed to photoprecipitate catalytically active platinum defined by the evolution of molecular hydrogen from hydrogen ions in aqueous solution.

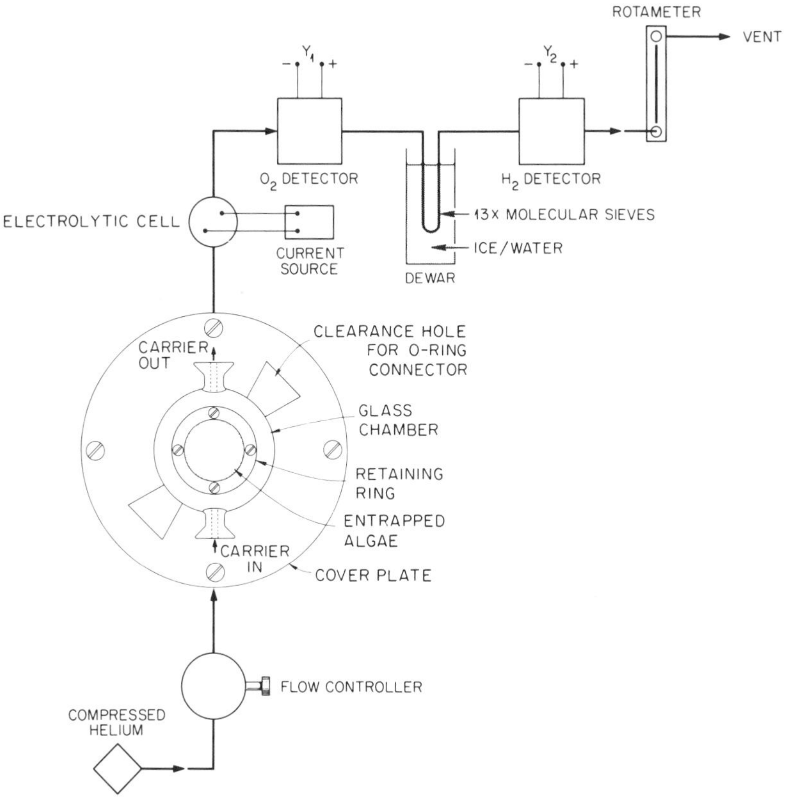

FIG. 1. Schematic illustration of measuring system. The reaction chamber used depended on whether a liquid suspension or entrapped film was studied. For example, in this illustration hydrogen and oxygen that is produced in the semisolid environment of the algal film diffuses into the gas phase adjacent to the film. The clear glass O-ring cover plate containing the entrapped film contains the photoproduced gases which are swept out of the chamber and down stream to oxygen and hydrogen sensors. Two smaller O-ring ports oriented vertically are the inlet and outlets for the helium carrier.

Hydrogen Evolution by Algal Water Splitting

An absolute measurement of thermodynamic conversion efficiency of light energy into Gibbs free-energy of molecular hydrogen requires two measurements: (1) the absolute rate of light absorbed by the algal sample and (2) the absolute rate of hydrogen photoproduction. Figure 1 is a schematic illustration of the chamber and flow system used to measure the light absorbed by the algae and to irradiate them during hydrogen and oxygen production. The main body of the chamber consists of a large glass O-ring cover plate pressing against a thin clear quartz backing plate. Two smaller O-ring connectors are used for inlet and outlet ports for gas flow to create a helium atmosphere for the algae. The algae are entrapped on filter paper and appressed to the thin quartz plate. The sensing element of a spectrally flat electro-optic radiometer (EG&G Model 550) is positioned behind the sensor and is used for absolute measurement of transmitted light. In this apparatus, the algae are entrapped in a well-defined circle of 36-mm diameter. Net light absorbed by the algae is determined by measuring the difference in transmitted light in the absence and presence of the algae. Measurement of the angular distribution of reflected and scattered radiation indicated that virtually all of the scattered radiation was forward-scattered and captured by the active area of the radiometer.

When the light is turned on, hydrogen and oxygen that are evolved in the semisolid environment of the algal film diffuse into the helium gas phase created by the helium carrier and are then swept downstream to oxygen and hydrogen detectors. Calibrations of the oxygen and hydrogen sensors were achieved with an electrolytic cell. Although the electrochemical oxygen sensor is linear over the range of gas phase concentrations studied, the tin oxide semiconducting hydrogen sensor is not. Calibration curves for both hydrogen and oxygen were constructed by observing steady-state d.c. deflections for sequential electrolysis currents of 5, 10, 25, 50 and 100 μA. Using a Hewlett-Packard 85 microcomputer, least squares fitting routines were used to generate analytical expressions for the calibration curves. It was determined empirically that good fits to the calibration data were obtained by fitting the oxygen data to a straight line and the hydrogen data to a rectangular hyperbola, $y = (\alpha x)/(1 + \beta x)$, where x is the rate of hydrogen generation, y is the sensor response and α and β are constants, the numerical values of which were determined by the least squares fitting routine. The accuracy of the gas sensors is $\pm 5\%$ or better. The accuracy for the overall efficiency measurement is estimated to be $\pm 15\%$ as is the statistical variability. The uniformity of the algal film was about 5%. This value was determined by scanning the fluctuation of transmitted light with a model 560B EG&G "Lite-Mike". Since the active area of this photodetector is only 0.051 cm^2 it is much smaller than the active area of the algal film. By sampling a small fraction of the transmitted light at numerous points behind the algal film the optical uniformity was determined.

RESULTS AND DISCUSSION

Platinized Chloroplasts

The time profiles of photoactivity are presented in Fig. 2 for three light-dark cycles of 2 hours each. Fig. 2 shows that the first irradiation period has a qualitatively different time profile than the subsequent two periods. In the first cycle, the oxygen profile underwent a transient gush (peaking at 7 μmol of 0_2 per hour) before settling down to steady state. The hydrogen rate climbed

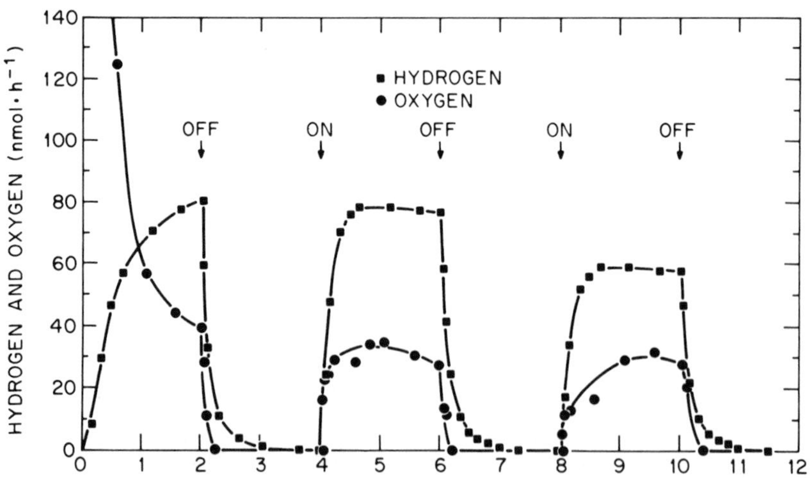

FIG. 2. Simultaneous photoproduction of hydrogen and oxygen by moistened platinized chloroplasts entrapped on filter paper. The ELH projector lamp, providing saturating illumination, was turned on at T = 0 and alternated with 2-hour on-off cycles as indicated. The peak oxygen rate was 7 μmol/hour. The area of the disk was 10.2 cm^2. The activity of the platinized chloroplasts was still measurable after 2 to 3 days of irradiation cycles at 10 to 20 percent of the initial rates. (From reference 7)

monotonically to steady state. The time required to reach 50 percent of steady state in the first cycle was ~25 minutes, whereas the corresponding time for the second and third cycles was 5 to 8 minutes.

The only material capable of sustained photoactivity was prepared by precipitating the $[Pt(Cl)_6]^{2-}$ ion in the presence of thylakoid membranes. Platinized chloroplasts prepared by precipitating the $[Pt(NH_3)_4]^{2+}$ ion in the presence of thylakoid membranes were incapable of any hydrogen photoactivity. These results suggest that $[Pt(NH_3)_4]^{2+}$ is repelled (either Coulombically or sterically) by the Photosystem I reduction site and that $[Pt(Cl)_6]^{2-}$ has a specific chemical affinity of, perhaps, an ion exchange-like nature for this site. It is known from several types of measurements, such as electrophoretic mobility studies, that thylakoid membrane surfaces bear a net negative electrostatic charge at neutral pH due to carboxyl groups [22]. However, the discrete nature of the proteins associated with the membrane prevents this charge from being uniformly distributed across the membrane surface. Photosystem I is capable of donating electrons to negatively charged electron acceptors such as $[Fe(CN)_6]^{3-}$ the classic Hill electron acceptor [23]. Moreover, as shown by Zweig and Avron [24] and Kok et al. [25], the methyl viologen (MV^{2+}) cation can be reduced by Photosystem I. Chow and Barber [26] have shown that MV^{2+} accumulates in the diffuse layer of the membrane interface. A comparison of the interactions of MV^{2+} and $[Pt(NH_3)_4]^{2+}$ with Photosystem I indicates that MV^{2+} can be reduced at a relatively large distance, whereas a smaller distance is required for platinum precipitation and catalytic activity from Photosystem I.

Platinum Catalyst Formation

In Fig. 3 the baseline at the left corresponds to the hydrogen sensor output in the dark with the complete mixture of chloroplasts and hexachloroplatinate ions. At the indicated time, the stroboscopic lamps begin flashing. By continously monitoring for the onset of hydrogen evolution and counting the number of flashes required to achieve that onset, a value, N, for the number of flashes can be determined for the first period of flashing. As indicated in Fig. 3, the rate of hydrogen evolution eventually reaches a steady state since a kinetic balance between the rate of photosynthetic electron transport and the catalytic rate of hydrogen evolution is achieved. Since hydrogen evolution in this system is derived from light activated photosynthetic water splitting, the baseline is restored when the light is turned off. Not indicated in Fig. 3 is the oxygen that is simultaneously coevolved with the hydrogen. It should be emphasized, however, that whereas a delay in the first onset of hydrogen evolution is observed with respect to the second onset, no such delay is observed for oxygen evolution. For each irradiation the onset of oxygen evolution appears within the response time of the apparatus. When the stroboscopic flash lamps are triggered for the second time, the onset of hydrogen is observed after N′ flashes. It can be seen in Fig. 3 that N = 1200 and N′ = 1000. The delay in the onset of hydrogen for the first interval of flash illumination is made up of instrumental response time delay plus photochemical delay of hexachloroplatinate converting to metallic platinum. However, delay in the onset of hydrogen for the second interval of illumination is made up of response time delay only, since the platinum catalyst has already been formed. From the foregoing discussion, it is clear that $\Delta N/4$ represents an upper limit for the minimum number of platinum atoms required to form a platinum metal catalyst as defined by the property of onset of hydrogen evolution. The values for N and N′ were determined by visual inspection of the event markers on the chart recording and comparing it with the reading of the electronic totalizer which recorded the number of flashes. It is estimated that the uncertainty of the position at which the break in the baseline occurs is ~25 flashes. The propagated error involved in combining two numbers N and N' with the stated uncertainty leads to an overall estimated uncertainty of about 35 flashes. This corresponds to an uncertainty of about ± 9 platinum atoms. Insoluble platinum was positively identified in the aqueous suspension by entrapping the platinized chloroplasts on filter paper, washing with distilled water, and monitoring the presence of retained platinum by X-ray fluorescence analysis. Control experiments in which the reaction mixture was held in the dark gave negative results when assayed for retained insoluble platinum. This confirmed the absence of any appreciable thermal reduction of hexachloroplatinate.

Apart from the differences in flash numbers corresponding to the onset of hydrogen evolution for the first and second irradiations in a given experiment, it will be further noted in Fig. 3 that the analytical shapes of the two traces are qualitatively different. Whereas the first trace has a pronounced sigmoidal character, the second has none. The data of Fig. 3 show an onset of hydrogen activity at approximately 50 atoms of platinum. Crystallographic data for metallic platinum indicate that it has a face-centered cubic structure with a nearest neighbor distance of 0.277 nm [27]. Fifty atoms of platinum, therefore, correspond to a cube with an edge dimension of about

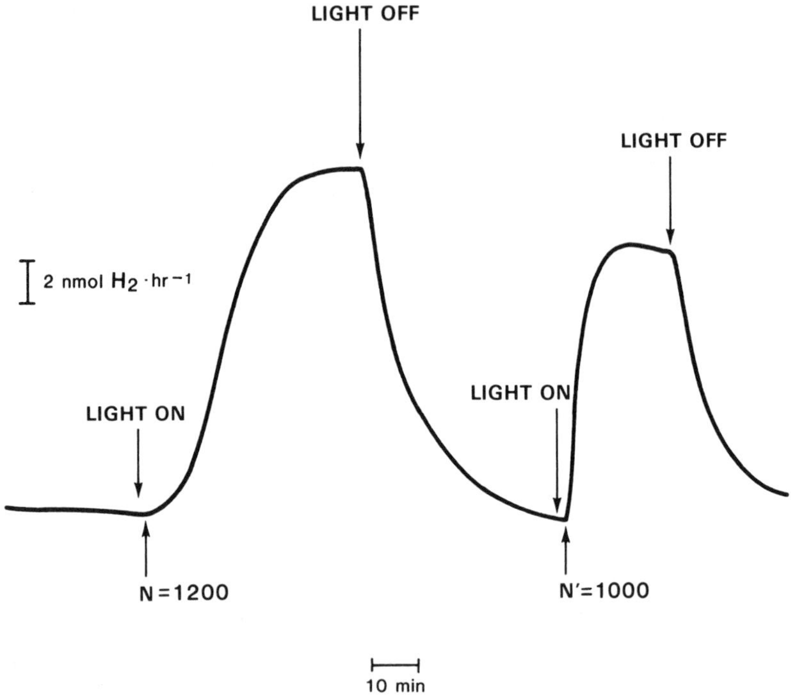

FIG. 3. Hydrogen-time profiles of the output of the hydrogen sensor as a function of flash illumination. At the indicated times, the chloroplast mixture was illuminated with saturating flashes of light at a frequency of 10 Hz. The decrease in steady-state rate during the second period of illumination is attributed to photodegradation of some of the photosynthetic units in the chloroplast preparation. The light source consisted of eight stroboscopic flash lamps, arranged symmetrically around the photoreactor and triggered simultaneously.

10Å. The sigmoidal character of the hydrogen time profile during the first interval of illumination suggests that although the onset of hydrogen evolution can be observed after the creation of 50 platinum atoms or less, a larger number is required to develop full catalytic activity. The inflection point of the sigmoid, which corresponds to about 500 atoms, suggests itself as a transition point between catalyst growth and catalyst stability. Five hundred platinum atoms correspond to a cube with an edge dimension of 22Å.

Therefore, an upper limit for the number of platinum atoms required to make an aqueous environment hydrogen-evolving platinum metal catalyst may be defined as follows: The onset of catalytic activity can be observed after the formation of 50 atoms (or less) whereas a stable catalyst can be formed with 500 (or less). These limits are based on the assumption of a quantum efficiency of unity. The actual value is probably close to this since the quantum efficiency of the primary process of photosynthesis (in the photosynthetic reaction centers) is close to unity [28]. Moreover, the charge on the hexachloroplatinate ion, $PtCl_6^{2-}$ is minus two. This negative charge gives it an electrostatic affinity to the Photosystem I-ferredoxin reduction binding site in the chloroplast membrane, which bears a net positive charge at and near neutral pH [29]. The saturation of the Photosystem I reduction sites is also aided by the relatively large stoichiometric excess of hexachloroplatinate ions to Photosystem I reaction centers. Since no water soluble electron carriers are present in this system, it follows that the hexachloroplatinate ions must be held in very close proximity to the Photosystem I reduction site.

These results imply relative stability of the one-electron reduction steps in the conversion of hexachloroplatinate ions to zero valent platinum. Such stability is reasonable based on data in thermodynamic tables which indicate that, on average, approximately 0.7 eV per electron is liberated in the conversion of platinum(IV) to zero valent platinum [30]. There is, moreover, an important precedent for stability of redox intermediates that are photogenerated at the photosynthetic membrane aqueous interface. These are the intermediates that are associated with the four-step oxidation of water to produce molecular oxygen [31].

These measurements can be performed because of the ability to satisfy several experimental requirements simultaneously, such as the conversion of hexachloroplatinate to platinum metal at pH 7 and room temperature, the electrostatic affinity of hexachloroplatinate ions to the ferredoxin binding site, and, perhaps most important, the fact that the chemical reductive steps that form the catalyst are the very ones that are needed to drive the reactions of the catalyst.

The new contribution of this result is the photochemical precipitation of catalytic platinum metal clusters by the photoreducing system of green plant photosynthesis. Since the structure and particle sizes of hydrogen-evolving platinum metal catalysts have been characterized authoritatively by several investigators, it is instructive to compare the results of the present work with that which has previously been performed. Toshima et al. have studied particle size effects of colloidal platinum catalysts for light-induced hydrogen evolution from water [32]. Their key conclusion was that maximum activity was obtained for colloid particle sizes of about 30Å in diameter. Kiwi and Grätzel [33] have studied catalytic hydrogen evolution as a function of platinum loading on titanium dioxide particles. The platinum particle size distributions were in the range of 5–30Å. Heller et al. [34] have prepared and characterized light-transmitting platinum films. One of their key conclusions is that, at the primary level, the monocrystalline stable platinum particles are approximately 50Å.

Algal Hydrogen Evolution

Table 1 summarizes the energy conversion efficiencies based on net absorbed photosynthetically active radiation. This efficiency was computed using Faraday's Law of Electrochemical Equivalence to determine the rate of hydrogen production by the electrolysis calibration cell and the associated Gibbs free-energy content ($\Delta G^\circ = 237$ kJ/mol) of hydrogen. The efficiency entries for Table 1

Table 1. Energy conversion efficiencies of green algae for hydrogen and oxygen production

Alga	Absorbed light ($\mu W/cm^2$)	Light on (no.)[a]	H_2 (nmol/h)	Efficiency (PAR) (%)[b]
Scenedesmus D_3	5.1	1	126	16
		2	181	23
C. reinhardtii (sup)	2.2	1	44	13
		2	54	16
		3	61	18
		4	64	19
		5	71	21
		6	71	21
		7	61	18
C. reinhardtii (UTEX 90)	8.4	1	78	6
		2	104	8
		3	104	8
C. moewusii	9.1	1	337	24
		2	309	22
		3	253	18

[a]The entries in this column correspond to the ordinal number of successive periods of illumination. The light was on for either a 3- or 4-h period, followed by an equal period of darkness.

[b]Conversion efficiency based on absorbed photosynthetically active radiation (PAR). Based on repeated measurements and calibrations, it is estimated that the experimental error in these measurements is, at most, ±15%. The efficiencies were computed for the rates of hydrogen evolution at the end of the period of illumination when the algae were in a steady (or nearly steady) state.

were computed for the actual conditions under which the experiments were run: hydrogen and oxygen simultaneously coevolved in the same volume and transported to downstream gas sensors by the helium carrier at relatively dilute concentrations. Energy expenditures for hypothetical processing steps such as separation and compression to multiatmosphere values, which will decrease effective efficiencies, are not included in Table 1.

All of the algae used for the experiments of Table 1 were grown photoautotrophically on minimal solution with atmospheric carbon dioxide as the sole carbon source. Therefore, all of the reductant that is expressed as molecular hydrogen is ultimately derived from water splitting. However, due to the presence of alternate electron acceptors [18] and endogeneous reductants [35, 36], the stoichiometric ratio of hydrogen to oxygen at any given moment is not necessarily equal to 2; this ratio is, however, usually close to 2. A photograph of the entrapped algal film is presented in Fig. 4. The *Scenedesmus* D_3 experimental data for the light ON-OFF cycles used to calculate the corresponding entries for Table 1 are illustrated in Fig. 5. Similar data were recorded for the other entries.

FIG. 4. *Scenedesmus* D_3 entrapped in filter paper. These immobilized algae are retained in a well-defined circular geometry that allows the measurement of the absolute amount of light that is absorbed. The number of monolayers is determined by the concentration of algal cells in the aqueous suspension from which the film is formed. The optical uniformity of the film is about 5% as determined by scanning the transmitted light with a small area photodiode.

Under light-limiting conditions, energy conversion efficiencies based on absorbed photosynthetically active radiation vary from 6 to 24% (Table 1). With increasing light intensity and under non-light-limiting conditions, however, net conversion efficiencies decrease. For example, at an incident light intensity of 100 mW/cm^2, conversion efficiencies are well below 1%. However, the range of efficiencies indicated in Table 1 implies that under appropriate experimental conditions a major fraction of the reductant that is generated by Photosystem I can be expressed as molecular hydrogen.

The kinetically limiting aspects of photosynthesis have been discussed by Kok [37] and Clayton [38]. Photosynthesis saturates with increasing light intensity because of the inability of thermally activated biochemical reactions to keep pace with the increasing rate of quantum excitation of the photosynthetic reaction centers at the higher intensities. As previously demonstrated, this kinetic

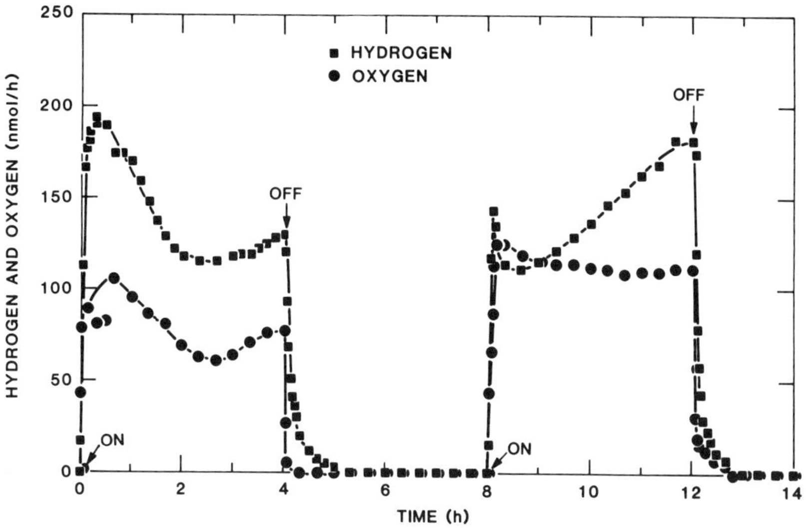

FIG. 5. Simultaneous photoevolution of hydrogen and oxygen from the anaerobically adapted green alga *Scenedesmus* D_3. The algae are irradiated with a projector lamp at normal incidence to the plane defined by the filter-paper entrapped algae. Two four-hour intervals of illumination are indicated. The values of hydrogen evolution used in the energy efficiency calculation were those at the end of each four-hour interval.

limitation expresses itself mechanistically in terms of the number of apparent functional photosynthetic units present in the algae [39]. The present results on energy efficiencies are completely consistent with the analysis of Kok [37] and Clayton [38] who have described a rational approach to overcome this inherent kinetic limitation of photosynthesis. By working with photosynthetic systems of smaller photosynthetic unit sizes (i.e., smaller optical cross sections), it should, in principle, be possible to preserve a kinetic balance between the rate of quantum excitation and the rates of thermally activated electron-transfer reactions at higher light intensities. For example, the structure of photosynthetic tissue is such that each reaction center is served by about 200 molecules of antenna chlorophyll. These antenna chlorophyll molecules do not participate directly in the photochemistry; they absorb light and deliver the energy to the reaction centers. If every quantum absorbed by the antenna in full sunlight were utilized photochemically at the reaction centers, electrons would be flowing through the complete chain from water to ferredoxin (or hydrogenase) at a rate of ~2000 per sec. However, the electron-transport chain can transport no more than about 200 electrons/sec, so that in full sunlight only 1/10 of the incoming quanta can be utilized. The prospect of increasing the rate at which the electron-transport system can operate is limited. However, if each reaction center were served by an antenna of only 20 chlorophyll molecules rather than 200, quanta would be delivered to the reaction centers at a rate of 200/sec in full sunlight rather than 2000/sec. The electron-transport machinery could then keep pace. The practical implication of preserving this kinetic balance is that the high conversion efficiencies which have been measured in this work at low incident light intensities will be preserved at higher incident light intensities. The logical upper limit for these higher intensities is, of course, the maximum value of terrestrial solar irradiance.

CONCLUSIONS

In conclusion, a new composite material capable of photosynthetically splitting water into molecular hydrogen and oxygen has been prepared and a new interfacial photochemical reaction at the photosynthetic membrane interface is reported. This reaction has been used to flash photoprecipitate platinum clusters and the number of flashes required to observe the onset of hydrogen evolution have been used to characterize the cluster size. While photoprecipitation of metal colloids at the photosynthetic membrane interface is obviously not a general technique for catalyst preparation it is, however, probably true that other metals (or combinations of metals) can be used to metallize chloroplasts thereby imparting a variety of catalytic properties to the composite material.

Based on absolute light absorption of visible polychromatic illumination in the low-intensity region of the light saturation curve, conversion efficiencies of 6 to 24% have been obtained. These values are the highest ever measured for hydrogen evolution by green algae.

ACKNOWLEDGEMENTS

The author thanks J. P. Eubanks, C. V. Tevault, and J. E. Thompson for technical support and P. S. Mattie for secretarial support. Part of this paper was previously published in reference 7, copyright 1985 by the American Association for the Advancement of Science. The energy conversion efficiency results for microalgae are in press in the *Biophysical Journal.* The results on photocatalytic flash precipitation are pending publication in *The Journal of Physical Chemistry.* The research described in this paper was supported by the Office of Basic Energy Sciences, Department of Energy and the Gas Research Institute. Oak Ridge National Laboratory is operated by Martin Marietta Energy Systems under contract DE-AC05-840R21400 with the Department of Energy.

REFERENCES

1. H. T. Brown and F. Escombe, *Proc. R. Soc. London Ser. B* **76,** 29–111 (1905).
2. O. Warburg, *Biochem. Z.* **100,** 230–270 (1919).
3. R. Emerson and W. Arnold, *J. Gen. Physiol.* **15,** 391–420 (1932).
4. R. Emerson and W. Arnold, *J. Gen. Physiol.* **16,** 191–205 (1932).
5. R. Hill, *Proc. Roy. Soc. Ser. B* **127,** 192–210 (1939).
6. J. Barber, *Plant, Cell and Environ.* **6,** 311–322 (1983).
7. E. Greenbaum, *Science* **230,** 1372–1375 (1985).
8. B. Kok, *Enzymologia* , 1–56 (1948–49).
9. S. J. Pirt, *Biotechnol. Bioeng.* **25,** 1915–1922 (1983).
10. A. C. Ley and D. C. Mauzerall, *Biochim. Biophys. Acta* **680,** 95–106 (1982).
11. L. N. M. Duysens in: *The Photochemical Apparatus: Its Structure and Function* (Brookhaven National Laboratory, 1959), pp. 10–25.
12. R. S. Knox, *Biophys. J.* **9,** 1351–1362.
13. J. R. Bolton, *Science* **202,** 705–711 (1978).
14. W. W. Parson, *Photochem. Photobiol.* **28,** 389–393 (1978).
15. R. T. Ross, R. J. Anderson, and T.-L. Hsiao, *Photochem. Photobiol.* **24,** 267–278 (1976).
16. H. Gaffron and J. Rubin, *J. Gen. Physiol.* **26,** 219–248 (1942).
17. P. F. Weaver, S. Lien, and M. Seibert, *Solar Energy* **24,** 3–45 (1980).
18. N. I. Bishop and L. W. Jones, *Curr. Top. Bioenerg.* **9,** 3–31 (1978).
19. E. Greenbaum, *Biotechnol. Bioeng. Symp.* **10,** 1–13 (1980).
20. S. G. Reeves and D. O. Hall, *Methods Enzymol.* **69,** 85–94 (1980).
21. E. Greenbaum, *Photobiochem. Photobiophys.* **8,** 323–332 (1984).
22. J. Barber, *Biochim. Biophys. Acta* **594,** 253–308 (1980).
23. A. Trebst, *Methods Enzymol.* **24,** 146–165 (1972).
24. G. Zweig and M. Avron, *Biochem. Biophys. Res. Commun.* **19,** 397–400 (1965).
25. B. Kok, H. J. Rurainski, and O. V. H. Owens, *Biochim. Biophys. Acta* **109,** 347–356 (1965).
26. W. S. Chow and J. Barber, *J. Biochem. Biophys. Methods* **3,** 173–185 (1980).

27. J. R. Anderson, **Structure of Metallic Catalysts,** (Academic Press, New York 1975) p. 447.

28. A. C. Ley and D. C. Mauzerall, *Biochim. Biophys. Acta* **680,** 936–943 (1983).

29. K. K. Colvert and D. J. Davis, *Arch. Biochem. Biophys.* **225,** 936–943 (1983).

30. G. Charlot, **Selected Constants: Oxidation-Reduction Potentials of Inorganic Substances in Aqueous Solution,** (Butterworths, London 1971) pp. 40–41.

31. T. J. Wydrzynski in: **Photosynthesis I: Energy Conversion by Plants and Bacteria,** Govindjee ed. (Academic Press, New York 1982) pp. 486–589.

32. N. Toshima, M. Kuriyama, Y. Yamada and H. Hirai, *Chem. Lett.,* No. 6, 793–796 (1981).

33. J. Kiwi and M. Grätzel, *J. Phys. Chem.* **88,** 1302–1307 (1984).

34. A. Heller, D. E. Aspnes, J. D. Porter, T. T. Sheng and R. G. Vadimsky, *J. Phys. Chem.* **89,** 4444–4452 (1985).

35. U. Klein and A. Betz, *Plant Physiol.* **61,** 373–377 (1978).

36. R. P. Gfeller and M. Gibbs, *Plant Physiol.* **77,** 509–511 (1985).

37. B. Kok in: **Proceedings of the Workshop on Bio-Solar Conversion,** M. Gibbs et al., eds. (Indiana University 1973) pp. 22–30.

38. R. K. Clayton in: **Chlorophyll–Proteins, Reaction Centers, and Photosynethetic Membranes,** J. M. Olson and G. Hind. eds. (Brookhaven National Laboratory, BNL/50530, 1977), pp. 1–15.

39. E. Greenbaum, *Science* **215,** 291–293 (1982).

ELECTRIC FIELD EFFECTS ON ELECTRON TRANSFER REACTIONS IN ISOTROPIC SYSTEMS

STEVEN G. BOXER, DAVID J. LOCKHART, AND STEFAN FRANZEN
Department of Chemistry,
Stanford University,
Stanford, California 94305 USA

ABSTRACT

The effects of electric fields on electron transfer reactions in an isotropic system are considered. Effects are expected on electron transfer kinetics and on the quantum yield of emission which competes with photoinduced electron transfer. The effects on free energy and electronic coupling are considered. Applications to the initial charge-separation step and $P^{+\cdot}Q^{-\cdot}$ recombination in photosynthetic reaction centers are presented.

INTRODUCTION

The rate of non-adiabatic electron transfer reactions can be expressed in the general form:

$$k_{et} = \frac{2\pi}{\hbar} |V|^2 FC \quad , \tag{1}$$

where V is the electronic interaction matrix element and FC is the thermally averaged Franck-Condon weighted density of states. V depends on the overlap of the orbitals on the donor and acceptor moieties; it depends strongly on the distance separating them, and to some extent, on the nature of the intervening orbitals. FC accounts for the dependence of the rate on the driving force (ΔG°_{et}) and on the reorganization energy for the reaction. In much of our current research we are using externally applied electric fields to manipulate the rates of electron transfer reactions. The field can affect both V and FC via different mechanisms and with different consequences. In the following we briefly describe the physical origin of these effects, starting with the FC term.

Effects of an Electric Field on FC

Most electron transfer reactions of interest for energy storage involve the separation of charge, while most wasteful reactions involve charge recombination, regenerating neutral molecules. When charge is separated, a state with an electric dipole moment is generated which can be

Photochemical Energy Conversion
James R. Norris, Jr. and Dan Meisel, Editors

quite substantial (4.8 Debye per Å for fully separated charges). In the presence of an electric field, the energy of such a state is changed depending on the orientation of the dipole in the field as shown in Figure 1. For electric fields which are easily generated in the laboratory (e.g. $1x10^6$ V/cm) these energy changes can be very substantial. For example a 50D dipole (full charge separated by 10Å) aligned with a $1x10^6$ V/cm field will have its energy decreased by 104 meV ($840cm^{-1}$) relative to its value at zero field. If such a dipolar state is the product of an electron transfer reaction whose precursor is neutral (or vice versa), the driving force for the electron transfer will be changed by this amount, and this leads to a change in the rate of the reaction. The specific relationship between rate and driving force for an electron transfer reaction has been the subject of enormous theoretical and experimental interest [1], and this relationship can be probed by studying the effects of an electric field on a particular reaction.

Consider for example the dependence of k_{et} on ΔG_{et} shown in Figure 2 [2]. This dependence is likely to be typical of many electron transfer systems. Three particular examples of zero-field driving force are indicated with the arrows in Fig. 2 to illustrate the effect of a field on the rate. Region I is the so-called normal region; region II is where the driving force is comparable to the reorganization energy; and region III is the so-called inverted region. The effect of temperature on such a curve is also illustrated in Figure 2B. Figure 3 shows the effect of an applied electric field on the rate for a dipole of 50D oriented either parallel or antiparallel to the field for each region and at each temperature. These are the largest effects that can be observed by this mechanism. There have been several very interesting studies of systems of this type using photosynthetic reaction centers (RCs) in lipid bilayers [3,4] or in Langmuir-Blodgett films [5-7].

We have addressed the possibility that electric field effects on electron transfer reactions can be measured in isotropic samples, for example with the donor-acceptor system of interest embedded in a solid polymer matrix with random orientation. In such cases the effect of an applied electric field will be to produce a spread in the driving force around the zero field value and a spread in the values of k_{et}. If, for example, the electron transfer rate were described by a single exponential at zero field, in the presence of the field a highly non-exponential decay would result. As can be seen from the simulations in Figure 4, the nature of the deviation from exponentiality can be quite characteristic of the shape of the k_{et} vs. ΔG_{et} curve near the value of ΔG_{et} at zero field. It is also apparent from Figure 4 that the magnitude of the deviation from exponentiality is expected to be quite small. This is because most of the dipoles in isotropic samples are oriented roughly perpendicular to the

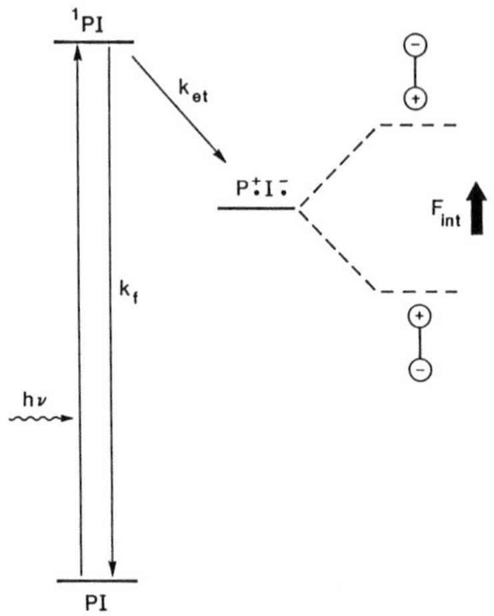

FIG. 1. General reaction scheme for electron transfer competing with fluorescence. The solid lines are schematic energy levels in zero electric field. For an isotropic immobilized sample in an applied electric field, the potential energy of the $P^{+}I^{-}$ state increases or decreases as shown.

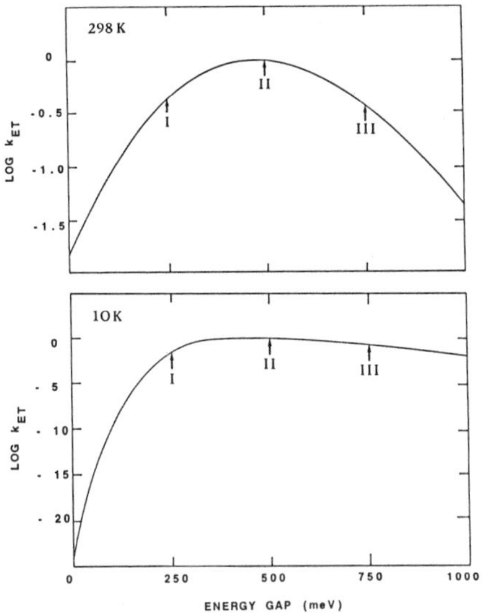

FIG. 2. Calculated value of the logarithm of the electron transfer rate as a function of the free energy change at 295K (upper panel) and 10K (lower panel). The theoretical curve represents the effect of Franck-Condon overlap of two modes in the donor and acceptor manifolds. The modes of frequency $\omega_1 = 1000 cm^{-1}$ and $\omega_2 = 50 cm^{-1}$ have coupling constants $S_1=1$ and $S_2=60$. The regions indicated by numerals I, II, and III are described in the text.

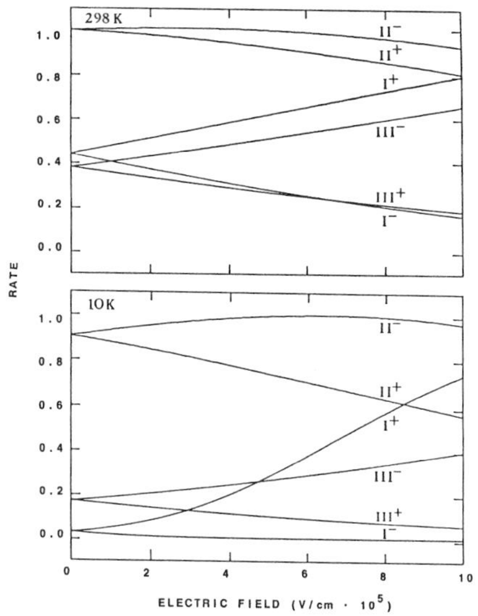

FIG. 3. Calculated effect of an electric field on the electron transfer rate for an *oriented* sample with a 50D dipole moment aligned with (+) or against (-) the field, for the three regions indicated in FIG. 2.

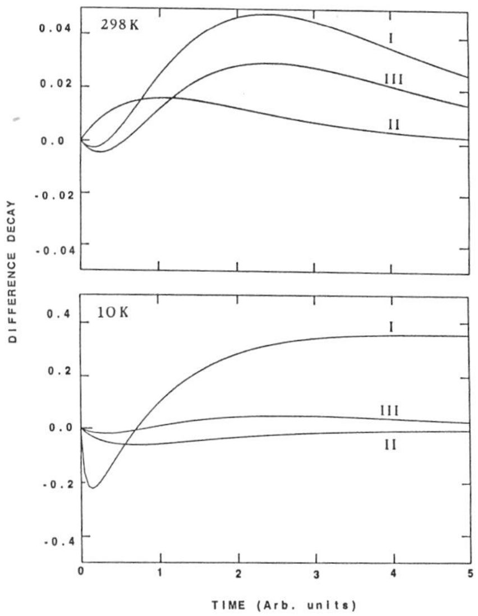

FIG. 4. Calculated effect of an electric field on the electron trnasfer rate for an *isotropic* sample shown as the difference between the ion-pair population in an applied electric field of 10^6V/cm and at zero field as a function of time. A 50D dipole moment was assumed. Regions are labeled in FIG. 2. Effects of the field on V are ignored in this figure and Fig. 3.

field and experience little change in ΔG_{et}. As a result, measurements of the effects of electric fields on reaction kinetics for isotropic samples will require the highest possible field, the largest possible electric dipole moment, and extremely good signal-to-noise.

A related approach, which is useful for electron transfer from photoexcited states, is to study the effect of the applied electric field on the emission quantum yield when this emission competes directly with electron transfer (for example, the fluorescence, rate constant k_f, in Figure 1). Since we are dealing with a donor-acceptor system immobilized in a matrix, we need only consider first order processes which compete with electron transfer and, in the simplest case, direct competition between emission and electron transfer. In the case of a singlet reaction, the quantum yield of fluorescence, Φ_f, is given by:

$$\Phi_f = \frac{k_f}{k_f + k_{et}} \tag{2}$$

Assuming that the radiative rate is independent of electric field and that k_{et} can be described by the classical Marcus theory, we can generate the curves in Figure 5 which show the dependence of Φ_f on ΔG_{et} (note that a similar shape would result for the dependence of k_{et} on ΔG_{et} shown in Fig. 2, however, it would not be symmetric; the classical Marcus curve is symmetric about $\lambda = \Delta G_{et}$). It is apparent that the *fractional* change in the quantum yield can be quite large depending on the shape of the curve and the zero field value of ΔG_{et}. Furthermore, for an isotropic, immobilized sample it is seen that the effect of an applied electric field is to increase the quantum yield of fluorescence, unless the difference between the zero-field value of ΔG_{et} and λ is quite large. Thus, even though some rates increase and others decrease in the applied field, as seen in Figures 3 and 4, the net effect on the quantum yield for an isotropic sample is usually to increase the fluorescence yield. The magnitude of the net increase in the emission quantum yield in an applied electric field depends on the exact shape of the k_{et} vs. ΔG_{et} curve around the zero-field value. Changes in emission quantum yields in an applied field can be measured with great sensitivity.

Because the rate of electron transfer depends on the orientation of the dipolar state relative to the electric field, the yield of any competing process, e.g. emission, will also depend on orientation. Thus, even when the sample is excited with unpolarized light, the emission will become polarized when an electric field is applied, and it is expected that the nature of the polarization will depend on the angle ζ_{et} between the transition dipole moment and the dipole moment of the state whose formation competes with the emission. We have recently discovered this effect and call it electric field induced fluorescence anisotropy. A detailed derivation describing the effect quantitatively can be found in the original paper[8]. We have shown that ζ_{et} can be determined by measuring the fluorescence polarization induced by the application of the electric field. Although it might seem that the specific k_{et} vs. ΔG_{et} curve is crucial to the degree of emission polarization, it can be shown that any dependence of k_{et} on ΔG_{et} which can be fit by a third order polynomial in the field perturbation over the relevant range of ΔG_{et} sampled by the field gives a simple expression for the change in fluorescence as a function of the desired internal angle, ζ_{et}, and the angle between the field direction and the direction of an analyzing polarizer, χ:

$$\Delta F_{et}(\chi) \;\; \alpha \;\; 5|\boldsymbol{\Delta\mu}_{\mathbf{et}}|^2 + (3\cos^2\chi - 1)\{3(|\boldsymbol{\Delta\mu}_{\mathbf{et}}|\cos\zeta_{et})^2 - |\boldsymbol{\Delta\mu}_{\mathbf{et}}|^2\} \qquad (3)$$

This expression is exactly analogous to the well-known expression for the orientation dependence of the Stark effect on absorption or fluorescence spectra in the absence of competing, electric field-dependent processes [9-12,15]. Thus, it is possible to determine the direction of the separation of charge which competes with emission relative to the emission transition dipole moment. If the direction of the emitting transition dipole moment is known in the molecular axis system then the direction of charge separation is determined (within a cone).

Throughout this discussion of the effects of a field on FC we have only considered effects on ΔG_{et}. In principle, the reorganization energy could also be affected, although we believe this is likely to be a less significant factor. For example, if solvent dipoles are not fully immobilized and become oriented in the field or if there are large induced dipoles, the field might affect the reorganization energy. Further, if the field were sufficiently large to perturb the wavefunction, there could be effects on the vibrational frequencies which are coupled to the electron transfer process. It is unlikely that the latter would be significant without there being large effects on V which are discussed in the following.

Effects of an Electric Field on V

The electronic coupling between the donor and acceptor depends on the overlap between the relevant orbitals on each. If the donor and acceptor are sufficiently separated, the overlap of the tails of the wavefunctions becomes very difficult to calculate accurately as the wavefunctions are optimized for properties related to their core where the majority of electron density is found. If the orbitals are highly polarizable, then the applied field can affect this overlap. We expect that this effect is likely to be small compared with the effect on ΔG_{et} described in the previous section, however, this remains to be demonstrated in model systems

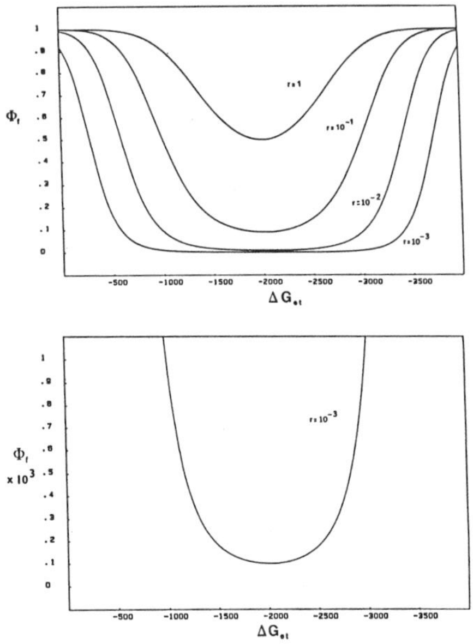

FIG. 5. The dependence of the fluorescence quantum yield, $\Phi_f = k_f/(k_f+k_{et})$, on ΔG_{et} using the classical Marcus expression [1] for k_{et} with λ=2000cm^{-1}. Plots are shown for different values of k_f relative to the maximum value of k_{et}, i.e. $r \equiv k_f/k_{et}$(max). k_f is assumed to be independent of field. The lower panel is an expansion of the upper plot for $r=10^{-3}$, a situation believed to be relevant to reaction centers.

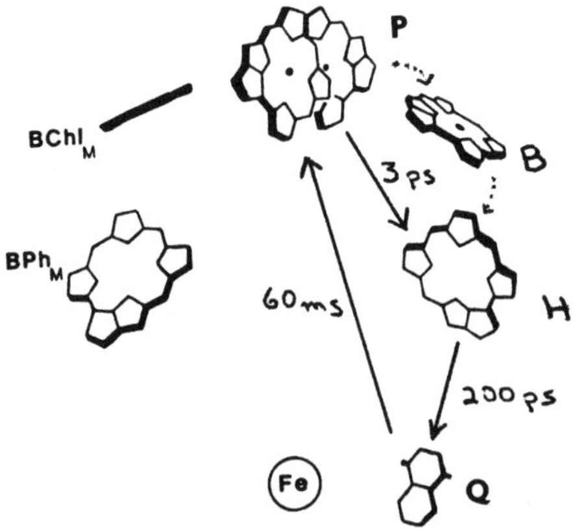

FIG. 6. Arrangement of chromophores in the x-ray structure of *Rps. viridis* reaction centers [19]. P is the special pair, B is the bridging bacteriochlorophyll monomer, H is the bacteriopheophytin acceptor, and Q is a quinone. The time constants of electron transfer steps are approximate and are at room temperature.

(experiments are in progress).

If the interaction between the donor and acceptor is mediated to a significant extent by the intervening medium, a superexchange mechanism, then the electric field can have a substantial effect on V. Consider a situation in which a donor excited state, ^1D, transfers an electron to an acceptor A, mediated by orbitals on the intervening medium M. The superexchange mechanism can be written by first considering the interaction between ^1D and the virtual state(s) $D^{+\cdot}M^{-\cdot}$. The interaction between this new state and the product state, $D^{+\cdot}A^{-\cdot}$, then determines V for the reaction [13]. The field can have a major effect because it affects the energy of the $D^{+\cdot}M^{-\cdot}$ virtual state. We have shown elsewhere [14] that the field will have little effect on V if the $D^{+\cdot}M^{-\cdot}$ state lies far above the ^1D state, but that it can introduce very large effects (comparable to the ΔG_{et} effects discussed above) if $D^{+\cdot}M^{-\cdot}$ is nearly degenerate with ^1D.

APPLICATIONS TO PHOTOSYNTHETIC REACTION CENTERS

To date most quantitative studies of electric field effects have been performed on photosynthetic RCs from bacteria. The arrangment of the chromophores in the RC is shown in Figure 6, from the x-ray structure solved by Deisenhofer and co-workers [19]. The reactive groups are: P, the dimeric electron donor or special pair of bacteriochlorophylls; B, a monomeric bacteriochlorophyll; H, a monomeric bacteriopheophytin; and Q, a quinone. The rates of electron transfer are indicated in the Figure. Previous studies of electric field effects have included examination of the $P^{+\cdot}Q^{-\cdot}$ recombination reaction in RCs oriented in a lipid bilayer [3,4] or in a Langmuir-Blodgett film [5,6]; effects have also been measured on the $P^{+\cdot}Q^{-\cdot}$ quantum yield [7]. It is useful to be able to vary the temperature so that thermally activated pathways can be avoided (note that a pathway which is activated at high temperature may also make a significant contribution at low temperature via superexchange, and this may affect the electric field effect as described in the previous section; this was not considered in [3-7]). In general it is very helpful to study the electric field effect on a single, elementary step, otherwise there is no obvious way to connect observables with theory. For this reason measurements of the $P^{+\cdot}Q^{-\cdot}$ quantum yield as a function of field are very difficult to interpret even in an oriented system [7]: this yield depends on the competition among at least four processes, $^1PHQ \rightarrow P^{+\cdot}H^{-\cdot}Q$; $P^{+\cdot}H^{-\cdot}Q \rightarrow P^{+\cdot}HQ^{-\cdot}$; $P^{+\cdot}H^{-\cdot}Q \rightarrow PHQ$; and singlet-triplet mixing and triplet recombination in $P^{+\cdot}H^{-\cdot}Q$ (chromophore B is left out for clarity, but is discussed further below). For this reason, we have measured the effect of an electric field at low temperature on: (1) the fluorescence to study the properties of the critical initial step, $^1PBHQ \rightarrow$ some dipolar state; and (2) the recombination reaction $P^{+\cdot}Q^{-\cdot} \rightarrow PQ$.

Experimental Methods

Experimental methods have been discussed previously in the context of absorption Stark effect studies of RCs [9-11]. Briefly, RCs are embedded in PVA films and the films are coated with semi-transparent electrodes, generally Ni. For absorption and fluorescence Stark effect measurements the samples are typically 30-80 μm thick; for studies of recombination the samples are spin-coated and are about 5 μm thick. For Stark effect measurements the samples are immersed in liquid N_2, whereas for kinetics measurements they are held against a copper block in a variable temperature (20-300K) closed cycle refrigerator. For the kinetics measurements the field is generated by a high-voltage amplifier and is applied following the excitation flash; for absorption and emission Stark effects the field is generated by a home-built AC high-voltage supply, and the signal is detected in a lock-in at the second harmonic. Absorption and emission electric field effects as a function of the experimental angle χ are measured as described elsewhere [8,9].

Stark Effect on Absorption

The absorption Stark effect spectrum has been measured for *Rb. sphaeroides*, *Rps. viridis* and *Rb.* capsulatus RCs. The spectrum of *Rb. sphaeroides* is shown in Figure 7. In all species the results are very similar: there is a large Stark effect (second derivative lineshape) for the special pair, with comparable magnitudes of the difference dipole moment $|\boldsymbol{\Delta\mu}_A|$ and values of the angle ζ_A between the transition moment and $\boldsymbol{\Delta\mu}_A$. This effect is much larger than for the monomer bands, whose magnitudes are similar in the RC to those for the pure monomeric chromophores [9,11]. Although the values of ζ_A for monomeric bacteriochlorophylls a and b are different [11], the value of ζ_A for the special pairs in the RCs are nearly identical and considerably larger than for the monomers (Fig.9B), despite the fact that *Rb. sphaeroides* and *Rb. capsulatus* contain a-type bacteriochlorophylls and *Rps. viridis* contains b-type. This suggests that the displacement of charge associated with excitation of the special pair is a characteristic of the structure rather than the individual chromophores (the RC structures are very similar), and that the nature of the charge displacement is quite different, e.g. a degree of charge separation between the monomers comprising P (i.e. mixing with pure charge-transfer states such as $P^{\dot{+}}P^{\dot{-}}$) or some charge separation involving the monomeric bacteriochlorophyll, B (i.e. mixing with pure charge-transfer states such as $P^{\dot{+}}B^{\dot{-}}$). The spectral range from 350 - 1300 nm has been surveyed for *Rb. sphaeroides* and no evidence has been found for new bands with a large $|\boldsymbol{\Delta\mu}_A|$

as might be expected if a direct transition to a pure charge-transfer state were being observed [11]. In addition, the Stark effect on the special pair Q_y absorption band for *Rb. sphaeroides* and *Rps. viridis* RCs is found to be homogeneous with respect to ζ_A, making it very unlikely that there is underlying structure originating from different electronic states or different degrees of mixing with charge-transfer states in either species. Similar results have been obtained by Losche and co-workers [15].

Stark Effect on Fluorescence

Fluorescence from ^1P in RCs is very weak because the initial electron transfer reaction is very fast [16,17]. Nonetheless, this fluorescence and the related stimulated emission have proven to be very useful for analyzing the initial charge separation [8,12]. In the absence of competing electron transfer, one expects the same second derivative electric field effect on the fluorescence as was observed in absorption, and as has been observed for the pure monomeric bacteriochlorophylls [8,12]. By contrast the fluorescence spectrum from *Rb. sphaeroides* RCs in an applied electric field shows an increase in intensity (i.e. the Stark effect has a zeroth-derivative lineshape) rather than the expected second derivative lineshape, as shown in Figure 8. It can be shown from this data that the value of $|\Delta\mu_F|$ can be no larger than $|\Delta\mu_A|$, and could be considerably smaller [12]. A reasonable explanation for the shape of the electric field effect was discussed in an earlier section: the applied field leads to a decrease in the quantum yield of the initial charge separation step and an increase in the quantum yield of the competing fluorescence (k_f is much smaller than k_{et} in this system). The data in Figure 8 are for quinone-depleted *Rb. sphaeroides* RCs which are used to avoid saturation of the sample due to slow $P^{+}Q^{-}$ recombination (about 1-2 orders of magnitude slower than ^3P decay). Very similar results have been obtained for Q-containing *Rb. sphaeroides*, *Rps. viridis*, and *Rb. capsulatus* (data not shown) RCs, so long as the light intensity used is kept low ($<10mW/cm^2$). For Q-containing samples of any species, high light intensity causes a change in the sign and angle dependence (see below) of the fluorescence Stark effect, an interesting new result which is discussed in detail elsewhere [14].

The angle dependence of the fluorescence Stark effect is strikingly different from the angle dependence in absorption, as shown in Figure 9. The angle ζ_{et} is much larger than ζ_A. For an ordinary Stark effect, we expect that ζ_F and ζ_A will be comparable as they both reflect properties of the difference between the same ground and excited electronic states. This is clearly demonstrated for bacteriochlorophyll a itself in Figure 9. By contrast, the relevant angle for the fluorescence Stark effect in the RC

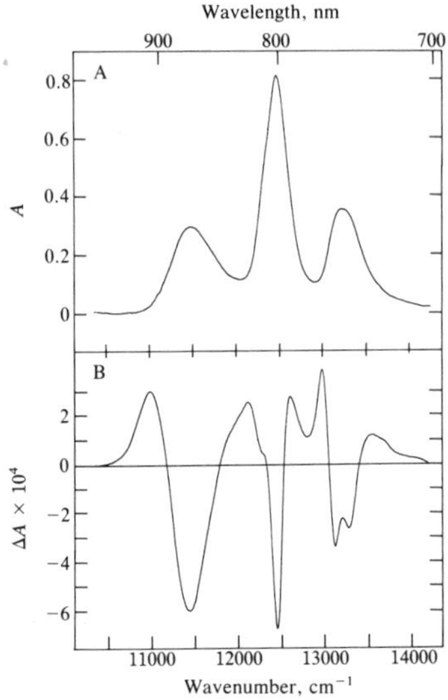

FIG. 7. Absorption (A) and Stark effect (B) spectra for *Rb. sphaeroides* RCs in a PVA film in the Q_y region at 77K (applied field 2.59×10^5V/cm).

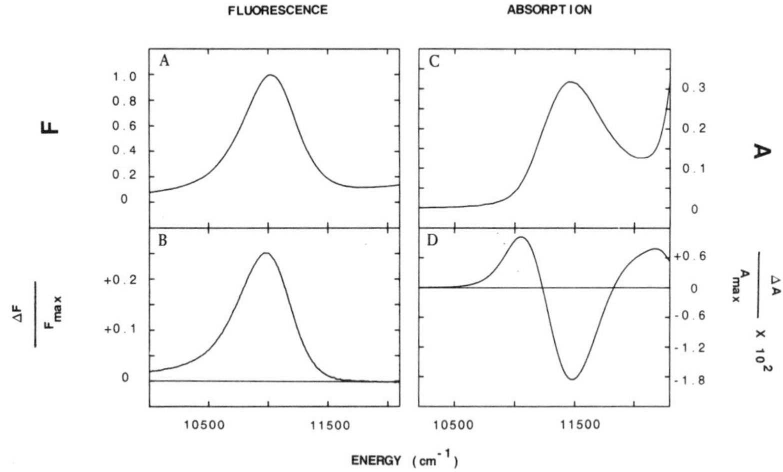

FIG. 8. (A) Fluorescence spectrum of Q-depleted Rb. sphaeroides RCs in PVA at 77K in the absence of a field, and (B) the change in fluorescence for the same sample in a field of 8.9×10^5V/cm. (C) Absorption spectrum of the Q_y transition of the special pair in Q-depleted RCs in PVA at 77K and (D) change in absorbance for the same sample in a field of 8.9×10^5V/cm.

depends on the direction of the dipolar product state of the electron transfer, $P^{+}\!\cdot B^{-}\!\cdot$ or $P^{+}\!\cdot H^{-}\!\cdot$, whose formation competes with fluorescence. The dipole moment associated with either of these product states is much larger than that of ^{1}P, so the difference dipole between the intial and final state of the electron transfer reaction, $\Delta\mu_{et}$, is given by: $\Delta\mu_{et} = \mu(P^{+}\!\cdot I^{-}\!\cdot) - \mu(^{1}P) \sim \mu(P^{+}\!\cdot I^{-}\!\cdot)$, where $I^{-}\!\cdot$ is $B^{-}\!\cdot$ or $H^{-}\!\cdot$. The relevant angle, ζ_{et}, between $\Delta\mu_{et}$ and the direction of the fluorescence transition dipole moment can be estimated by combining the results of single crystal absorption measurements, fluorescence polarization measurements, and the x-ray structure coordinates, making reasonable assumptions discussed in detail elsewhere [8]. The results are shown schematically in Figure 10. It is seen that the data in Figure 9 are incompatible with a reaction mechanism in which formation of $P^{+}\!\cdot B^{-}\!\cdot$ competes with fluorescence.

The analysis in the previous paragraph is based on the assumption that the electric field only affects ΔG_{et} (i.e. the FC term in Eqn. 1) and not the electronic coupling, V. This is a reasonable assumption for the analysis up to this point because it is unlikely that superexchange plays a significant role in the hypothetical step, $^{1}PB \rightarrow P^{+}\!\cdot B^{-}\!\cdot$, which we have just ruled out. On the other hand, the alternative initial step, $^{1}PBH \rightarrow P^{+}\!\cdot BH^{-}\!\cdot$, is unlikely to have a rate of about $5\times10^{12}s^{-1}$ for a direct electron transfer over such a large distance. Thus, the bridging chromophore B, may play a significant role in mediating electron transfer between ^{1}P and H, as has been argued by several investigators [13,18]. In principle the observed angle ζ_{et} and the magnitude of the fluorescence enhancement in the applied field contain information on the magnitude of the superexchange contribution. As discussed at the beginning, the energy of the $P^{+}\!\cdot B^{-}\!\cdot$ virtual state is crucial to this analysis. It can be shown [14] that the angle dependence of ΔF_{et} (Figure 9) as a function of temperature should be sensitive to the importance of superexchange. Measurements of this are in progress.

Electric Field Effect on the $P^{+}\!\cdot Q^{-}\!\cdot$ Recombination Kinetics

The effects of an applied field on the $P^{+}\!\cdot Q^{-}\!\cdot \rightarrow PQ$ recombination reaction at 60K are shown in Figure 11. The experimental difference decays are almost identical between 20-80K; the decay kinetics at zero field is also nearly temperature independent in this region. At higher temperature (data not shown) the effect of an electric field is significantly different. This suggests that parameters which govern the electron transfer reaction are not independent of temperature as is often assumed. As can be seen from the shape of the difference decays at low temperature the $P^{+}\!\cdot Q^{-}\!\cdot$ recombination reaction lies somewhere between region II and region III, by comparison with the calculated curves in Figures 2 and 4.

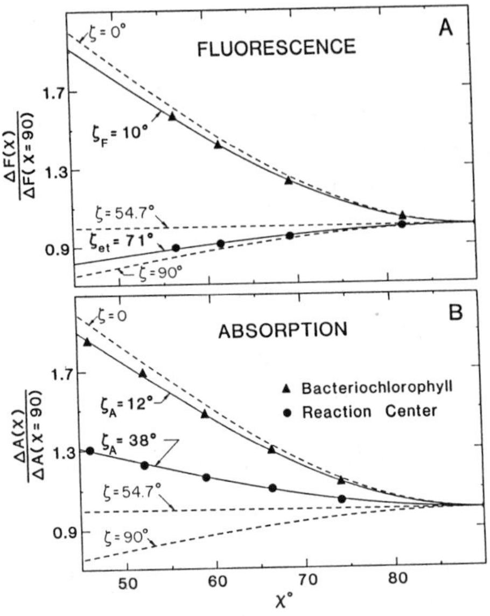

FIG. 9. (A) Dependence of $\Delta F_{et}(\chi)/\Delta F_{et}(\chi{=}90)$ on the experimental angle χ for Q-depeleted *Rb. sphaeroides* RCs and bacteriochlorophyll a in PMMA. (B) Dependence of $\Delta A(\chi)/\Delta A(\chi{=}90)$ on χ for the Q_y absorption band of *Rb. sphaeroides* RCs and bacteriochlorophyll a in PMMA. Fits are to Eqn. 3.

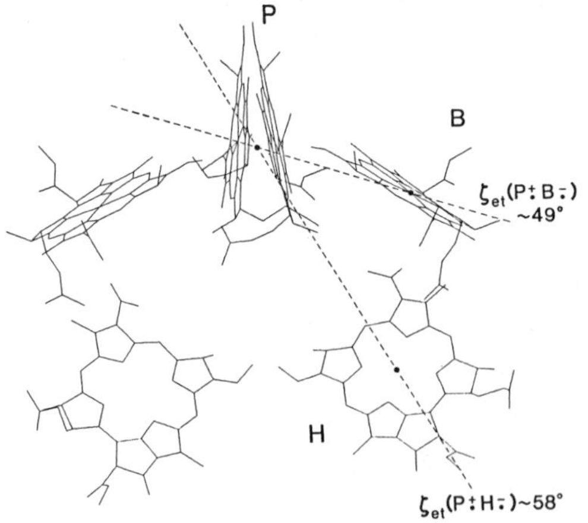

FIG. 10. Arrangement of chromophores in the RC as in Fig. 1 from a different projection, with directions between hypothetical $P^{+}B^{-}$ and $P^{+}H^{-}$ dipole moments and the transition dipole moment of P.

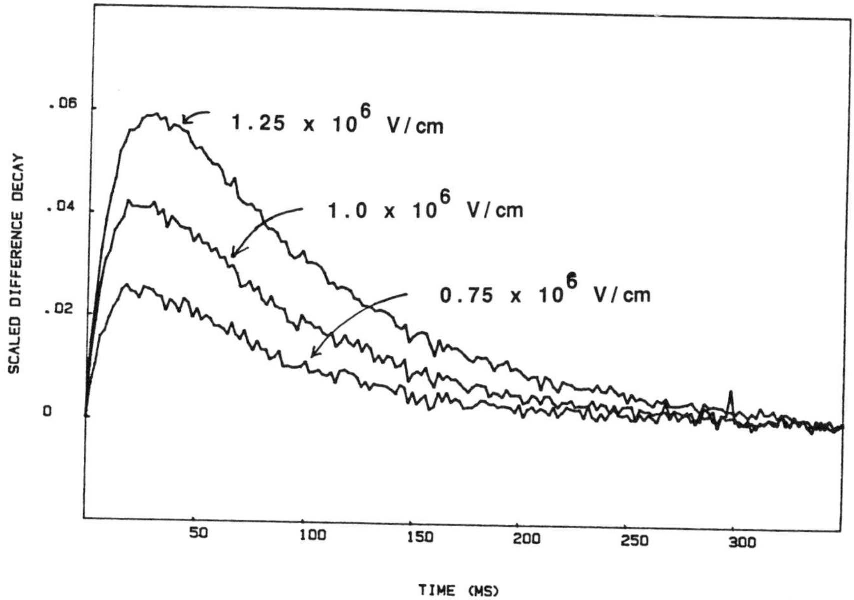

FIG. 11. Difference decay (decay with the field on minus decay with the field off) for the $P^{+}\cdot Q^{-}\cdot$ decay in *Rb. sphaeroides* RCs at 60K measured at 868nm. The data has not been corrected for the electric field effect on absorption at this wavelength. See Figs. 2 and 4 for calculations related to this data.

ACKNOWLEDGMEMENTS: This work was supported in part by the National Science Foundation and the Gas Research Institute.

REFERENCES

1. R.A. Marcus and N. Sutin, Biochim. Biophys. Acta 767, 265 (1987).
2. S.G. Boxer, R.A. Goldstein, and S. Franzen in: Photoinduced Electron Transfer, M.A. Fox and M. Chanon, Eds. (Elsevier, Amsterdam) in press (1988).
3. A. Gopher, Y. Blatt, M. Schonfeld, M.Y. Okamura, G. Feher, and M. Montal, Biophys. J., 48, 311 (1985).
4. T. Arno, A. Gopher, M.Y. Okamura and G. Feher, Biophys. J. Abstracts 53, 270a (1988).
5. Z.D. Popovic, G.J. Kovas, P.S. Vincett, and P.L. Dutton, Chem. Phys. Lett. 116, 405 (1985).
6. Z.D. Popovic, G.J. Kovas, P.S. Vincett, G. Alegria and P.L. Dutton, Biochim. Biophys. Acta 851, 38 (1986).
7. Z.D. Popovic, G.J. Kovas, P.S. Vincett, G. Alegria and P.L. Dutton, Chem. Phys. 110, 227 (1986).
8. D.J. Lockhart, R. Goldstein, and S.G. Boxer, J. Chem. Phys., in press (August 1988).
9. D.J. Lockhart and S.G. Boxer, Biochem. 26, 664 (1987).
10. S.G. Boxer, D.J. Lockhart and T.R. Middendorf, Spr. Proc. Phys. 20, 80 (1987).
11. D.J. Lockhart and S.G. Boxer, Proc. Natl. Acad. Sci. 85, 107 (1988).
12. D.J. Lockhart and S.G. Boxer, Chem. Phys. Letts. 144, 243 (1988).
13. M.E. Michel-Beyerle, M. Plato, J. Deisenhofer, H. Michel, M. Bixon and J. Jortner, Biochem. Biophys. Acta 932, 52 (1988).
14. D.J. Lockhart and S.G. Boxer, submitted.
15. M. Losche, G. Feher, and M.Y. Okamura, Proc. Natl. Acad. Sci. 84, 7537 (1987).
16. N.W. Woodbury, M. Becker, D. Middendorf, and W.W. Parson, Biochem. 24, 7516 (1985).
17. J.-L. Martin, J. Breton, A.J. Hoff, A. Migus and A. Antonetti, Proc. Natl. Acad. Sci. 83, 957 (1986).
18. A. Warshel, S. Creighton, and W.W. Parson, J. Phys. Chem. 92, 2696 (1988).
19. J. Deisenhofer, O. Epp, K. Miki, R. Huber and H. Michel, Nature 318, 618 (1985).

STRUCTURAL, THEORETICAL AND EXPERIMENTAL MODELS OF PHOTOSYNTHETIC ANTENNAS, DONORS AND ACCEPTORS

K. M. BARKIGIA,* L. CHANTRANUPONG,* L. A. KEHRES,** K. M. SMITH** and J. FAJER*. *Department of Applied Science, Brookhaven National Laboratory, Upton, New York 11973; ** Department of Chemistry, University of California, Davis, California 95616.

ABSTRACT

Theoretical calculations, based on recent x-ray studies of bacterial reaction centers, suggest that the light-absorption properties of the special pair phototraps in bacteria are controlled by the interplanar spacing between the bacteriochlorophyll subunits that constitute the special pairs. The calculations offer attractively simple explanations for the range of absorption spectra exhibited by photosynthetic bacteria.

The wide range of (bacterio)chlorophyll skeletal conformations revealed by x-ray diffraction studies raise the intriguing possibility that different conformations, imposed by protein constraints, can modulate the light-absorption and redox properties of the chromophores *in vivo*. Electron-nuclear double resonance data obtained for the primary acceptors in green plants suggest specific substituent orientations and hydrogen bonding that may help optimize the orientations of the acceptors relative to the donors.

In vitro studies provide working models for the unusual chlorophyll-based antennas of green photosynthetic bacteria. The latter light-adapt by forming oligomeric self-aggregates with different peripheral substituents that range from ethyl to propyl, isobutyl and neopentyl. These aggregates offer appealing guidelines for the synthesis of self-assembled biomimetic antenna systems.

INTRODUCTION

Recent structural and experimental results have unveiled the molecular architecture used by photosynthetic organisms to carry out light harvesting and charge separation [1]. We examine here some of the structural factors that help determine light-absorption properties, redox potentials, orientations and vectorial electron flow *in vivo*, and suggest some guidelines for the construction of synthetic biomimetic models based on structural and theoretical considerations.

RESULTS AND DISCUSSION

1. Absorption spectra of the primary donors in photosynthetic bacteria.

In isolated bacterial reaction centers (R.C.), the lowest energy absorption bands, attributed to the primary donors, range from 840-870 nm in bacteria containing bacteriochlorophylls (BChl) *a* to 960 nm in those comprised of BChls *b* [1]. The spectral features of RCs have been treated theoretically by a variety of methods ranging from exciton theory to quantum mechanical techniques [2]. We extend here Zerner's INDO method [3] to consider the effects of the interplanar spacing between the BChl subunits that constitute the special pair donors, and to treat specifically the peripheral substituents that differentiate BChls *a* and *b*.

Photochemical Energy Conversion
James R. Norris, Jr. and Dan Meisel, Editors

Thompson and Zerner [4] used INDO calculations and preliminary x-ray data from the Rhodopseudomonas viridis RC to show that the overlap of the two BChls *b* that comprise the special pair results in a "supermolecule" with a significantly red-shifted low energy absorption band. (A shift of ~3600 cm^{-1} was calculated relative to the monomeric BChls). However, examination of the early x-ray data used reveals some unreasonably close contacts between the two chromophores which are no longer present in more recent refinements. We have thus repeated the calculations using the new x-ray data for the Rps. viridis special pair at 2.8Å resolution in which the average spacing between the overlapping rings I of the BChls is 3.3 rather than 3.1Å [5]. The axial histidines and the major peripheral substituents such as the 9-keto and 2-acetyl groups were included in the calculations but, to minimize computation time, the ethylidene of ring II was replaced by a methylene group, and all other peripheral substituents were replaced by hydrogens. INDO calculations using these skeletons and the "new" Rps. viridis coordinates still yield significant red-shifts for the dimer at 3.3Å separation: the average shift relative to the monomers is 1050 cm^{-1} with a calculated absorption maximum at 854nm. (A calculation using a full set of peripheral substituents yields a maximum at 870 nm). Although agreement with experiment is not perfect, the trend is certainly correct.

Interestingly, conversion of the ring II substituents characteristic of BChl *b* to those of BChl *a* has little effect on the predicted shifts. Since the optical spectra of monomeric BChls *a* and *b* differ little in solution [6], it is perhaps not surprising that identically constructed dimers of the two chromophores should yield similar results.

The earlier calculations of Thompson and Zerner offer a possible explanation for the differences observed in vivo between organisms containing BChls *a* and *b*. Since tight dimers with interplanar separations of ~3.1Å yield large red shifts, it is intuitively obvious that increasing the distances between the monomers will decrease the "supermolecule" effects in the dimers. Indeed, calculations for BChl *a* dimers at 3.4, 3.5 and 3.6Å separations result in progressively smaller red shifts relative to the monomers: 760, 530 and 360 cm^{-1}, respectively. These correspond to blue shifts of 20, 36 and 48 nm compared to the maximum calculated for the dimer at 3.3Å. These results suggest that the differences in the spectra of the special pairs of Rps. viridis (BChls *b*, λ max = 960 nm) and Rhodobacter sphaeroides (BChls *a*, λ max = 870 nm) do not arise from intrinsic differences between BChls *b* and a. Instead, the different maxima in vivo may simply arise from different interplanar spacings in the two organisms. Calculations for Rb. sphaeroides predict a spacing larger by 0.2-0.3Å than that of Rps. viridis, a value consistent with the current x-ray distance of ~3.5Å [7,8]. In addition, the red shifts observed at high pressures or on cooling the reaction centers [1] may be due in part to contractions of the protein that result in smaller gaps within the special pairs. Similarly, the small variations in absorption maxima for organisms that contain the same chromophores (840-870 nm for BChls *a*) may arise from small differences in the spacing induced by the protein environment.

Clearly, other factors may also influence the spectral properties of the chromophores in the protein. Likely among these are the polarity of neighboring protein residues, the orientations of the acetyl groups and conformational variations of the BChl skeletons imposed by ligands, hydrogen bonding and steric constraints. We discuss the consequences of such possible conformational variations next.

2. Implications of structural variations for redox, light-absorption properties and vectorial electron flow.

Recent structural data for (bacterio)chlorophylls and chlorins as isolated molecules and in proteins demonstrate the skeletal flexibility of the chromophores that can be imposed by crystal packing and/or protein constraints [5,9-13]. As examples, we present here single crystal x-ray data for a homologous series of methyl bacteriopheophorbides d, derived from the antenna chlorophylls of the green photosynthetic bacterium Chlorobium vibrioforme, that illustrate the crystallographically significant conformational variations possible for the same chlorophyll skeleton.

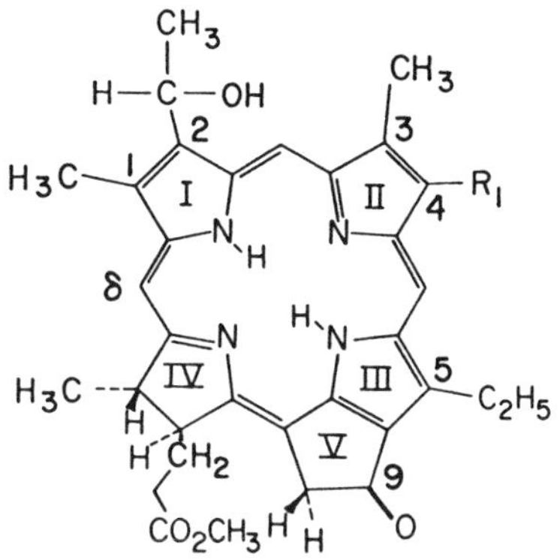

Fig. 1. Structural formula of methylbacteriopheophorbide d. R_1 can be ethyl, n-propyl, isobutyl or neopentyl.

Smith and Bobe [14] have demonstrated that green bacteria light-adapt by sequential alkylation of the antenna chlorophylls to yield a series of homologues with different substituents at position 4 of ring II (see structure in Fig. 1). Single crystal x-ray diffraction of metal-free derivatives (pheophorbides) from Chlorobium vibrioforme yield the following results [12]: a) The substituents range from ethyl to n-propyl, isobutyl and neopentyl. b) Removal of the magnesium affords a series of dimeric and higher aggregates that are hydrogen bonded via the 2-(1-hydroxyethyl) group to the 9-keto group or to the keto group of the propionic acid side chain of ring IV, depending on the method of crystallization. c) Significant conformational variations are found for the same pheophorbide skeleton, depending on the aggregation. The deviations of the pyrrole and exocyclic rings from the plane defined by the four nitrogens range between ±0.5Å, and are illustrated in Fig. 2.

Comparable variations have also been noted, albeit at lower resolution, for the BChls a in the BChl antenna protein of Prosthecochloris aestuarii [13] and the BChls b that comprise the special pair of the Rps. viridis RC protein [5].

We now consider theoretically the possibility that such conformational variations can affect the highest occupied (HOMOS) and lowest unoccupied (LUMOS) molecular orbital levels of the chromophores and thereby modulate their redox potentials and light-absorption properties. We first test the concept by demonstrating that a sterically distorted porphyrin of known structure exhibits experimental optical and redox properties in solution consonant with theory, and extend the calculations, using crystallographic data for the Rps. viridis primary donor, to show that a redox asymmetry is possible in the BChl subunits that comprise the special pair of that bacterium.

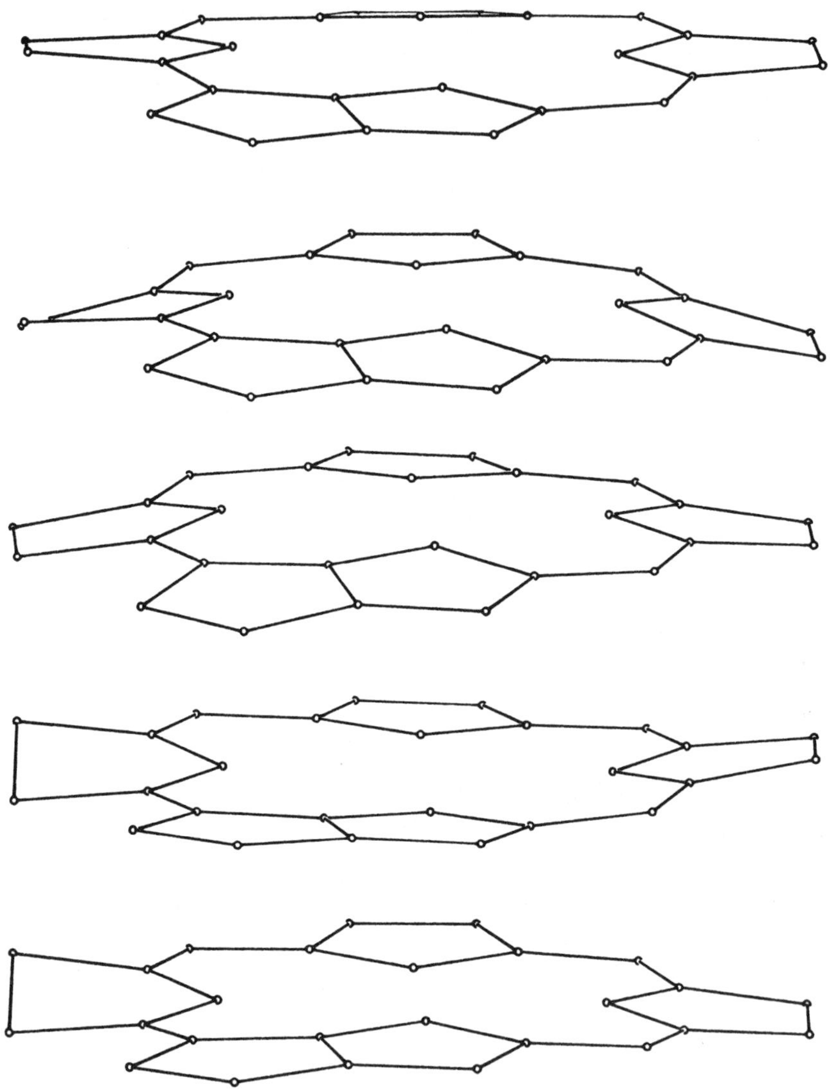

Fig. 2. Edge-on views of the different conformations assumed by the methyl bacteriopheophorbides d in single crystals. The deviations of the outer atoms from the planes defined by the four nitrogens range between plus and minus 0.5Å. Shown, from top to bottom are: 4-propyl, 5-ethyl; 4-isobutyl, 5-ethyl; 4-neopentyl, 5-ethyl; and the two crystallographically independent molecules in the unit cell of 4,5-diethyl methylbacteriopheophorbide d.

Preliminary single crystal x-ray data show zinc tetraphenyl octaethyl porphyrin (ZnTPOEP, 5,10,15,20-tetraphenyl, 2,3,7,8,12,13,17,18-octaethylporphinato zinc (II)) to be severely saddle shaped (Fig. 3) with the β carbons of adjacent pyrrole rings displaced by ~±1Å relative to the plane of the four nitrogens. NMR data establish that the puckered conformation is retained in CH_2Cl_2 solution [15]. Remarkably, the first absorption band of the compound is shifted to 637 nm compared with λ max of 586 nm for Zn tetraphenyl porphyrin (ZnTPP) or 569 nm for Zn octaethyl porphyrin (ZnOEP) [16]. Also noteworthy, the oxidation halfwave potential of ZnTPOEP in CH_2Cl_2 has decreased to +0.47V (vs SCE) compared to those of ZnTPP and ZnOEP, $E_{1/2}$ = 0.75V and 0.63V, respectively, while the reduction potentials in tetrahydrofuran are as follows: $E_{1/2}$ = -1.54, -1.35, and -1.63V, for ZnTPOEP, ZnTPP and ZnOEP, respectively.

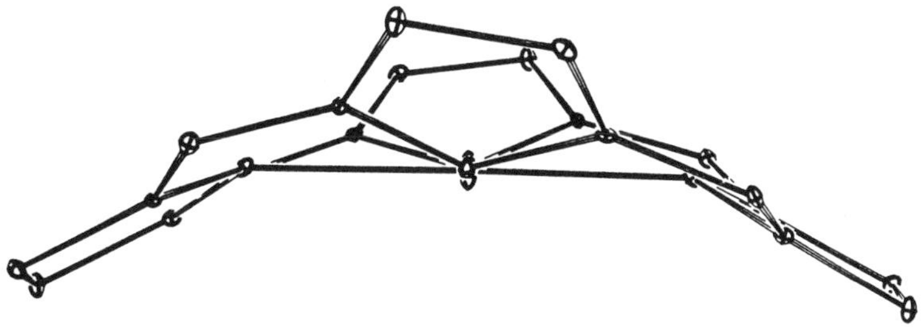

Fig. 3. Edge-on view of the skeleton of ZnTPOEP that illustrates the severe saddle shape of the macrocycle. The deviations of the peripheral carbons from the plane defined by the four nitrogens are ~±1Å.

INDO calculations predict the experimental trends with calculated red shifts of 1900 cm^{-1} for a conformational change from a planar Zn porphyrin to one with the saddle structure of ZnTPOEP to be compared with observed shifts of 1370 cm^{-1} and 1880 cm^{-1} relative to ZnTPP and ZnOEP, respectively [16]. The calculations also predict that the puckered porphyrin will be easier to oxidize by 0.12 eV whereas reduction is insensitive to the distortion. Similar calculations for puckered chlorins and bacteriochlorins predict red shifts of 1200 and 820 cm^{-1}, respectively.

Extension of the calculations to the different conformations observed for the BChls b that comprise the special pair of Rps. viridis also predicts different optical and redox properties. For the BChl b associated with the L protein subunit, λ max = 807 nm with calculated energies of -2.030 and -5.850 eV for the LUMO and HOMO, respectively, and for the BChl associated with the M subunit, λ max = 762 nm with energies of -1.920 and -5.880 eV for the LUMO and HOMO.

The above combination of experimental and theoretical results clearly suggests that the conformational variations observed in vitro and in vivo can provide a mechanism for altering optical and redox properties. Such effects, in conjunction with additional modulations induced by neighboring protein residues may thus combine to cause the observed asymmetry in the triplet [17] and oxidized donor [18] of Rps. viridis and to the vectorial electron flow that occurs in the reaction center [19].

In addition to providing simple and reasonable explanations for the properties of the photosynthetic chromophores, the above results may have useful consequences for controlling the redox and light-absorption properties of synthetic biomimetic models: changing the spacing between dimers or aggregates, or crowding the molecules provides readily available synthetic means for predictably changing redox potentials and optical spectra.

3. Protein effects on conformations and orientations.

As noted above, the protein pockets into which the chromophores fit and neighboring residues provide obvious means of altering the conformations and properties of the pigments. We address now the possibility that the protein environment also helps control the orientation of the macrocycles and their substituents and thereby optimizes the relative orientations of donors and acceptors to facilitate electron transfer. Considerable attention has been paid to the fact that a glutamic acid is properly situated to hydrogen bond the 9-keto group of the bacteriopheophytin (BPheo) acceptors in Rps. viridis and in Rb. sphaeroides [5,7,8]. (This residue is also present in photosystem II (PS) of green plants.) It has therefore been suggested [5,20] that the hydrogen bond serves to alter the redox properties of the BPheo and thereby helps to differentiate between the two possible electron transport branches of the RCs revealed by the x-ray structures [5,7,8]. However, theoretical calculations indicate that hydrogen bonding changes the reduction potential of BPheo by only 50mV or less [19,21]. This effect seems rather small to be the major factor that determines the path of electron flow. Rather, we would suggest that such bonding helps to lock the BPheo in a position optimal for electron transfer.

A combination of theoretical calculations, model studies and electron-nuclear double resonance (ENDOR) results for the pheophytin (Pheo) a acceptor in PS II tends to support the existence of a hydrogen bond in that acceptor and, in addition, suggests specific orientations of the 2-vinyl substituent that would further immobilize the molecule in the binding site.

The electron transport chain of PS II mirrors that of photosynthetic bacteria in that pheophytins and quinones act as early electron acceptors that can be trapped and studied by electron paramagnetic resonance techniques [22]. In frozen media, the 1- and 5- methyl groups of the Pheo anion in PS II are readily detected and exhibit proton hyperfine coupling constants of 4.6-4.7 and 12.5-13.0 MHz, respectively, to be compared with constants of 5.4-5.5 and 9.9-10.5 MHz for $Pheo^-$ in vitro [22,23]. (The assignment of the 5- methyl group was confirmed by selective deuteration [22]). Iterative extended Hückel calculations, using crystal coordinates of methylpheophorbide a, indeed predict substantial hyperfine constants for the 1- and 5- methyl groups of $Pheo^-$ in a ratio of ~1:2, as observed in vitro [22]. The calculations also indicate that if the 2-vinyl substituent were rotated to lie perpendicular to the plane of the Pheo, the unpaired spin densities would decrease by 24% at the 1-CH_3 and increase by 8% at the 5-CH_3.

This prediction can be tested experimentally by examining the anion of methyl mesopheophorbide a in which the 2-vinyl group normally found in Pheo a is replaced by an ethyl group, a situation analogous to rolling the vinyl group out of plane and out of conjugation with the Pheo π system. The 1- and 5- methyl coupling constants of meso $Pheo^-$ are found to be 4.2 and 11.2 MHz, respectively, for a decrease of 22% at the 1- position and an increase of 10% at the 5- position, in excellent agreement with the calculations.

If, in addition to rotating the 2-vinyl group out of plane, a protein residue is assumed to hydrogen bond to the 9-keto group, the calculated spin densities, relative to the "canonical" x-ray coordinates of Pheo a, are predicted to decrease by 24% at the $1-CH_3$ and increase by 16% at the $5-CH_3$. The experimentally observed differences between $Pheo^-$ in vitro and in vivo are a decrease of 15% at the $1-CH_3$ and an increase of 25% at the $5-CH_3$, in reasonably good agreement with the calculations. The conclusion, therefore, is that the protein environment imposes specific spatial orientations on the chromophore by a combination of hydrogen bonding and orientation of substituents. Such control would therefore optimize the relative orientation of donors and acceptors presumably required to carry out efficient and rapid charge separation. At the same time, these constraints provide a mechanism for altering the conformations of the pigments which, in turn, can modulate both the light-absorption properties and redox potentials of the chlorophylls, as discussed above.

4. Self-assembled arrays of Chlorobium chlorophylls: models for synthetic and in vivo antennas.

We have suggested above that the 2-vinyl group of Pheo a helps to anchor the molecule in the protein. We discuss now the role of a different substituent, 2-(1-hydroxyethyl), found at the same position in Chlorobium chlorophylls, the chromophores that act as antennas in green and brown photosynthetic bacteria. The latter are particularly intriguing because they appear to straddle green plants and purple bacteria on an evolutionary scale. The organisms contain both BChl a and chlorophyll derivatives, named Chlorobium chlorophylls as a class, and BChls c, d and e depending on their substituents. (The BChl nomenclature is misleading: the compounds are derivatives of pyrochlorophylls a and b in which the 2-vinyl group has been hydrated to a hydroxyethyl function. A structure of BPheo d is shown in Fig. 1. BChls c and e incorporate δ-methyl groups between rings I and IV and, in addition, BChl e has a formyl group at position 3 of ring II [12].)

The unusual hydroxyethyl group that characterizes all the Chlorobium chlorophylls has naturally attracted attention as to its function in the antennas of green and brown bacteria. Smith et al. [24] demonstrated that BChls c, d and e aggregate in low dielectric solvents such as hexane to yield species with large optical red shifts (~1300-1700 cm^{-1}) compared to the spectra of the monomers. The effect is illustrated in Fig. 4 in which an aggregate of BChl d (λ max = 728 nm) is back titrated with methanol to yield the solvated monomer with λ max = 656 nm. A similar trend is observed if zinc is substituted for magnesium: the monomer absorbs at 648 nm and the aggregate at 716 nm. Under comparable conditions, pyrochlorophyll derivatives without the 2-(1-hydroxyethyl) substituent do not show the red shift. NMR data in cyclohexane reveal only broad featureless spectra that suggest an oligomeric rather than a dimeric structure. A comparison of the absorption spectra found for the aggregated BChls c, d and e with those of bacteria containing the same chromophores reveals a close analogy between the aggregates in vitro and in vivo (λ max = 748, 728 and 708 nm versus 750, 730 and 720 nm, respectively). It seems probable therefore that the red shifts observed in vivo are caused by aggregation. A likely function of the 2-(1-hydroxyethyl) variation in these organisms is thus to promote such aggregation and its concomitant modulation of the light-absorption properties of their antennas.

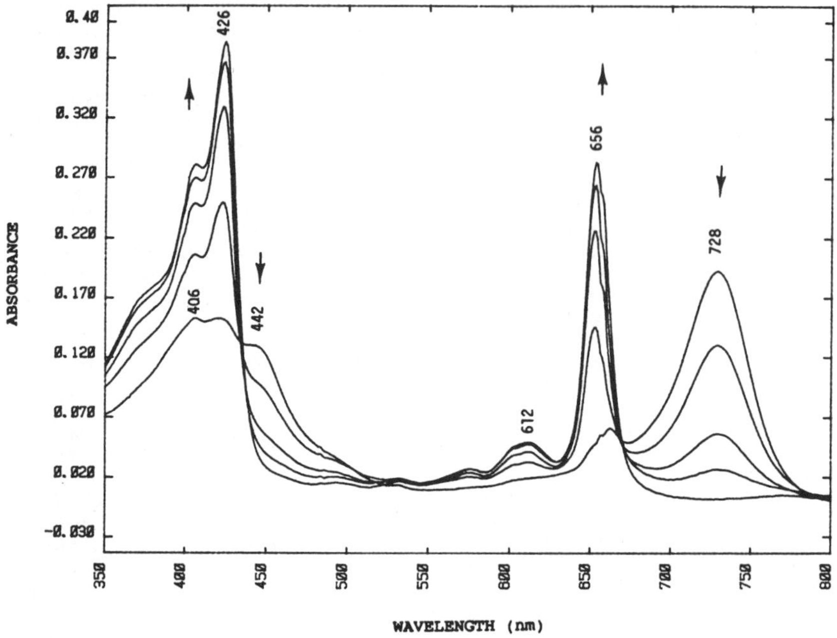

Fig. 4. Spectrophotometric titration of a solution of BChl d in hexane/dichloromethane (200:1). uL aliquots of methanol were added to dissociate the aggregate (λ max = 728 nm) to the monomeric form (λ max = 656 nm).

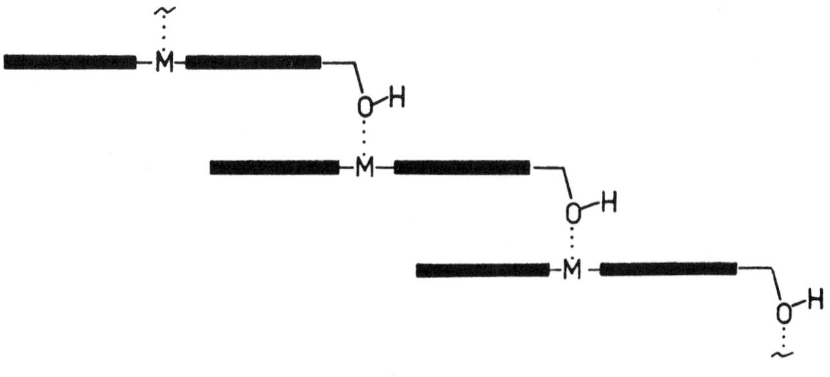

Fig. 5 Schematic representation for the oligomeric aggregates in hexane. The porphyrin planes need not be strictly coplanar and the 9-keto group may participate but is not essential for the aggregation to occur [24].

A minimal and simple model that accounts for the observed aggregation is shown in Fig. 5. (Variations of this model have been proposed by other groups but most invoke the same type of interaction [25].) Since green bacteria can light-adapt by changing substituents on ring II [14], it is possible, in addition, that the bulkier substituents alter the spacing between the chromophores or the tilt angle between the planes to further control the absorption maxima.

Finally, we note that BChls d dimerize in solvents of higher dielectric such as chloroform with an optical maximum at 672 nm that shifts to 656 nm on disaggregation [26]. Clearly, Chlorobium chlorophylls offer appealing models for the synthesis of biomimetic aggregates in which simple synthetic manipulations of peripheral substituents can induce self-assembly of the chromophores ranging from dimers to oligomers with significantly different and predictable optical characteristics.

ACKNOWLEDGMENTS We thank Dr. J. Deisenhofer for the Rps. viridis special pair coordinates, Dr. D. A. Goff for isolating the methylbacteriopheophorbides, Dr. M. W. Renner for the redox data for the zinc compounds, and Drs. L. K. Hanson and M. C. Zerner for useful discussions. This work was supported by the U.S. Department of Energy, Division of Chemical Sciences, under contract DE-AC02-76CH00016, at BNL and by the National Science Foundation, grant CHE-86-19034, at UCD.

REFERENCES

1. For recent reviews, see D. E. Budil, P. Gast, C. H. Chang, M. Schiffer, and J. R. Norris, Ann. Rev. Chem. 38, 561 (1987). C. Kirmaier and D. Holten, Photosyn. Res. 13, 225 (1987). H. Zuber, Photochem. Photobio. 42, 821 (1985).
2. For a survey of recent literature, see L. K. Hanson, Photochem. Photobio. 47, in press (1988).
3. J. Ridley and M. C. Zerner, Theor. Chim. Acta 32, 111 (1973), 42, 223 (1976). M. C. Zerner, G. Loew, R. Kirchner, U. Mueller-Westerhoff, J. Am. Chem. Soc. 102, 589 (1980).
4. M. A. Thompson and M. C. Zerner, J. Am. Chem. Soc. 110, 606 (1988).
5. H. Michel, O. Epp, and J. Deisenhofer, EMBO J. 5, 2445 (1986). J. Deisenhofer (private communication).
6. J. Fajer, M. S. Davis, D. C. Brune, L. D. Spaulding, D. C. Borg, and A. Forman, Brookhaven Symp. Bio. 28, 74 (1976).
7. J. R. Norris and M. Schiffer, private communication. C. H. Chang, D. Tiede, J. Tang, U. Smith, J. R. Norris, and M. Schiffer, FEBS Lett. 205, 82 (1986).
8. J. P. Allen, G. Feher, T. O. Yeates, H. Komiya and D. C. Rees, Proc. Nat'l. Acad. Sci. USA, 84, 5730 (1987).
9. H. C. Chow, R. Serlin, and C. E. Strouse, J. Am. Chem. Soc. 97, 7230 (1975). R. Serlin, H. C. Chow, and C. E. Strouse, ibid, 97, 7237 (1975).
10. C. Kratky and J. D. Dunitz, Acta Crystallogr., Sect. B. B32, 1586 (1975), ibid, B33, 545 (1977), J. Mol. Biol. 113, 431 (1977).
11. K. M. Barkigia, J. Fajer, K. M. Smith, and G. J. B. Williams, J. Am. Chem. Soc. 103, 5890 (1981). K. M. Barkigia, J. Fajer, C. K. Chang, and R. Young, ibid, 106, 6457 (1984).
12. K. M. Smith, D. A. Goff, J. Fajer, and K. M. Barkigia, J. Am. Chem. Soc. 104, 3747 (1982), ibid, 105, 1674 (1983). J. Fajer, K. M. Barkigia, E. Fujita, D. A. Goff, L. K. Hanson, J. D. Head, T. Horning, K. M. Smith, and M. C. Zerner in: Antennas and Reaction Centers of Photosynthetic Bacteria, M. E. Michel-Beyerle, Ed. (Springer-Verlag, Berlin, 1985) pg. 324.

13. D. E. Tronrud, M. F. Schmid, and B. M. Matthews, J. Mol. Biol. 188, 443 (1986).
14. K. M. Smith and F. W. Bobe, J. Chem. Soc. Chem. Commun. 276 (1987).
15. M. W. Renner and J. Fajer, unpublished results.
16. J. Fajer, D. C. Borg, A. Forman, R. H. Felton, L. Vegh, and D. Dolphin, Ann. NY Acad. Sci. 206, 349 (1973).
17. J. R. Norris, D. E. Budil, H. L. Crespi, M. K. Bowman, P. Gast, C. P. Lin, C. H. Chang, and M. Schiffer in: Antennas and Reaction Centers of Photosynthetic Bacteria, M. E. Michel-Beyerle, ed. (Springer-Verlag, Berlin, 1985) pg. 147.
18. W. Lubitz, F. Lendzian, M. Plato, K. Mobius, and E. Trankle in: Antennas and Reaction Centers of Photosynthetic Bacteria, pg. 164.
19. M. E. Michel-Beyerle, M. Plato, J. Deisenhofer, H. Michel, M. Bixon, and J. Jortner, Biochim. Biophys. Acta 932, 52 (1988).
20. D. F. Bocian, N. J. Boldt, B. W. Chadwick, and H. A. Frank, FEBS Lett. 214, 92 (1987).
21. L. K. Hanson, M. A. Thompson, and J. Fajer, in: Progress in Photosynthesis Research, J. Biggins, ed. (Martines Nijhoff, Amsterdam, 1987) pg. 311. L. K. Hanson, M. A. Thompson, M. C. Zerner, and J. Fajer in: The Photosynthetic Bacterial Reaction Center. Structure and Dynamics, J. Breton and A. Vermeglio, eds. (Plenum, New York, 1988) in press.
22. J. Fajer, M. S. Davis, A. Forman, V. V. Klimov, E. Dolan, B. Ke, J. Am. Chem. Soc. 102, 7143 (1980). A. Forman, M. S. Davis, I. Fujita, L. K. Hanson, K. M. Smith, and J. Fajer, Israel J. Chem. 21, 265 (1981). J. Fajer, I. Fujita, M. S. Davis, A. Forman, L. K. Hanson, and K. M. Smith, Adv. Chem. Ser. 201, 489 (1982).
23. W. Lubitz, M. Plato, G. Feher, R. A. Isaacson, and M. Y. Okamura, Biophys. J. 53, 67a (1988).
24. K. M. Smith, L. A. Kehres, and J. Fajer, J. Am. Chem. Soc. 105, 1387 (1983).
25. D. C. Brune, G. H. King, and R. E. Blankenship in: Organization and Function of Photosynthetic Antennas, H. Scheer and S. Schneider, eds. (Walter de Gruyter & Co., Berlin, 1988) in press, and references therein.
26. R. J. Abraham, K. M. Smith, D. A. Goff, and F. W. Bobe, J. Am. Chem. Soc. 107, 1085 (1985).

Interpretation of the Electric Field Sensitivity of the Primary Charge Separation in Photosynthetic Reaction Centers

C. C. Moser, G. Alegria, M. R. Gunner and P. L. Dutton
Department of Biochemistry and Biophysics
University of Pennsylvania
Philadelphia, Pennsylvania USA 19067

ABSTRACT

Popovic et al. [1] have shown that the efficiency of light induced charge separation in a monolayer of oriented photosynthetic reaction center (RC) is sensitive to electric field. With the data provisionally interpreted as indicating a sensitivity of the picosecond electron transfer to field induced free energy changes of the charge separated states, current models for the light induced picosecond electron transfer in the RC from bacteriochlorophyll dimer ($BChl_2$) to bacteriopheophytin assisted by an interposed bacteriochlorophyll monomer (BChl) are challenged to fit the data. A model presuming a simple field sensitive Franck-Condon overlap between $BChl_2$ and BChl based on single classical vibration coupled to electron transfer can only model the data if the negative free energy change does not match the reorganization energy. An even simpler model presuming a rapid field sensitive thermal equilibrium between $BChl_2$ and the nearby BChl fits equally well. However, a model involving superexchange electronic coupling between the excited $BChl_2$ and the BChl with or without similar field dependent Frank-Condon overlap factors provides the best fits to the data and suggests that the relevant dielectric constant of the RC may be about 1 with an $BChl_2$-BChl energy gap of about 100 meV.

1. Popovic, A.D., Kovacs, G.J., Vincett, P.S., Alegria, G., and Dutton, P.L. (1986), Biochim. Biophy. Acta, 851, 38-48.

INTRODUCTION

Although recent X-ray crystallography [1-3] of the bacterial photosynthetic reaction center (RC) [4-6] has revealed the arrangement of the redox centers, it has not made the mechanism of the remarkably efficient psec electron transfer reaction from the bacteriochlorophyll dimer ($BChl_2$) to the bacteriopheophytin (BPh) anymore obvious. For example, a bacteriochlorophyll monomer (BChl) is interposed between the $BChl_2$ and the electron transfer active BPh (see figure 1), but its role is completely uncertain. Thus the challenge has now turned away from structure towards the development of an electron transfer theory that can explain the observed kinetics and can be tested by generating predictions for those relatively few modifications of the $BChl_2$/BChl/BPh system that are experimentally possible. The most straightforward modification is temperature. All theories must accomodate the relatively weak temperature dependence [7,8]. Chemical modification can be accomplished by reduction of the BChl monomer to a BPh on the "inactive" branch generating recently observed spectral kinetic changes [9], or by site directed mutagenisis, which is expected to generate insight into the role of individual amino acids along the electron transfer pathway.

Photochemical Energy Conversion
James R. Norris, Jr. and Dan Meisel, Editors

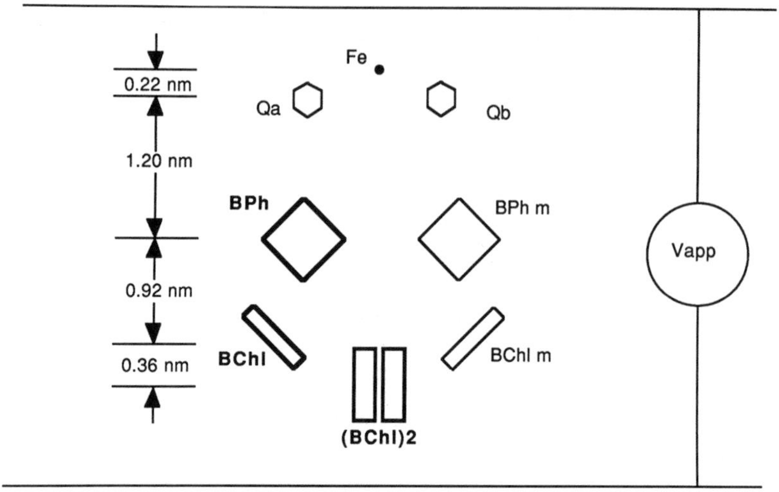

Figure 1: The arrangement of the redox centers in the reaction center with respect to the external electrodes. Center to center monolayer normal distances are given. Vapp represents the applied voltage.

It is also possible to manipulate chromophore energetics with external electric fields. The application of a field of hundreds of mV/nm across an oriented sample should dramatically change the energies of participants in the initial charge separation (see figure 2). Popovic et al. [10] have already shown that external electric fields applied to asymmetrically oriented, spectrally dichroic thin films of RC deposited from a Langmuir-Blodgett trough modulate the efficiency of the light induced charge separation between $BChl_2$ and Qa. While Popovic et al. [10] monitored the reaction yield at the relatively slow time scale (msec) at which Qa reduction is complete, there is reason to believe that the field induced failure of electron transfer occurs during the initial psec charge separation. Because the transmembrane electron transfer distances from $BChl_2$ to BPh and from BPh to Qa are similar (see figure 1), the field induced free energy changes are likely to be similar. However, the relative field induced free energy change for the BPh to Qa reaction will be less because the zero field free energy drop is much greater [11,12]. Furthermore, figure 3 shows experiments that change the free energy gap of the BPh to Qa reaction by substitution of synthetic quinones for native ubiquinone have shown that this reaction rate is not dramatically free energy dependent [13,14]. In addition, there is some evidence that the decay of $BChl_2$+BPh- to the ground state is relatively insensitive to field [15]. Experiments are in progress in this laboratory to directly examine the field effect on the electron transfer kinetics down to the psec time scale. Using the provisional interpretation that the field induced failure occurs during the initial charge separation, we find the Popovic et al. data imposes significant constraints to the various theories describing the reaction.

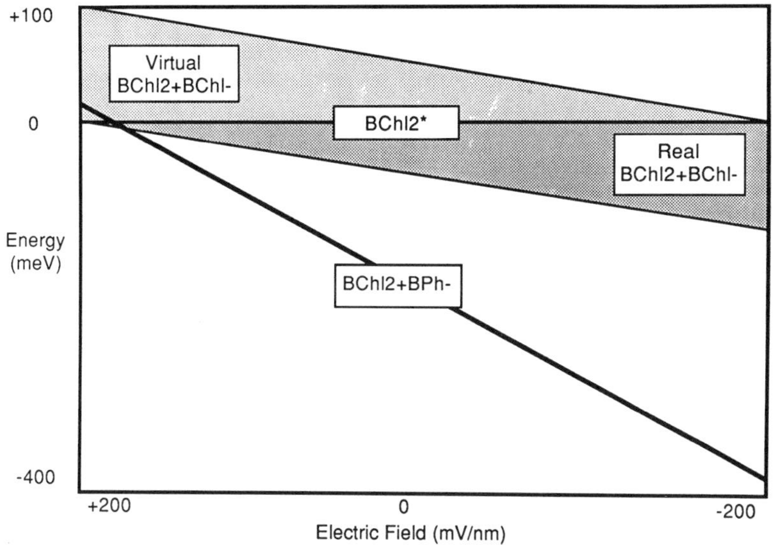

Figure 2: The effect of external field on the relative energies of the redox centers involved in picosecond electron transfer: $BChl_2$* the excited dimer; $BChl_2$+BChl- the hypothetical monomer reduced intermediate in electron transfer as a virtual state (above $BChl_2$* in energy) or a real state (below $BChl_2$* in energy); $BChl_2$+BPh- the photoreduced bacteriopheophytin state. The latter two states represent dipolar states which vary in energy with field.

Popovic et al. [10] related the electrical signal developed by a flash activated RC film (R) to three fundamental parameters: Φ, the efficiency of forward electron transfer relative to decay from the excited state; δ, the fraction of the reaction centers oriented such that the field assists the overall electron transfer minus the fraction in the opposite orientation; and α, the electrical signal expected from fully oriented reaction centers.

(1) $$R(E) = \alpha \{ (1+\delta)\,\Phi(E) - (1-\delta)\,\Phi(-E) \}$$

δ was previously estimated by assuming Φ approached 100% at high positive fields, an assumption we choose not to make in this paper. We estimate δ from the magnitude of the electrical response at zero field compared to the density of active reaction centers per unit area in the film as assayed by optical means. Using a parallel plate capacitor model for the voltage generated by the light activated RC with the maximum voltage at zero field of 29 mV, an active RC density of 1.6×10^{16} molecules/m^{-2}, a membrane normal distance from $BChl_2$ to Qa of 2.5 nm (from the X-ray crystal structure) and an estimated dielectric constant [16,17] of 3 we estimate δ to be about 0.14 with some uncertainty due to errors in measurement and possible mosaic spread of the RC orientation. α can be calculated from equation 1 using the response at zero applied field.

(2) $$\alpha = R(0) / 2\, \delta\, \Phi(0)$$

Popovic et al. assumed a RC dielectric constant value of 3, similar to the dielectric constant of the bulk of the sample capacitor insulation. However, the dielectric properties of proteins is uncertain and may vary in both space [18-20] and time. In addition, it is not clear that the macroscopically derived concept of dielectric constant applies on the level of the psec charge separation. Nevertheless, we retain the convenient concept of dielectric constant and account for any dielectric dependence by introducing a multiplier ϵ_r describing the variation of this dielectric constant from the originally assumed value of 3. Thus the above δ and α are multiplied by ϵ_r to give δ' and α', while the intraprotein intensity of electric field, E, is divided by ϵ_r to give E'.

Φ can be related to the rate of forward electron transfer kET and all other rates of decay from the excited state (kd) by the relation [10]

(3) $$\Phi = kET/(kET + kd).$$

where we expect kET to be dependent and kd to be relatively independent of the electric field because of the relatively large and small dipoles involved respectively. If we express kET relative to the zero field rate kET(0) and use the observation that $\Phi(0)$ at zero field is 0.98 [21], then

(4) $$\Phi(E') = (1 + (1-\Phi(0))\ kET(0)/kET(E')\ )^{-1}$$

The present paper will examine several of the electron transfer theories that have been developed and applied to the $BChl_2$ to BPh electron transfer [7,8,21-28] by considering the field dependences of the models and the constraints imposed by a fit to the Popovic et al. [10] data. We will restrict our examination to those models which assume an assisting role for the intervening BChl and are consistent with the observation that significant concentrations of the reduced monomer are not apparent on a psec time scale (see for example [8,29-31]).

The first and simplest model that we consider does not give an explicit field dependence for the forward electron transfer kinetics from $BChl_2$ to BChl, but instead assumes a field sensitive, free energy dependent equilibrium between $BChl_2$* and $BChl_2$+BChl- which is followed by a slower electron transfer to BPh- [9,32]. Because of the required rapidity (sub-psec) of the equilibrium, this model suggests a strong coupling between the donor and acceptor states may be involved, so that the electron transfer may be adiabatic.

Other models explicitly describe the rate dependence of the initial electron transfer beginning with the golden rule (see eg. [21,22]) and are traditionally associated with weak
coupling and non-adiabatic electron transfer:

(5) $$kET = (2\,\pi/\,\hbar)\ V^2\ FC$$

where V is the rate dependence on the electronic coupling and FC the rate dependence on the nuclear coupling refered to as Franck-Condon overlap intergrals. The second model introduces the imposed field dependence only through FC factors. We consider the relatively simple semi-classical model of Marcus [33], recognizing that more sophisticated models involving a more complete quantum mechanical approach are available [34,35]. A third model introduces a field dependence through the electronic coupling term. However, because the direct orbital overlap between $BChl_2$ and BPh over an edge to edge distance of 1.06 nm is likely to be quite small [36], we consider an indirect coupling (superexchange) through a virtual reduced BChl. Finally, we will consider a model in which both superexchange electronic coupling and FC nuclear coupling are field sensitive.

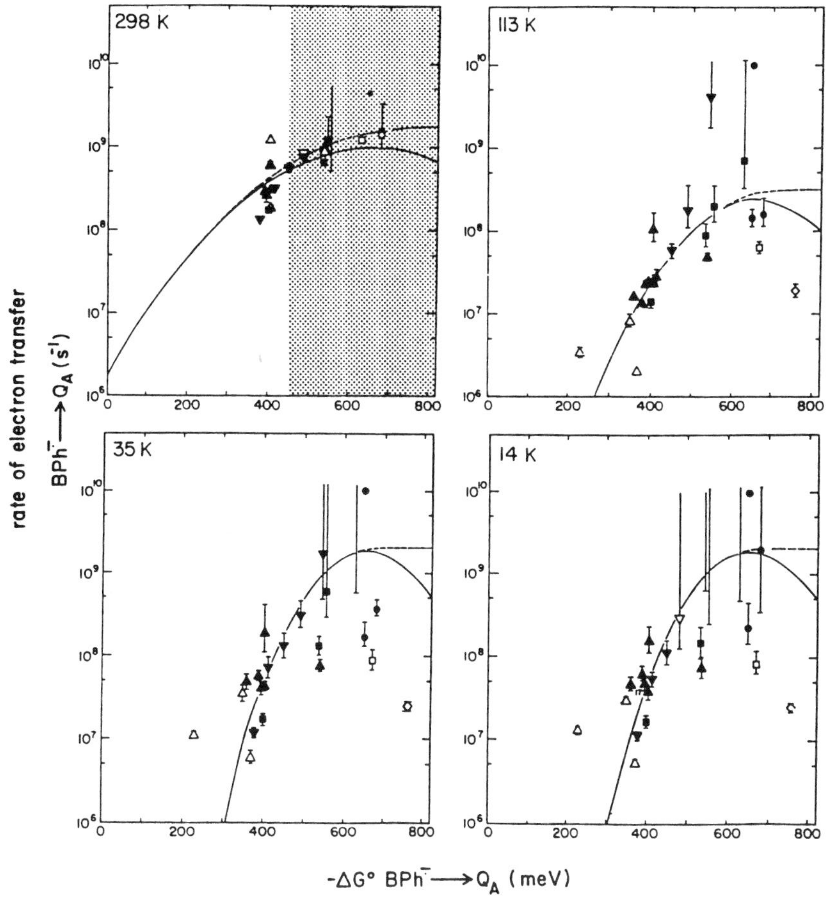

Figure 3: The dependence of the BPh to Qa electron transfer rate on changes in the free energy of the reaction by quinone substitution at four temperatures. The shaded region in the room temperature panel represents the approximate free energy range shown in figure 2 and emphasizes the relatively small free energy dependence under these conditions. Also shown are theoretical rates expected from a model in which free energy dependent Franck-Condon overlap factors are treated quantum mechanically with coupling to a single mode (solid line) at which $h\omega = 15$ meV and $\lambda = 600$ meV, or coupling to an additional mode (dashed line) $h\omega = 200$ meV and $\lambda = 200$ meV (see reference 13).

RESULTS AND DISCUSSION

A: The Equilibrium Model:

The simplest theory supposes that the rate of electron transfer is dependent on a rapid equilibrium between $BChl_2$* and $BChl_2$+BChl- dictated by a thermal distribution reflecting the field dependent free energy gap between these states. Thus the ratio of the forward and reverse electron transfer rates between $BChl_2$* and $BChl_2$+BChl- (k_1/k_{-1}) is related to the free energy at zero applied field ($-\Delta G$) minus the electrostatic energy drop of the $BChl_2$+BChl- dipole component normal to the membrane (μ), with a value derived from the structure of 0.36 e-nm, in the dielectric constant corrected applied field (E')

(6) $$(k_1/k_{-1}) = \exp[(\Delta G+\mu E')/kT]$$

Assuming equilibrium is established faster than the rate of electron transfer from BChl- to BPh (k_2) then the overall rate will be

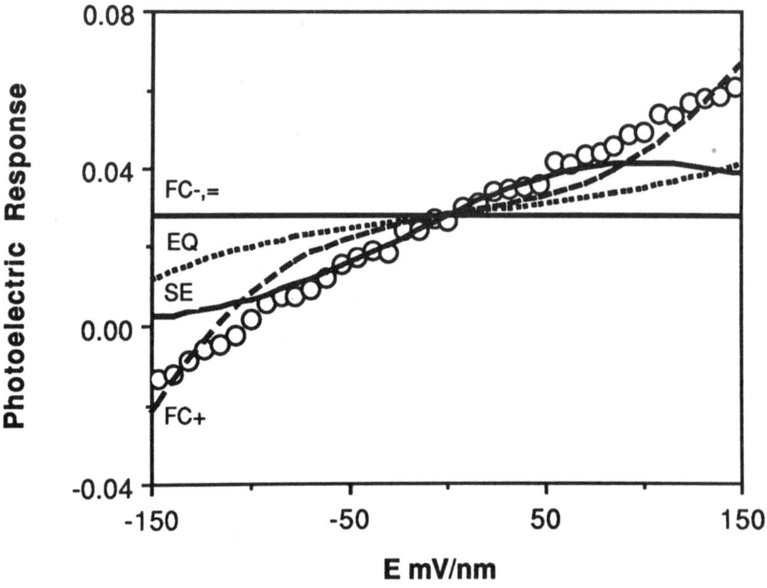

Figure 4: Examples of the theoretical photoelectric response as a function of applied field maintaining the RC dielectric constant at 3. The original Popovic et al. [10] data is presented as open circles. The equilibrium model (EQ) is shown as a fine dashed line. The models in which Franck-Condon factors are field sensitive and $-\Delta G = \lambda$ (FC=) with λ set at 50 meV, and that in which $-\Delta G$ does not equal λ and is exothermic (FC-) with λ set at 125 meV and ΔG = -100 meV are nearly superimposable (solid horizontal line). The Franck-Condon sensitive model with endothermic ΔG = 200 meV and λ best fit at 95 meV is shown as the coarse dashed line (FC+). The superexchange model with electronic factors only field sensitive is shown as a curved solid line with the virtual state energy gap at 12 meV.

(7) $$kET = k_2 (k_1/k_{-1})$$

Note that the rate will be temperature dependent as $\Delta G + \mu E$ departs from zero. Normalizing kET relative to the rate at zero applied field (kET(0)), removes the dependence on ΔG and k_2.

(8) $$kET(E')/kET(0) = \exp[\mu E'/kT].$$

Substituting this expression in (1) and (3) gives a field dependent electrical response of

(9) $$R(E') = \alpha'\{(1+\delta')(1+(1-\Phi(0))\exp[-\mu E'/kT])^{-1} - (1-\delta')(1+(1-\Phi(0))\exp[\mu E'/kT])^{-1}\}.$$

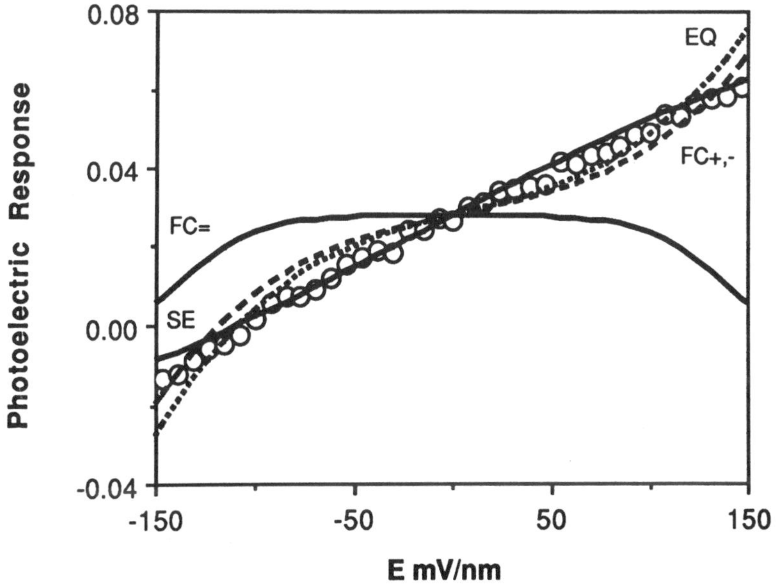

Figure 5: Examples of the theoretical photoelectric response as a function of applied field assuming variable RC dielectric constant less than 3. The imposed electric field strength axis is calibrated for a dielectric constant of 2 or 1 (upper and lower numbers respectively). The original Popovic et al. [10] data is presented as open circles. The equilibrium model (EQ) is shown as a fine dashed line with dielectric constant of 2. The model in which Franck-Condon factors are field sensitive and $-\Delta G = \lambda$ (FC=), with λ set at 50 meV to show the symmetric curvature of the trace, is a solid line. The models in which $-\Delta G$ does not equal λ and is exothermic (FC-) with λ set at 125 meV, ΔG = -100 meV and RC dielectric constant of 1 is nearly superimposable with the endothermic model (FC+) with ΔG = 200 meV and λ best fit at 275 meV and RC dielectric constant of 2 (coarse dashed line). The superexchange model with electronic factors only field sensitive is shown as a curved solid line with the virtual state energy gap at 96 meV and RC dielectric constant of 1.

With a RC dielectric constant of 3, the equilibrium model does not follow the data (see figure 4). However, if the dielectric constant is lowered to 2, this simple model is moderately successful at following the trend of the data (see figure 5).

B: Field Dependence of Only Franck-Condon Factors:

A mechanism that involves electron transfer via a $BChl_2$+BChl- intermediate to $BChl_2$+BPh- may have an electric field dependence due to Franck-Condon factors. If the intermediate is of lower energy than the excited dimer, such a mechansism will be a sequential electron transfer in which the formation of the intermediate is followed by a rapid electron transfer to BPh- such that little BChl- is built up, as required by rapid kinetic data [30,31]. A relatively simple expression for the Franck-Condon factors developed by Marcus [33] assumes a simple harmonic oscillator energy vs. nuclear coordinate description, where in the high temperature limit:

(10) $$F = (4\pi\lambda kT)^{-1/2} \exp(-(\Delta G+\lambda)^2/4\lambda kT)$$

ΔG here refers to the energy difference between the lowest energy states of $BChl_2$* and $BChl_2$+BChl-, and λ reflects the reorganization energy in passing between equilibrium conformations of the reactants and products.

As the initial charge separation is relatively temperature insensitive, the Marcus theory suggests that λ is approximately equal to $-\Delta G$ under native conditions, although more sophisticated theories required to treat electron transfer by a quantum mechanical rather than a semi-classical calculation of the Franck-Condon overlap integrals shows clearly that temperature insensitivity may occur when $-\Delta G$ in not equal to λ (see figure 3).

Using the semi-classical Marcus theory in which $-\Delta G$ at zero field equals λ yields a FC factor at a given imposed field strength F(E') relative to that at zero imposed field F(0) of:

(11) $$F(E')/F(0) = \exp[-(\mu E')^2/4\lambda kT]$$

an expression symmetric with respect to positive and negative imposed electric fields, that cannot be fit to the data no matter what RC dielectric constant is chosen (see figures 3 and 4).

However, if $-\Delta G$ at zero field is not set equal to λ, then

(12) $$F(E')/F(0) = \exp[-\{(\Delta G+\lambda-\mu E')^2-(\Delta G+\lambda)^2\}/4\lambda kT]$$

(13) $$\Phi(\pm E') = [1+(1-\Phi(0))\exp[\{(-\Delta G-\lambda\pm\mu E')^2-(\Delta G+\lambda)^2\}/4\lambda kT]]^{-1}$$

If we substitute these expressions into equation (1), and use an exothermic ΔG with a dielectric constant of 3, the result is a fit nearly as poor as the model in which $-\Delta G = \lambda$ (see figure 4). However, if the dielectric constant is dropped to 1, then the model follows the data somewhat better than the equilibrium model. Interestingly, if we use a model in which the ΔG is endothermic, that is the intermediate state is above the excited dimer in energy, then a model in which only FC factors are sensitive to field will follow the data moderately well for both a dielectric constant of 3 (figure 4) and below (figure 5). However, in this case the $BChl_2$+BChl- would no longer correspond to a real electron transfer intermediate or a severe zero field temperature dependence of the rate might be expected.

C: Superexchange Model:

The superexchange model specifically accomodates a $BChl_2$+BChl- state that is higher in energy than $BChl_2$* [7,27,28,36,37]. The superexchange electronic contribution to the electron transfer rate depends directly on the square of the electronic overlap between $BChl_2$ and BChl and between BChl and BPh and inversely on the square of the energy gap (ΔE) between the $BChl_2$* and $BChl_2$+BChl- states at the lowest energy nuclear coordinates of $BChl_2$*. The imposed electric field will profoundly influence the relative energies of the $BChl_2$* and dipolar $BChl_2$+BChl- state, while the electronic overlap terms will be much less dependent on field. Thus the field dependence of the electronic contribution will be:

$$(14) \qquad kET(E')/kET(0) = [V(E')/V(0)]^2 = [\Delta E/(\Delta E-\mu E')]^2 = [1/(1-\mu E'/\Delta E)]^2.$$

Substitution into expression (1) and (3) gives:

$$(15) \qquad R(E') = \alpha' \{(1+\delta')(1+(1-\Phi(0))(1-\mu E'/\Delta E)^2)^{-1} - (1-\delta')(1+(1-\Phi(0))(1+\mu E'/\Delta E)^2)^{-1}\}.$$

Such a model has a fair fit to the data with a dielectric constant of 3 (figure 4) but a much improved fit with a dielectric constant of 1. Under such circumstances the virtual state energy gap appears to be about 100 meV.

D: Superexchange with Field Dependent Electronic and Franck-Condon Factors:

Model D presents a more general extension of the superexchange model C by adding field dependent Franck-Condon factors to the electron coupling dependent reaction rate.
For $-\Delta G = \lambda$,

$$(16) \qquad kET(E)/kET(0) = [1/(1-\mu_B E/ \epsilon_B \Delta E)]^2 \exp[-(\mu_H E/\epsilon_H)^2/4\lambda kT]$$

while for $-\Delta G$ not equal to λ

$$(17) \quad kET(E)/kET(0) = [1/(1-\mu_B E/\epsilon_B \Delta E)]^2 \exp[-\{(\Delta G+\lambda-\mu_H E/\epsilon_H)^2-(\Delta G+\lambda)^2\}/4\lambda kT]$$

The electric field dependence of the FC factors enters into the ΔG expression by adding the term $\mu_H E/\epsilon_H$, where μ_H here is the membrane normal projection of $BChl_2$+BPh-, estimated on a center to center basis as 1.28 e-nm, and ϵ_H is the dielectic constant between $BChl_2$ and BPh; μ_B and ϵ_B refer to the charge separation or $BChl_2$+BChl-. Substitution into equation (1) generates a model which successfully follows the data although the uncertainty of the data may not justify the introduction of the additional parameters to the electronic coupling only model.

CONCLUSIONS

Any of the 4 electron transfer models presented is capable of following the general trend of signal vs. imposed electric field of the Popovic et al. experiment, except a model in which only the Frank-Condon overlap is field sensitive (a sequential electron transfer model) in which $-\Delta G = \lambda$ and the Franck-Condon factors are based on a classical harmonic oscillator approximation. The fits of all other models are improved if the dielectric constant is less than 3 (figure 5). Such values may not be unreasonable, especially considering that the dielectric constant is a macroscopically defined concept which may be modified on the short time and distance scales of this experiment. Both the equilibrium and FC factor models display inflections which are not apparent in the data and result in conspicuous deviations at the extrema of positive and negative fields. The superexchange model, on the other hand, follows the smooth response of signal vs. field somewhat better, and for this reason the superexchange model may be preferred. Under circumstances in which the RC dielectric constant is 1, the virtual state energy gap would be about 100 meV.

More sophisticated models than those examined here have been used successfully to fit the considerably more complete data available for the slower electron transfers in the RC. For example, simultaneous ΔG and temperature variations of the electron transfer rates from BPh to Qa (figure 3) and from Qa back to the $BChl_2$ [34] required the use of a complete quantum mechanical treatment of the Franck-Condon factors that allows coupling to multiple modes [34,35]. We will examine more sophisticated models as field dependent experiments progress and delimit the expected field induced rate changes. The appearance of a field induced stabilization of BChl monomer would obviously clarify the validity of the different models. In addition, the field induced ΔG change should permit us to eliminate and native matching of $-\Delta G$ and λ and reveal a field induced temperature dependence of the electron transfer rate permiting us to calculate the strength of any modes coupled to the electron transfer.

REFERENCES

1 Deisenhofer, J., Epp, O., Miki, K., Huber, R., and Michel, H., Nature (London) 318, 618-624 (1985).

2 Chang, C.H., Tiede, D., Tang, J., Smith, U., Norris, J. and Schiffer, M., FEBS Lett., 205, 82-86 (1986).

3 Allen, J.P., Feher, G., Yeates, T.O., Komiya, H., and Ress, D.C., Proc. Natl. Acad. USA, 84, 5730-5734 (1987).

4 Sistrom, W.R. and Clayton, R.K. (eds.) The Photosynthetic Bacteria, (Plenum Press, New York 1978).

5 Okamura, M.Y., Feher, G., and Nelson, N. in: Photosynthesis Energy Conversion in Plants and Bacteria, Govindjee, ed. (Academic Press, London 1982) pp. 195-272.

6 Dutton, P.L in: Encyclopedia of Plant Physiolology, Vol. 19, Staehelin, L.A. and Arntzen, C.J., eds., (Springer-Verlag, Berlin 1986) pp. 197-223.

7 Woodbury, N.W., Becker, M., Middendorf, D., and Parson, W.W. Biochemistry 24, 7516-7521 (1985).

8 Fleming, G.R., Martin, J.L., and Breton, J., Nature 333, 190-192 (1988).

9 Chekalin, S.V., Matveetz, Y.A., Shkuropatov, Y.A., Shuvalov, V.A., and Vartzev, A.P., FEBS Lett., 216, 245-248 (1987).

10 Popovic, A.D., Kovacs, G.J., Vincett, P.S., Alegria, G., and Dutton, P.L., Biochim. Biophy. Acta, 851, 38-48 (1986).

11 Woodbury, N.W.T, and Parson, W.W., Biochim. Biophys. Acta, 767, 345-361 (1984).

12 Arata, H., and Parson, W.W., Biochim. Biophys. Acta 638, 201-209 (1981).

13 Gunner, M.R., and Dutton, P.L., Submitted to J. Amer. Chem. Soc, (1988).

14 Woodbury, N.W., Parson, W.W., Gunner, M.R., Prince, R.C., and Dutton, P.L., Biochim. Biophys. Acta. 851, 6-22 (1986).

15 Cruz, J.M.R., Master's Thesis, University of Toronto, Canada (1986).

16 Momany, F.A., McGuire, R.F., Burgess, A.W., and Scheraga, H.A., J. Phys. Chem. 79, 2361-2381 (1975).

17 Van Krevelen, D.W., and Hoftzyer, P.J., in: Properties of Polymers, 2nd Edn. (Elsevier, Amsterdam 1976) p. 231.

18 Churg, A.K., and Warshel, A., Biochemistry 25, 1675-1681 (1986).

19 Rees, D.C., J. Mol. Biol. 141, 323-326 (1980).

20 Rogers, N.K., Moore, G.R., and Sternberg, M.J., J. Mol. Biol. 182, 613-616 (1986).
21 Wraight, C.A., and Clayton, R.K., Biochim. Biophys. Acta 333, 246-260 (1973).
22 Jortner, J., J. Am. Chem. Soc., 102, 6676-6685 (1980).
23 Martin, J.L., Breton, J., Hoff, A.J., Migus, A., and Antonetti, A., Proc. Natl. Acad. Sci. USA 83, 957-961 (1986).
24 Shuvalov, V.A., Klevanik, A.V., Sharkov, A.V., Matveetz, Y.A., and Kryukov, P.G., FEBS Lett. 91, 135-139 (1978).
25 Haberkorn, R., Michel-Beyrle, M.E., and Marcus, R.A., Proc. Natl. Acad. Sci. USA 76, 4185-4189 (1979).
26 Kirmaier, C., Holten, D., and Parson, W.W., Biochim. Biophys. Acta 810, 33-42 (1985).
27 Ficher, S.F., and Scherer, P.O., J. Chem. Phy. 115, 151-158 (1987).
28 Jortner, J., Proc. NATO Workshop, Cadarache, (1987).
29 Martin, J.L., Breton, J., Hoff, A.J., Migus, A., and Antonetti, A., Proc. Natl. Acad. Sci. USA 83, 957-961 (1986).
30 Breton, J., Martin, J.L., Petrich, J., Migus, A., and Antonetti, A., FEBS Lett. 209, 37-43 (1986).
31 Martin, J.L., Fleming, G., and Breton, J., Proc. NATO Workshop, Cadarache, (1987).
32 Dutton, P.L., Alegria, G., and Gunner, M.R., Proc. NATO Workshop, Cadarache, (1987).
33 Marcus, R.A., J. Chem. Phys. 24, 966-978 (1956).
34 Gunner, M.R., Robertson, D.E., and Dutton, P.L., J. Chem. Phys. 90, 3783-3795 (1985).
35 Devault, D., Q. Rev. Biophys. 13, 387-564 (1980).
36 Jortner, J., and Bixon, M., in: Proceedings of the Philadelphia Conference on Protein Structure, (Academic Press, New York 1986).
37 McConnel, H.M., J. Chem. Phys. 35, 508-515 (1961).

THE CRYSTAL STRUCTURE OF THE PHOTOSYNTHETIC REACTION CENTER FROM RHODOPSEUDOMONAS VIRIDIS

Johann Deisenhofer and Hartmut Michel
Max-Planck-Institut fuer Biochemie
D-8033 Martinsried, FRG

INTRODUCTION

The primary event in photosynthesis is the absorption of a light quantum, followed by the transfer of an electron across a cell membrane. The light driven charge separation happens with high speed and a quantum efficiency of near unity; it is performed by a membrane-bound complex of proteins and pigments, the photosynthetic reaction center (RC). In the RC the excited state of a "primary electron donor" is deactivated by the transfer of an electron to an electron acceptor via intermediate carriers. It is one of the functions of the surrounding proteins to keep the electron donor, intermediate carriers and acceptors in a fixed geometry such that electron donor and acceptor are near opposite surfaces of the photosynthetic membrane. After transfer of an electron a major part of the energy of the absorbed light is stored in the form of difference in electric potentials across the membrane ("membrane potential") and redox energy in the form of the reduced electron acceptor.

The best known reaction centers are those from the purple photosynthetic bacteria (for reviews see e.g. Feher & Okamura, 1978, Hoff, 1982, Parson, 1987, Kirmaier & Holten, 1987). Most of them contain three protein subunits which are called H (heavy), M (medium) and L (light) subunits according to their apparent molecular weights as determined by sodium dodecylsulphate acrylamide gel electrophoresis. In addition, RCs from several purple photosynthetic bacteria, including Rps. viridis, contain a tightly bound cytochrome molecule.

The cytochrome subunit of the Rps. viridis RC has four heme groups covalently linked to the protein. Photosynthetic pigments are four

Photochemical Energy Conversion
James R. Norris, Jr. and Dan Meisel, Editors

bacteriochlorophyll-bs (BChl-bs), two bacteriopheophytin-bs (BPh-bs), one menaquinone ("primary quinone" or "Qa"), one non-heme-iron and one ubiquinone ("secondary quinone" or "Qb"). The RCs from most of the other purple photosynthetic bacteria contain BChl-a instead of BChl-b, BPh-a instead of BPh-b, and a second ubiquinone instaed of the menaquinone.
The light driven charge separation process in RCs of purple bacteria can be summarized as follows: The primary electron donor is a pair of BChl molecules with absorption bands extending to the near infrared region (about 960nm for Rps. viridis). Absorption of a photon by the primary donor is followed within about 3ps by the transfer of an electron to one of the BPhs. From there the electron is transferred within about 200ps to the Qa molecule; it ends up at the Qb molecule after another 0.1ms. After having received two electrons and two protons, the Qb molecule leaves the RC, and is replaced from a quinone pool in the membrane. In Rps. viridis the photooxidized primary donor is reduced within about 270ns by the bound cytochrome.

The RC from Rps. viridis was one of the first integral membrane proteins which could be crystallized (Michel, 1982). The RCs retained their photochemical activity in the crystalline state (Zinth et al., 1983). The crystals were suitable for X-ray structure analysis at atomic resolution. In a study at 3Å resolution, using phases determined from multiple isomorphous replacement experiments with heavy atom compounds, the arrangement of the major prosthetic groups (Deisenhofer et al., 1984), and the folding of the protein subunits (Deisenhofer et al.,1985) were determined. The amino acid sequences of the RC's protein subunits were derived from the corresponding gene sequences (Michel et al., 1985; Michel et al., 1986a; Weyer et al., 1987); information from X-ray diffraction and from sequencing together allowed precise model building, and a detailed description of pigment-protein interactions in the RC (Michel et al., 1986b). The model of the RC was further improved by crystallographic refinement, first at 2.9Å resolution, later at 2.3Å resolution (Deisenhofer, Epp, Michel, in preparation).

THE RC STRUCTURE

The structural studies resulted in an atomic model of the Rps. viridis RC, a simplified representation of which is shown in figure 1. The membrane spanning core of the RC is formed by the subunits L and M, hydrophobic polypeptides of 273 and 323 amino acids, repectively. Large portions of these subunits are folded in a similar way; outstanding elements of secondary structure are 5 membrane spanning helices in each subunit. The arrangement of L and M shows a remarkable local 2-fold symmetry. On both

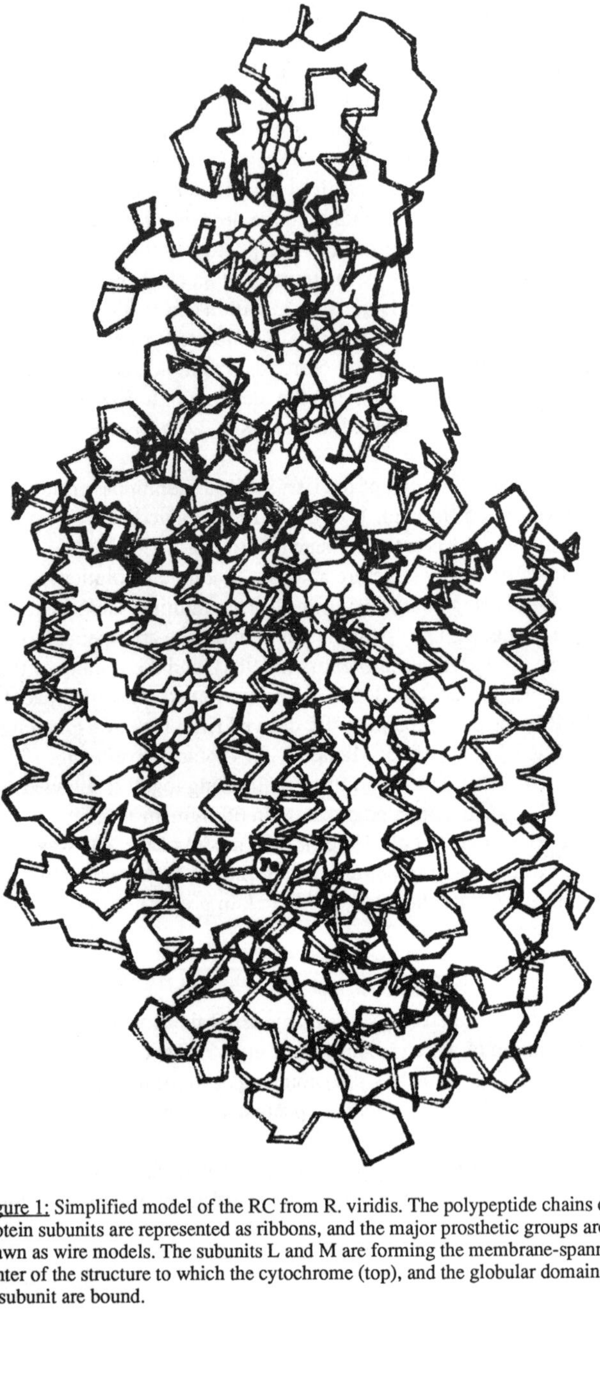

Figure 1: Simplified model of the RC from R. viridis. The polypeptide chains of the protein subunits are represented as ribbons, and the major prosthetic groups are drawn as wire models. The subunits L and M are forming the membrane-spanning center of the structure to which the cytochrome (top), and the globular domain of the H-subunit are bound.

sides of the membrane a peripheral subunit is noncovalently attached to the L-M complex: The 4-heme cytochrome with 336 amino acids at the periplasmic surface, and the globular part of the H-subunit at the cytoplasmic surface. The N-terminal segment of the H-subunit forms a membrane spanning helix; in total, this subunit consists of 258 amino acids. Imbedded into the L-M complex, and exhibiting largely the same local 2-fold symmetry, is the group of pigments involved in the primary charge separation: 4 BChl-b, 2 BPh-b, a menaquinone, and the non-heme iron. These pigments form 2 symmetrically arranged branches, both originating from the tightly interacting pair of BChl-bs, the special pair, near the periplasmic surface of the membrane. Each branch leads via an accessory BChl-b and a BPh-b to a quinone binding site near the membrane's cytoplasmic surface. Only the binding site of Qa, the primary quinone acceptor, is fully occupied by a menaquinone molecule; part of the Qb molecules (ubiquinone) appear to have been lost during isolation or crystallization of the RC. The non-heme iron is bound by 4 histidines and one glutamic acid; it sits on the local twofold axis between both quinone binding sites.

During crystallographic refinement a number of ordered constituents of the RC structure were discovered which, due to large phase errors and lack of resolution, could not be recognized during the early stages of the X-ray analysis. These new features include a long, bent molecule near one of the accessory BChl-bs, which most probably represents the carotenoid found by chemical analysis in the RC (Sinning, Michel, Eugster, unpublished). Another new feature in the refined model is the Qb head group, found with low occupancy in the Qb binding site which had been identified from symmetry considerations, and from binding studies with the inhibitors o-phenanthroline and terbutryn. About 200 bound water molecules were located, some of them in positions that indicate probable functional importance. A few electron density features, too big to be water, were interpreted as bound ions, most probably sulfate. The list of ordered solvent is completed by one molecule of the detergent LDAO; the vast majority of the detergent in the crystal is disordered. In a low resolution neutron diffraction study the detergent regions in the crystal could be localized (Roth, Lewitt-Bentley, Deisenhofer, Michel, to be published).

The improved accuracy of the refined atomic coordinates allowed to detect significant deviations of the pigment arrangement from local twofold symmetry. These deviations may help to explain the functional asymmetry of the pigments in the core of the RC (Michel-Beyerle et al., 1988).

RELATED STRUCTURES

Recently, RCs from Rb. sphaeroides could also be crystallized, and the photochemical activity of the crystalline RCs could be demonstrated (Allen & Feher, 1984), Chang et al., 1985, Gast & Norris, 1984). The model of the Rps. viridis RC (without the cytochrome subunit) was used as a starting point for the crystal structure analysis of RC from Rb. sphaeroides (Allen et al., 1986, Chang et al., 1986). These studies confirmed the close structural similarity between RCs from Rps. viridis and Rb. sphaeroides.

Sequence comparisons, making use of structural information on RCs from purple bacteria led to the hypothsis that the proteins D1 and D2 of photosystem II from green plants and cyanobacteria correspond to subunits L and M, respectively (see e.g. Michel & Deisenhofer, 1988). Thus, similarly folded protein subunits, arranged with an approximate twofold symmtry, seem to be a frequent structural motif in photosynthetic reaction centers.

References:

Allen, J. and Feher, G. (1984) Proc. Natl. Acad. Sci. USA **81**, 4795-4799

Allen, J.P., Feher, G., Yeates, T.O., Rees, D.C., Deisenhofer, J., Michel, H., and Huber, R. (1986) Proc. Natl. Acad. Sci. USA **83**, 8589-8593

Chang, C.H., Schiffer, M., Tiede, D., Smith, U., Norris, J. (1985) J. Mol. Biol. **186**, 201-203

Chang, C.-H., Tiede, D., Tang, J., Smith, U., Norris, J. and Schiffer, M. (1986) FEBS-Lett. **205**, 82-86

Deisenhofer, J., Epp, O., Miki, K., Huber, R. and Michel, H. (1985) Nature **318**, 618-624.

Deisenhofer, J., Epp, O., Miki, K., Huber, R. and Michel, H. (1984) J. Mol. Biol. **180**, 385-398.

Feher, G. and Okamura, M.Y. (1978) In Clayton, R.K. and Sistrom, W.R. (eds), The Photosynthetic Bacteria, pp. 349-386, Plenum Press, New York.

Gast, P. and Norris, J.R. (1984) FEBS-Lett. **177**, 277-280

Hoff, A.J. (1982) In Fong, F.K. (ed.) Molecular Biology, Biochemistry, and Biophysics, Vol. 35, pp. 80-151, 322-326, Springer, Berlin.

Kirmaier, C., and Holten, D. (1987) Photosynthesis Research **13**, 225-260

Michel, H. (1982) J. Mol. Biol. **158**, 567-572.

Michel, H., Weyer, K.A., Gruenberg, H. and Lottspeich, F. (1985) EMBO J. **4**, 1667-1672.

Michel, H., Weyer, K.A., Gruenberg, H., Dunger, I., Oesterhelt, D. and Lottspeich, F. (1986a). EMBO J. **5**, 1149-1158.

Michel, H., Epp, O. and Deisenhofer, J. (1986b) EMBO J. **5**, 2445-2451.

Michel-Beyerle, M.E., Plato, M., Deisenhofer, J., Michel, H., Bixon, M., Jortner, J. (1988) Biochim. Biophys. Acta **932**, 52-70.

Parson, W.W. (1987) in Amesz, J. (ed.) New Comprehensive Biochemistry Vol. 15: Photosynthesis, pp. 43-61, Elsevier, Amsterdam

Weyer, K.A., Lottspeich, F., Gruenberg, H., Lang, F., Oesterhelt, D. and Michel, H. (1987) EMBO J. **6**, 2197-2202.

Zinth, W., Kaiser, W. and Michel, H. (1983) Biochim. Biophys. Acta **723**, 128-131

MOLECULAR ORGANIZATION IN PHOTOSYSTEM II REACTION CENTER

KIMIYUKI SATOH
Department of Biology, Faculty of Science, Okayama University, Okayama 700, Japan

ABSTRACT

The primary photochemistry occuring in the photosystem II is thought to involve a charge separation between the primary donor chlorophyll(s) of P-680 and a pheophytin acceptor. Recent comparative molecular biological evidence has been suggested that the D1 and D2 proteins of photosystem II form the site of charge separation, in a similar manner to the L and M subunits forming the purple bacterial reaction center whose structure has been determined by x-ray crystallographic analysis. The proposal has recently been substantiated by a successful isolation of a pigment-protein consisting of D1 and D2 proteins and cytochrome b-559 and exhibiting an efficient charge separation between P-680 and pheophytin *a*. The chemical, photochemical, and spectroscopical characteristics of this complex with special reference to the organization of components on the donor side of photosystem will be discussed in this talk.

INTRODUCTION

Photosynthesis is the process whereby light energy is converted into chemical energy and conserved in the form of ATP and NAD(P)H, which can be used as the source of free energy and reducing power in the biosynthesis of organic molecules. Photosynthesis occurs in a wide variety of organisms, including both prokaryotes and eukaryotes, *e.g.*, photosynthetic bacteria, cyanobacteria, algae and higher plants. The prokaryotic photosynthesis of purple and green bacteria has only a single photosystem and uses reductants other than water (*e.g.*, H_2, H_2S or reduced organic compounds) to provide the reducing equivalents. In contrast, the photosynthesis of cyanobacteria, algae and higher plants has two photosystems, *i.e.*, photosystem I (PSI) and photosystem II (PSII). The realization of the latter photosystem in organisms made possible the use of the earth's huge water reservoir as electron donor and, at the same time, led to the creation of an aerobic atmosphere. The oxygen thereby formed is a powerful oxidant which permits an energetically highly efficient nutrient turnover in biological systems (respiration).

Previously, PSI was regarded as phylogenically equivalent to the bacterial photosystem and PSII as having been acquired in the course of evolution in order to adapt to the shortage of environmental organic molecules and reducing compounds [1]. However, the belief had to change when it was found that there exist striking similarities between PSII and the purple bacterial photosystem in the sequence of electron acceptors as shown in Fig. 1. The resemblaces in the electron transfer chain can be summarized as follows [2-4]: (1) The primary electron acceptor is one of the two (bacterio)-pheophytin molecules present in the reaction center in perpendicular to the plane of membrane which then reduces a quinone acceptor (Qa) in a few hundred ps; when Qa is reduced, the pheophytin anion accumulates under steady-state illumination (photoaccumulation): (2) Two quinone acceptors exist, Qa and Qb, functioning in series; Qa is a one-electron carrier, which has a redox potential of about -200 mV and is firmly bound to proteins, whereas Qb is a two-electron carrier weakly bound in the Q and QH_2 forms, but more strongly in

Photochemical Energy Conversion
James R. Norris, Jr. and Dan Meisel, Editors

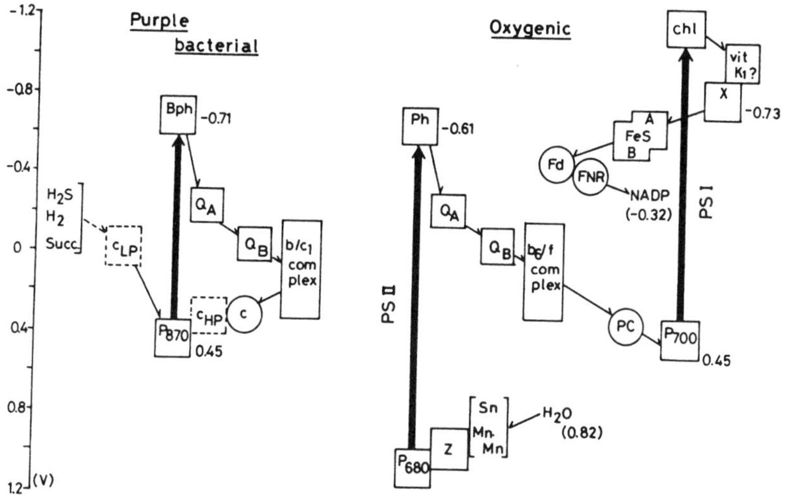

FIG. 1. Photosynthetic electron transfer systems (see text, modified from C.A. Wraight, 1982).

the Q and QH_2 forms, but more strongly in the Q^- form: (3) A ferrous-Fe interacts with both Qa and Qb, and the electron transfer from Qa to Qb is efficiently inhibited by atrazine type herbicides which act by competitive displacement of Qb: (4) The Qb delivers electrons to the cytochrome b/c_1 (b_6/f)-complex. Although these similarities do not fully extend into the details, e.g., the properties of the iron atom in PSII is different in the reactivity [5] and the molecular environment [6] from those in the purple bacterial photosystem, these data are sufficient enough to predict the presence of similar special relationships, supported by a similar proteinaceous environment, of the electron acceptors in both photosystems.

The donor side of PSII in the redox scheme, however, is not comparable with that in purple bacteria. Redox-potential of the primary donor, P-680, is approximately 600-800 mV more positive than that of the equivalent component of purple bacteria. In addition, PSII has an array of high-potential electron transfer chain which leads to the water cleavage and this part is truely unique to PSII reaction center.

PROTEIN SUBUNITS OF PSII

The reactions of PSII take place in a highly organized matrix which is made up of several classes of molecules which cooperate in fulfilling complementary roles: architectural support, light absorption, energy and electron transfer including the primary charge separation. Recent biochemical investigations [7,8] have identified and characterized a pigment-protein complex responsible for the photochemical reactions of oxygenic photosystem (PSII), which is consisted of 6 chloroplast-encoded protein subunits of hydrophobic character (PSII core complex), as shown in TABLE I. The genes for these proteins have been identified and sequenced and thus the primary structure have been predicted based on the nucleotide sequence of the genome [8-11]. The genes psbB and psbC code for the two largest subunits of about 50 and 40 kDa, respectively, which can be isolated in association with chlorophyll a and beta-carotene in a stoichiometric ratio [12,13]. The genes psbA and psbD code for the two medium-sized subunits of about 30 kDa, which have often been called D1 and D2, respectively, because of their diffuse migration patterns on SDS-polyacrylamide gel electrophoresis [9,14]. The genes psbE and psbF code for the alpha- and beta-subunits, respectively, of a b-type cytochrome (b-559) of unknown function [15]. It is now clearly established that the electron transfer from water to the quinone acceptor(s) can basically be performed in this pigment-protein consisted of the above mentioned 6 proteins, although a few more peripheral proteins are implicated in the stabilization and optimization under physiological conditions [16-18].

TABLE I. Protein subunits of PSII core complex.

Subunit	Approx. size (kDa)	Gene	Prosthetic group
α	50	psbB	chlorophyll a, β-carotene
β	40	psbC	chlorophyll a, β-carotene
γ (D-1)	30	psbA	chlorophyll a, pheophytin a
δ (D-2)	30	psbD	β-carotene, plastoquinone-9 ?
(α)	9	psbE	cytochrome b-559 heme
(β)	4	psbF	

"P-680 APOPROTEIN"

Until recently nearly all available experimental evidence seemed to indicate that the largest subunit of about 50 kDa (psbB gene product) of the PSII core complex is the site of primary photochemistry and thus the name "P-680 apoprotein" has been given to this subunit in the literature [19]. However, this was based only on the following circumstantial evidence: (1) It appeared that the psbB gene product is present in all photoactive subfractions of PSII so far isolated [20-22]: (2) A genetically engineered mutant depleted in the psbB gene product was totally inactive in the PSII activity [23]: (3) An isolated chlorophyll-protein consisting of the 50 kDa component (CP-47) exhibits a chracteristic emission band of F-695 which was postulated to originate from the primary pheophytin acceptor of PSII [24,25]: (4) The 50 kDa component has shown to be the specific binding site for azido-plastoquinone in photoaffinity labeling experiments [26], which may represent the binding site of plastoquinone functioning in the primary reactions in PS II.

HOMOLOGY OF AMINO ACID SEQUENCES OF PSII AND BACTERIAL REACTION CENTERS

Recently the notion that the 50 kDa component is the site of charge separation in PSII, however, has been challenged on the basis of comparative molecular biological evidence. As shown in Fig. 2, a weak, but significant amino acid sequence homology, as deduced from the nucleotide sequences of the chloroplast genome, has been documented to exist between the D1 and D2 protein of PSII and the L and M subunits of the purple bacterial reaction center [27,28], the structure of which has recently been determined by x-ray crystallographic analysis [29,30]. Although the sequence homology is only about 10-15 %, it is significant in the region of putative transmembrane helices predicted on the basis of hydropathy plots and immunological studies [27, 31,32]. It is remarkable that the amino acids (such as glycine and proline) conserved between L, M, D1 and D2 subunits are found at, or close to, the ends of the alpha-helices at the outer or the inner surfaces of the photosynthetic membranes. Furthermore, the homology is striking for the amino acids of structural importance in the co-factor binding in the purple bacterial reaction centers such as the histidine ligands to the magnesium atoms of the "special pair" bacterio-chlorophyll (His-173 and His-200 in L and M subunits, respectively, of *R. viridis*) and to the ferrous-Fe in the quinone (Qa and Qb)-iron acceptor complex (His-190 and His-230 in L, and His-217 and His-264 in M subunits of *R. viridis*). Tryptophan, which forms part of the

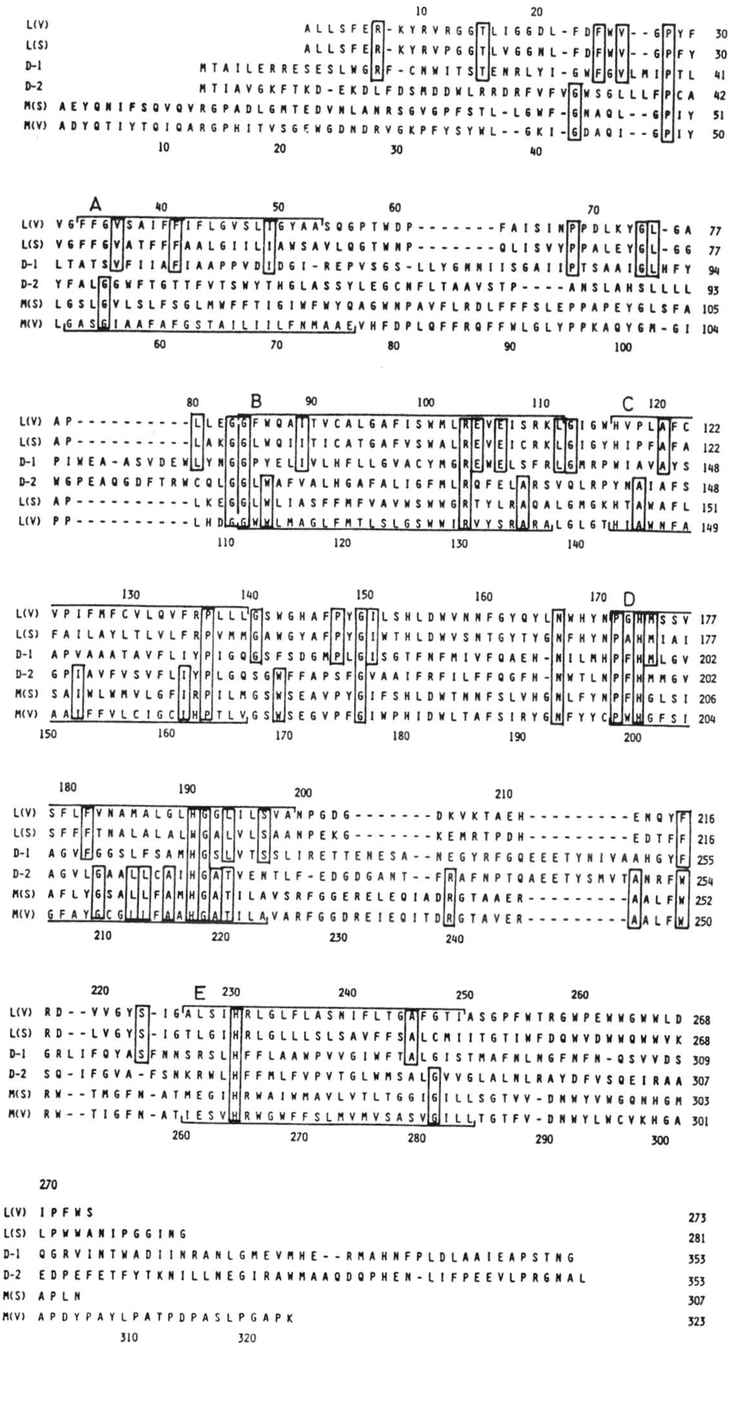
L(V) ALLSFER-KYRVRGGTLIGGDL-FDFWV--GPYF 30
L(S) ALLSFER-KYRVPGGTLVGGNL-FDFWV--GPFY 30
D-1 MTAILERRESESLWGRF-CNWITSTENRLYI-GWFGVLMIPTL 41
D-2 MTIAVGKFTKD-EKDLFDSMDDWLRRDRFVFVGWSGLLLFPCA 42
M(S) AEYQNIFSQVQVRGPADLGMTEDVNLANRSGVGPFSTL-LGWF-GNAQL--GPIY 51
M(V) ADYQTIYTQIQARGPHITVSGEWGDNDRVGKPFYSYWL--GKI-GDAQI--GPIY 50
A
L(V) VGFFGVSAIFFIFLGVSLIGYAASQGPTWDP-------FAISINPPDLKYGL-GA 77
L(S) VGFFGVATFFFAALGIILIAWSAVLQGTWNP-------QLISVYPPALEYGL-GG 77
D-1 LTATSVFIIAFIAAPPVDIDGI-REPVSGS-LLYGNNIISGAIIPTSAAIGLHFY 94
D-2 YFALGGWFTGTTFVTSWYTHGLASSYLEGCNFLTAAVSTP----ANSLAHSLLLL 93
M(S) LGSLGVLSLFSGLMWFFTIGIWFWYQAGWNPAVFLRDLFFFSLEPPAPEYGLSFA 105
M(V) LGASGIAAFAFGSTAILIILFNMAAEVHFDPLQFFRQFFWLGLYPPKAQYGM-GI 104
B C
L(V) AP----------LLEGGFWQAITVCALGAFISWMLREVEISRKLGIGWHVPLAFC 122
L(S) AP----------LAKGGLWQIITICATGAFVSWALREVEICRKLGIGYHIPFAFA 122
D-1 PIWEA-ASVDEWLYNGGPYELIVLHFLLGVACYMGREWELSFRLGMRPWIAVAYS 148
D-2 WGPEAQGDFTRWCQLGGLWAFVALHGAFALIGFMLRQFELARSVQLRPYNAIAFS 148
L(S) AP----------LKEGGLWLIASFFMFVAVWSWWGRTYLRAQALGMGKHTAWAFL 151
L(V) PP----------LHDGGWWLMAGLFMTLSLGSWWIRVYSRARALGLGTHIAWNFA 149
D
L(V) VPIFMFCVLQVFRPLLLGSWGHAFPYGILSHLDWVNNFGYQYLNWHYNPGHMSSV 177
L(S) FAILAYLTLVLFRPVMMGAWGYAFPYGIWTHLDWVSNTGYTYGNFHYNPAHMIAI 177
D-1 APVAAATAVFLIYPIGQGSFSDGMPLGISGTFNFMIVFQAEH-NILMHPFHMLGV 202
D-2 GPIAVFVSVFLIYPLGQSGWFFAPSFGVAAIFRFILFFQGFH-NWTLNPFHMMGV 202
M(S) SAIWLWMVLGFIRPILMGSWSEAVPYGIFSHLDWTNNFSLVHGNLFYNPFHGLSI 206
M(V) AAIFFVLCIGCIHPTLVGSWSEGVPFGIWPHIDWLTAFSIRYGNFYYCPWHGFSI 204
L(V) SFLFVNAMALGLHGGLILSVANPGDG-------DKVKTAEH---------ENQYF 216
L(S) SFFFTNALALALHGALVLSAANPEKG-------KEMRTPDH---------EDTFF 216
D-1 AGVFGGSLFSAMHGSLVTSSLIRETTENESA--NEGYRFGQEEETYNIVAAHGYF 255
D-2 AGVLGAALLCAIHGATVENTLF-EDGDGANT--FRAFNPTQAEETYSMVTANRFW 254
M(S) AFLYGSALLFAMHGATILAVSRFGGERELEQIADRGTAAER----------AALFW 252
M(V) GFAYGCGILFAAHGATILAVARFGGDREIEQITDRGTAVER----------AALFW 250
E
L(V) RD--VVGYS-IGALSIHRLGLFLASNIFLTGAFGTIASGPFWTRGWPEWWGWWLD 268
L(S) RD--LVGYS-IGTLGIHRLGLLLSLSAVFFSALCMIITGTIWFDQWVDWWQWWVK 268
D-1 GRLIFQYASFNNSRSLHFFLAAWPVVGIWFTALGISTMAFNLNGFNFN-QSVVDS 309
D-2 SQ-IFGVA-FSNKRWLHFFMLFVPVTGLWMSALGVVGLALNLRAYDFVSQEIRAA 307
M(S) RW--TMGFN-ATMEGIHRWAIWMAVLVTLTGGIGILLSGTVV-DNWYVWGQNHGM 303
M(V) RW--TIGFN-ATIESVHRWGWFFSLMVMVSASVGILLTGTFV-DNWYLWCVKHGA 301
L(V) IPFWS 273
L(S) LPWWANIPGGING 281
D-1 QGRVINTWADIINRANLGMEVMHE--RMAHNFPLDLAAIEAPSTNG 353
D-2 EDPEFETFYTKNILLNEGIRAWMAAQDQPHEN-LIFPEEVLPRGNAL 353
M(S) APLN 307
M(V) APDYPAYLPATPDPASLPGAPK 323

FIG. 2. The amino acid sequences of the D1 and D2 proteins of spinach with those of the L and M subunits of purple bacterial reaction centers; Rps. viridis (L(V) and M(V)) and Rb. sphaeroides R-26 (L(S) and M(S)). Amino acids common to all six or three subunits are indicated. The position of the transmembrane helices in Rps. viridis reaction center is bracketted (A-E) above and below the sequences of L and M subunits (previous page).

binding site of Qa, is conserved between D2 and M subunit (Tyr-250 in M-subunit of R. viridis). A phenylalanine is found in the equivalent position in the binding site of Qb and is conserved between D1 and L subunit (Phe-216 in L-subunit of R. viridis). When this phenylalanine in D1 mutates to tyrosine, the PSII reaction center becomes resistant to atrazine [33].

The structural and genetic evidence documented above is in agreement with the proposal that the heterodimer consisting of D1 and D2 proteins forms the site of primary photochemistry, in a similar manner to the L and M subunits forming the reaction center of purple bacteria [27,28], although no direct biochemical evidence has not been presented.

ISOLATION AND CHARACTERIZATION OF PSII REACTION CENTER

The first convincing evidence for the presence of D1 and D2 proteins in association with pigments was provided by a successful isolation of a pigment-protein complex consisting of D1, D2 and cytochrome b-559 [34]. The complex was isolated from PSII enriched membranes of spinach after solubilization with Triton X-100 followed by an ion-exchange chromatography using DEAE-Toyopearl. Antibodies raised against each of the constituent polypeptides of the core complex, i.e., against psbB gene product, psbC gene product, cytochrome b-559, psbA gene product expressed in E. coli, or against a synthetic oligopeptide corresponding to the specific 7 amino acid residues of D2 protein, were utilized to convince that the isolated pigment-protein does not contain any small amount of the 50 kDa component, but consisted only of D1 and D2 proteins and cytochrome b-559 [34,35].

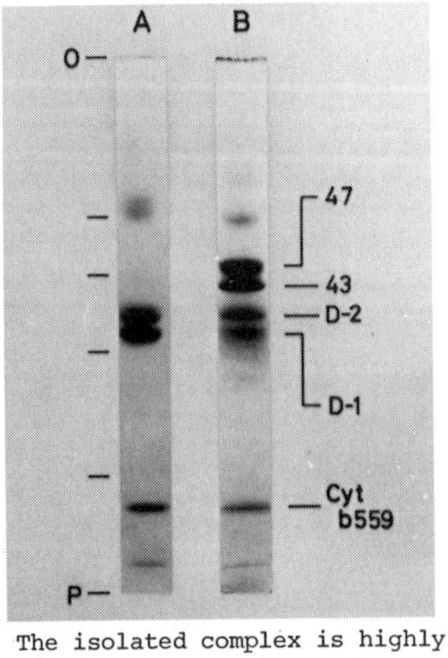

FIG. 3. SDS-polyacrylamide gel electrophoresis of PSII reaction center complex (lane A). For comparison, the profile for the PSII core complex is also shown (lane B). Tick marks on the left side indicate the positions of the analyzing gel (o), 68, 45, 25, and 12.4 kDa, respectively, from the top, and the pigment front (p).

The isolated complex is highly enriched in pheophytin a, which acts as the primary acceptor, in a molar ratio of 2 pheophytin a to 4-5 chlorophyll a, similar to the bacterio-chlorophyll/bacterio-pheophytin ratio in purple bacterial reaction centers [2]. The complex contains virtually no plastoquinone-9, the counterpart of which is present in the isolated bacterial reaction centers [2,29,30]. The complex contains 1-2 cytochrome b-559 per 2 pheophytin a and one non-heme Fe (Barber, personal communication) but no manganese atom which plays an essential role in the water cleavage.

The pigment-protein is totally inactive in the photooxidation of water (oxygen evolution) and the photoreduction of 2,6-dichlorophenol indophenol by 1,5-diphenylcarbazide. However, the complex exhibits a reversible absorbance change upon steady-state illumination as measured in the presence of dithionite and methylviologen (Fig. 4). The characteristic negetive peaks at 422, 515, 545 and 682 nm and a positive peak at 450 nm in the light-induced spectrum as well as their kinetic behavior coincide well with those of photoaccumulation of reduced pheophytin a in the photosystem II reaction center as reported by Klimov et al. [36], although the negative peak position at the red region is 2-3 nm blue-shifted. Using milimolar difference absorption coefficient of 32 for pheophytin a - pheophytin a^- at the red maximum [37], the absorption change upon saturating actinic illumination was estimated to correspond to about one pheophytin a molecule photoreduced out of two chemically estimated molecules in the complex.

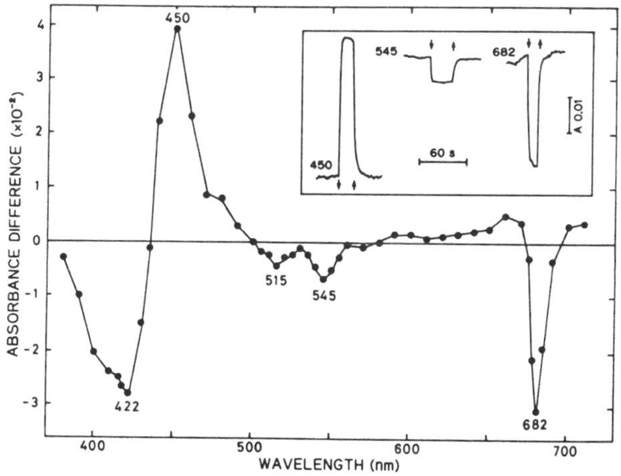

FIG. 4. Photoaccumulation of reduced pheophytin in PSII reaction center.

Primary PSII reactions have been studied by flash absorption spectroscopy with ps and ns time resolution [38,39]. Fig. 5 shows the absorption difference spectrum of the pigment-protein in the region 430-685 nm at 4 ns after a non-saturating 695 nm flash, and around 545 nm also after 24 ns (inset, dashed line). These spectra show features attributable to the reduction of pheophytin a. In particular, the negative band at 545 nm (inset) and the positive band at 450 nm strongly suggest that the same absorption changes which are observed upon photoaccumulation of reduced pheophytin a (Fig. 4) also contribute to this difference spectrum. The kinetic measurement indicated that the rise took place within the 25 ps flash duration; the decay was exponential with a best-fitting lifetime of 32-36 ns [38,39]. This is remarkably longer than 2-4 ns radical pair lifetimes reported for PSII, but not unexpectedly; both presence of Q_a^- and that of antenna chlorophyll a might cause a decrease of the observed lifetime [40]. The quantum yield of pheophytin reduction upon flash excitation was estimated to be substantially higher than 65 % [38,39].

The formation of ^{3}P-680 as a product of radical pair recombination with a yield of about 23 % at 276 K (80 % at 10 K) was detected by flash absorption spectroscopy [39]. The EPR spectrum of the light-induced triplet state in the pigment-protein has also been observed and characterized [41]. The spectrum (Fig. 6) exhibits the polarization AEE AAE (where A and E refer to absorption and emission, respectively) that is indicative of a radical pair precusor. The values of the zero-field splitting parameters D and E were obtained from the spectrum:

$$D = 287 \times 10^{-4} cm^{-1}, \quad E = 43 \times 10^{-4} cm^{-1}$$

These values are in agreement with those obtained for PSII core complex [41] and also with those reported previously [42].

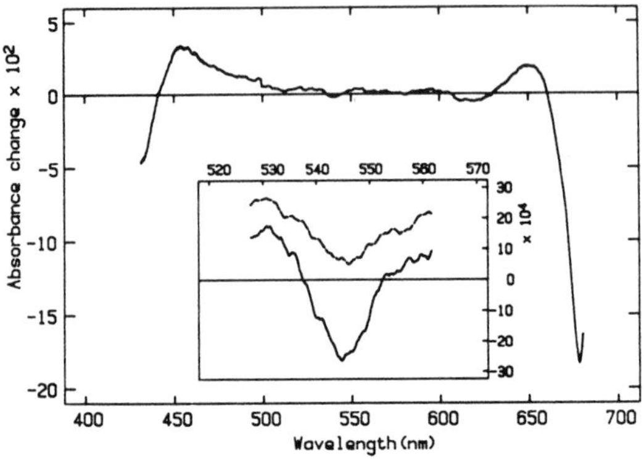

FIG. 5. Absorbance difference spectra of PSII reaction center 4 ns after excitation by a 25 ps, 695 nm pulse. (Inset), the 545 nm region enlarged; the upper line shows the difference spectrum at 24 ns after excitation.

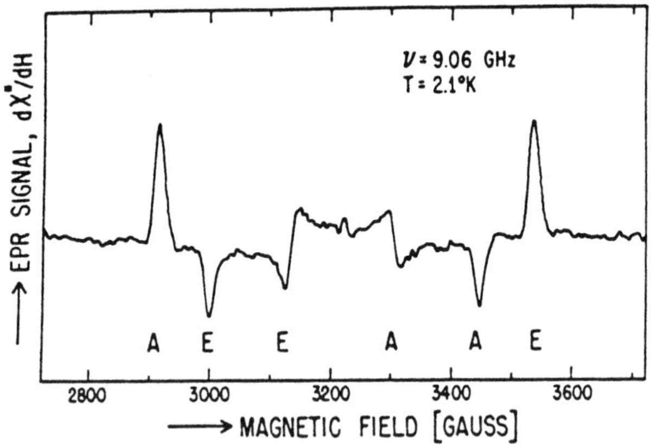

FIG. 6. EPR spectrum of the light-induced triplet in PSII reaction center complex. A and E refer to absorption and emission. Light source: Leitz 500 W projector with 650 nm filter (BW 50 nm), microwave power 10 μW, field modulation 100 kHz, 5 Gauss p.t.p.

The low temperature absorption spectrum of the pigment-protein complex shows two bands in the Q_y region located at 670 and 680 nm as shown in Fig. 7 [43]. On the basis of its absorption maximum and orientation the latter component is attributed in part to P-680 and in part to pheophytin a. The presence of carotenoid is indicated by the absorption shoulders around 460-490 nm which are due to beta-carotene. One of the striking feature of the absorption spectrum is the presence of a strong absorption peak at about 415 nm, which is due to gamma-bands of cytochrome b-559 and pheophytin a.

The F-695 nm emission band was absent in the low temperature emission spectrum [43], which indicates that it does not originate from the reaction center pheophytin of PSII as earlier proposal [25].

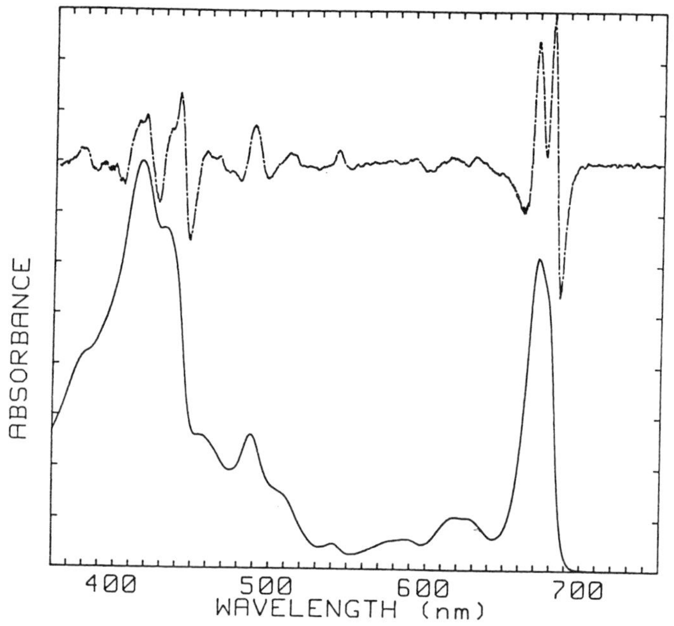

FIG. 7. Absorption (lower) and the second derivative (upper) spectra of PSII reaction center at 77 K.

A major conclusion reached from these analyses is that the pigment-protein complex consisting of D1 and D2 proteins and cytochrome b-559, but completely devoid of psbB gene product (50 kDa component), contains the site of primary charge separation in PSII, although the stabilization of the separated charges is not attained in this complex because of the absence of plastoquinone. The functional role of CP-47 which emits F-695 fluorescence [25] might be a specific antenna to the reaction center, since this component carries substantial amounts of chlorophyll a and beta-carotene, but exhibits no photochemical activity. This component could also be involved in the plastoquinone binding since isolated reaction center does not contain any plastoquinone molecule. The functional role of cytochrome b-559 in the reaction center complex is still a matter of conjection, although a photochemical reaction of this cytochrome in the isolated complex has been reported [44].

IDENTITY AND ORGANIZATION OF COMPONENTS ON THE DONOR SIDE OF PSII

A. P-680, primary donor

Resonance Raman spectroscopy of chlorophyll a moiety in the isolated PSII reaction center in the CC double-bond region (1552 and 1615 cm^{-1}) and the keto CO stretching region (1672 cm^{-1}) has indicated that the Mg atom is 5-coordinated and the keto CO is hydrogen-bonded [45].

The effect of an applied electric field on the optical absorption spectrum (Stark effect) of PSII reaction center in polyvinyl alcohol films at 77 K was measured [46]. The Stark spectrum (Fig. 8, lower) shows two bands with minima at 680 and 670 nm and has a shape similar to the second derivative (middle) of the absorption spectrum (upper and Fig. 7). As mentioned above, the band at 680 nm has contributions from the primary donor P-680 and pheophytin acceptor. The change in the dipole moment, Δμ, between the ground and excited states obtained from the ratio of the Stark and second derivative spectra were 0.8 and 1.5 Debye for the 680 and 670 (accessory chlorophyll) nm peaks, respectively. The angles between respective Δμs and transition moments were 40° and 20°. These results differ significantly from the situation in purple bacterial reaction centers where the primary donor, a bacterio-chlorophyll dimer, has a much larger Δμ than its monomeric bacterio-chlorophylls [47,48]. This suggest that P-680 is a monomeric chlorophyll species, although a symmetric dimer cannot excluded.

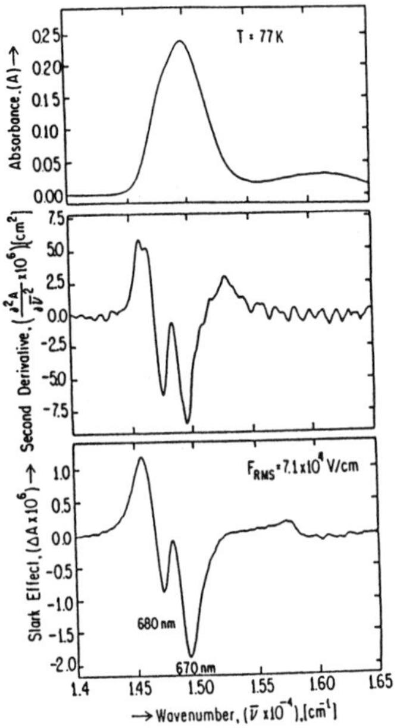

FIG. 8. Absorption spectrum (upper), second derivative (middle) and Stark spectrum of the PSII reaction center at 77 K.

A marked orientation dependence of the ^{3}P-680 EPR signal(Fig. 6) in oriented multilayers of PSII reaction center was observed [49], as previously reported for PSII preparations [50]. The large outer Z peaks were maximal when the oriented layers were in perpendicular to the magnetic field, while both the X and Y peaks were maximal when the layers were oriented parallel to the magnetic field. Assuming a monomeric structure for ^{3}P-680 as suggested by Stark effect measurements, the orientation data indicate that the chlorophyll ring plane is in the plane of the membrane. The results are markedly different from those obtained for the analogous signal in purple bacteria, the actual structure has now been elucidated by x-ray crystallographic analysis [29,30].

Two observations reported above exemplify that the structure of P-680 is somehow different from that of pueple bacterial counterpart, in spite of the fact that the histidine ligands in L and M subunits are preserved in the D1 and D2 proteins. The essential functional difference between the two types of reaction center is the oxidizing power generated at the photoactive pigments. It could be speculated that this problem is simply solved by exchanging chlorophyll a for the bacterio-chlorophyll dimer, because the first excited singlet state of chlorophyll is energetically higher by about 400 mV than that of bacterio-chlorophyll and because there is no large difference between the redox potential for the reduction of pheophytin and bacterio-pheophytin (see Fig. 1). However, the utilization of chlorophyll a is not a sufficient condition, as is shown by the example of chlorophyll a in PSI reaction center (P-700). Therefore, the nature of the protein matrix must be of central relevance for the redox potential of P-680.

B. Electron donor(s) to P-680

On the donor side of P-680, there exist two redox components, Z and D (not shown in Fig. 1), of unknown chemical entity. The Z serves as the secondary electron donor, whereas the function of the latter component which usually is in the D^+ state is unknown. This component is not directly involved in the main-path of electron transport to P-680. However, it does have the capacity to interact with s-states (Sn in Fig. 1., n = 0-4) of Mn cluster of oxygen-evolving system [51-53]. The oxidized form of the two components gives rise to an identical EPR spectrum, but with different stability: Z exhibits a rapidly decaying photoreversible signal (signal $II_{f(vf)}$) whereas D^+ exhibits a dark-stable signal (signal II_s). Arguments have been presented to suggest that Z^+ and D^+ are radicals of quinone molecules tightly bound to the PSII proteins.

Previous iodo-labeling experiment has shown that the D2 protein is exclusively iodinated in the dark, whereas the iodination of D1 protein takes place only in the light as a result of photooxidation by PSII reaction center [54]. The chemical entity responsible for the activation of iodide in the dark was suggested to be the dark-stable EPR species (signal II_s, D^+) which could be tyrosine residue on D2 protein. This idea has recently been substantiated by a selective deutration experiment using an amino acid auxotrophs, Synechocystis [55]. It was found that the normal 20 Gauss linewidth of the g = 2.0046 EPR signal is narrowed to 7 Gauss in cells fed with deutrated tyrosine. Taking into account that D^+ can interact with P-680 and s-states, a tyrosine residue at position 160 on D2 protein which is close to the putative P-680 binding site was considered to be the likely candidate for D [54,55]. This possibility has recently been demonstrated by two independent experiments [56,57], where they have applied the technique of site-directed mutagenesis. Using a transformable strain of Synechocystis, they mutated tyrosine 160 of D2 protein to phenylalanine and found that the resultant mutant lacks the EPR signal II_s indicative of D^+. One of the most important conclusion derived from these experiments is that Z is almost certainly the symmetrically related tyrosine residue on D1 protein as has been suggested by iodo-labeling experiments [54,58]; the common sequence between D1 and D2 proteins around the tyrosine residues in question (Tyr-161 of both D1 and D2 proteins of spinach, see Fig. 2) is -VFLIYPXGQ- (X = I or L).

The identification of the iodinated residue(s) on D1 protein has been conducted by enzymatic and chemical cleavages [59]. The conclusion from the analysis was that the tyrosine 147 and/or tyrosine 161 located on the third transmembrane helices (C-helices in Fig. 1) is the site of iodination in D1 protein. The geometry consideration further suggested that the tyrosine 161 is highly probable, since this residue is located at distances from the thylakoid membrane surface equivalent to that expected for P-680, assuming histidines 198 on D1 and D2 proteins form ligands for the chlorophyll(s), in the two-dimentional model of D1 prtein proposed by Trebst [31] (Fig. 9). Proteins have an obvious role of maintaining the co-factors at well defined positions and probably in suitable environment (acidity, polarity etc.) in reaction centers. In this case, however, proteins seems to be more directly involved in the electron transfer.

It is interesting to note that there is a symmetrical arrangement of co-factors on the donor side, as in the acceptor side which has been so clearly shown for purple bacterial reaction centers [29,30]. It is also interesting that the similarity of the environments of the two tyrosines of D1 and D2 can give rise to dramatically different kinetics of Z and D.

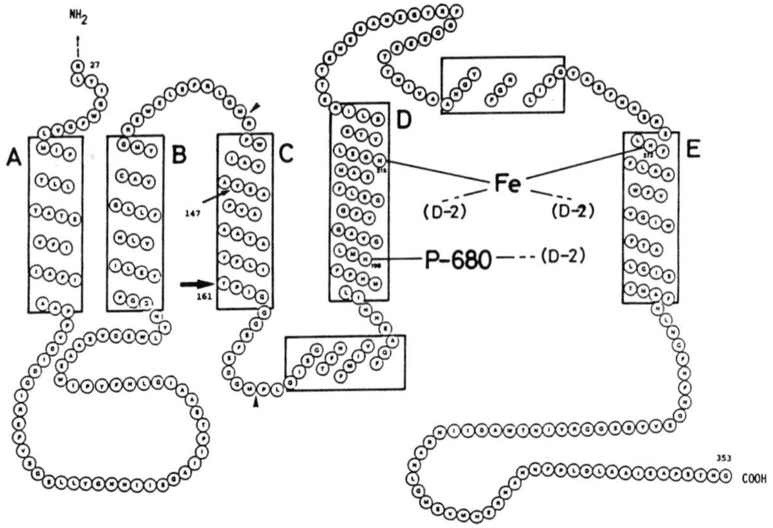

FIG. 9. The folding model of Dl protein of PSII reaction center (after Trebst [31]). Possible iodination sites are indicated by arrows (see text for further explanation).

REFERENCES

1. D.I. Arnon, M. Losada, F.R. Whatley, H.Y. Tsujimoto, D.O. Hall and A.A. Horton, Proc. Nat. Acad. Sci. 47, 1314-1334 (1961).
2. M.Y. Okamura, G. Feher and N. Nelson, in: Photosynthesis, Govindjee, ed. (Academic Press, N.Y. 1982) pp. 195-272.
3. P. Mathis and A.W. Rutherford, in: Photosynthesis, J. Amesz, ed. (Elsevier, Amsterdam 1987) pp. 63-96.
4. P. Mathis, in: Progress in Photosynthesis Research, J. Biggins, ed. (Martinus Nijhoff Publishers, Dordrecht 1987) pp. 151-160.
5. V. Petrouleas and B.A. Diner, Biochim. Biophys. Acta 849, 264-275 (1986).
6. A.W. Rutherford, in: Progress in Photosynthesis Research, J. Biggins, ed. (Martinus Nijhoff Publishers, Dordrecht 1987) pp. 277-283.
7. K. Satoh, in: The Oxygen Evolving System of Photosynthesis, Y. Inoue, A.R. Crofts, Govindjee, N. Murata, G. Renger and K. Satoh, eds. (Academic Press, Tokyo 1983) pp. 27-38.
8. K. Satoh, Photochem. Photobiol. 42, 845-853 (1985).
9. G. Zurawski, H.J. Bohnert, P.R. Whitfeld and W. Bottomley, Proc. Nat. Acad. Sci. 79, 7699-7703 (1982).
10. J. Alt, J. Morris, P. Westhoff and R.G. Herrmann, Curr. Genet. 8, 597-606 (1984).
11. J. Morris and R.G. Herrmann, Nucleic Acids Res. 12, 2837-2850 (1984).
12. P. Delepelaire and N.-H. Chua, Proc. Nat. Acad. Sci. 76, 111-115 (1979).
13. M. Fujitani and K. Satoh, in preparation.
14. J.-D. Rochaix, M. Dron, M. Rahire and P. Malnoe, Plant Mol. Biol. 3, 363-370 (1984).
15. W.A. Cramer, S.M. Theg and W.R. Widger, Photosynth. Res. 10, 393-403 (1986).
16. T. Ono and Y. Inoue, FEBS Lett. 168, 281-286 (1984).
17. N. Murata and M. Miyao, Trends Biochem. Sci. 10, 122-124 (1985).
18. X.-S. Tang and K. Satoh, FEBS Lett. 179, 60-64 (1985).
19. K. Satoh, Physiol. Plant. 72, 209-212 (1988).

20. E.L. Camm and B.R. Green, Biochim. Biophys. Acta 724, 291-293 (1983).
21. C. de Vitry, F.-A. Wollmann and P. Delepelaire, Biochim. Biophys. Acta 767, 415-422 (1984).
22. K. Satoh, FEBS Lett. 204, 357-362 (1986).
23. W.F.J. Vermaas, J.G.K. Williams, A.W. Rutherford, P. Mathis and C.J. Arntzen, Proc. Nat. Acad. Sci. 83, 9474-9477 (1986).
24. H.Y. Nakatani, B. Ke, E. Dolan and C.J. Arntzen, Biochim. Biophys. Acta 765, 347-352 (1984).
25. J. Breton, FEBS Lett. 147, 16-20 (1982).
26. K. Satoh, S. Katoh, R. Dostatni and W. Oettmeier, Biochim. Biophys. Acta 851, 202-208 (1986).
27. A. Trebst, Z. Naturforsch. 41c, 240-245 (1986).
28. H. Michel and J. Deisenhofer, in: Encyclopedia Plant Physiol. NS, Vol. III, A.C. Staehelin and C.J. Arntzen, eds. (Springer-Verlag, Berlin 1986) pp. 371-381.
29. J. Deisenhofer, O. Epp, K. Miki, R. Huber and H. Michel, Nature 318, 618-624 (1985).
30. J.P. Allen, G. Feher, T.O. Yeates, H. Komiya and D.C. Rees, Proc. Nat. Acad. Sci. 84, 5730-5734 (1987).
31. A. Trebst, Z. Naturforsh. 42c, 742-750 (1986).
32. R.T. Sayre, B, Andersson and L. Bogorad, Cell 47, 601-608 (1986).
33. J.M. Erickson, M. Rahire, J.-D. Rochaix and L. Mets, Science 228, 204-207 (1985).
34. O. Nanba and K. Satoh, Proc. Nat. Acad. Sci. 84, 109-112 (1987).
35. K. Satoh, Y. Fujii, T. Aoshima and T. Tado, FEBS Lett. 216, 7-10 (1987).
36. V.V. Klimov, A.V. Klevanik, V.A. Shuvalov and A.A. Krasnovsky, Dokl. Acad. Nauk SSSR 236, 241-244 (1977).
37. I. Fujita, M.S. Davis and J. Fajer, J. Am. Chem. Soc. 100, 6280-6282 (1978).
38. R.V. Danielius, K. Satoh, P.J.M. van Kan, J.J. Plijter, A.M. Nuijs and H.J. van Gorkom, FEBS Lett. 213, 241-244 (1987).
39. Y. Takahashi, O. Hansson, P. Mathis and K. Satoh, Biochim. Biophys. Acta 893, 49-59 (1987).
40. O. Hansson, J. Duranton and P. Mathis, Biochim. Biophys. Acta 932, 91-96 (1988).
41. M.Y. Okamura, K. Satoh, R.A. Isaacson and G. Feher, in: Progress in Photosynthesis Research, J. Biggins, ed. (Martinus Nijhoff Publishers, Dordrecht 1987) pp. 379-381.
42. A.W. Rutherford, D.R. Paterson and J.E. Mullet, Biochim. Biophys. Acta 635, 205-214 (1981).
43. R.J. van Dorssen, J. Breton, J.J. Plijter, K. Satoh, H.J. Gorkom and J. Amesz, Biochim. Biophys. Acta 893, 267-274 (1987).
44. T. Nishikata, M. Fujitani and K. Satoh, in preparation.
45. M. Fujiwara, H. Hayashi, M. Tasumi, M. Kanaji, Y. Koyama and K. Satoh, Chem. Lett. 2005-2008 (1987).
46. M. Losche, K. Satoh, G. Feher and M.Y. Okamura, Biophys. J. in press.
47. D. Lockhardt and S. Boxer, Biochemistry 26, 2664-2668 (1987).
48. M. Losche, G. Feher and M.Y. Okamura, Proc. Nat. Acad. Sci. 84, 7537-7541 (1987).
49. A.W. Rutherford and K. Satoh, in preparation.
50. A.W. Rutherford, Biochim. Biophys. Acta 807, 189-201 (1985).
51. A. Kawamori, J. Satoh, T. Inui and K. Satoh, FEBS Lett. 217, 134-138 (1987).
52. S. Styring and A.W. Rutherford, Biochemistry 26, 2401-2405 (1987).
53. J.H.A. Nugent, C. Demetriou and C.J. Lockett, Biochim. Biophys. Acta 894, 534-542 (1987).
54. Y. Takahashi, M. Takahashi and K. Satoh, FEBS Lett. 208, 347-351 (1986).
55. B.A. Barry and G.T. Babcock, Proc. Nat. Acad. Sci. 84, 7099-7103 (1987).
56. R.J. Debus, B.A. Barry, G.T. Babcock and L. McIntosh, Proc. Nat. Acad. Sci. 85, 427-430 (1988).

57. W.F.J. Vermaas, A.W. Rutherford and O. Hansson, Proc. Nat. Acad. Sci. in press.
58. M. Ikeuchi and Y. Inoue, FEBS Lett. 210, 71-76 (1987).
59. Y. Takahashi and K. Satoh, in preparation.

PANEL DISCUSSION ON PHOTOSYNTHESIS AND BIOMIMETIC SYSTEMS

Communicated by L. K. Patterson[1] (Panel Chairman)

Panel Members: Maryln Gunner[2], D. Gust[3], H. Levanon[4], V. Krishnan[5], V. Ya. Shafirovich[6]

The attention of this panel focused on electron transfer from the excited state chromophore to a primary (or secondary) acceptor and on the efficiency of back reaction which annuls charge separation. The discussion covered these processes in: (a) the photosynthetic reaction center protein; (b) covalently bound donor-acceptor systems designed to both optimize the initial yield and retard the back reaction; (c) model assemblies which utilize molecular organization to achieve optimal charge separation and stabilization. Additionally, the use of time resolved EPR to characterize intermediates in the charge-transfer pair was described, and the consideration of covalently bound donor-acceptor systems was extended to questions of chromophore orientation effects on energy transfer.

Modified Reaction Centers

Advances in biochemical techniques have made it possible to isolate reaction center proteins for study and, further, to alter their composition in order to establish thermodynamic as well as kinetic characteristics of various steps in the charge separation process as well as in back reactions. For example, reaction center s (e.g., *Rb sphaeroides*) may be isolated and the native quinone replaced by various other quinone species whose electron transfer properties in homogeneous

---

[1]Radiation Laboratory, University of Notre Dame, Notre Dame, IN 46556
[2]Department of Biochemistry & Biophysics, University of Pennsylvania, Philadelphia, PA 19104
[3]Department of Chemistry, Arizona State University, Tempe, AZ 85287-1604
[4]Department of Physical Chemistry, The Hebrew University, Jerusalem 91904, Israel
[5]Indian Institute of Science, Dept. of Inorganic and Physical Chemistry, 560012 - Bangalor, Karnatana, India
[6]Institute of Chemical Physics, USSR Academy of Sciences, 142432 Chernogolovka, USSR

Published 1989 by Elsevier Science Publishing Co., Inc.
Photochemical Energy Conversion
James R. Norris, Jr. and Dan Meisel, Editors

media are known. To determine the environmental effects of the protein on the quinone function, $E_{1/2}$ for the Q_a/Q_a^- couple have been measured for numerous quinones. When compared to values in solvent, it was shown that no clear correlation could be found between $E_{1/2}$ in the protein and in solvent, which would apply to all classes of quinones considered. Such findings establish a significant dependence of quinone function on protein environment. With these modified systems it has been possible to compare $-\Delta G$ with rate constants for both forward and back reactions for various quinone. This work establishes that while the processes with the various quinones generally follow Marcus theory, the native quinones are "optimized" in both directions with regard to predictions of the theory. The significance of an optimized back reaction remains a puzzle.

The application of fast EPR methods, namely time-resolved cw diode detection (~300 ns time resolution), or pulsed microwave electron spin echo and fast Fourier transform (FFT) spectroscopies (~10 ns time resolution), have opened new possibilities in the study of fast photochemical processes in which transient paramagnetic states and/or species are involved. In particular, electron transfer reactions in photosynthesis are being extensively examined. Particular attention is being focused o model systems in which the photochemical path leads to charge separation ending at a quinone constituent. While ultrafast optical spectroscopies are most appropriate to follow the primary electron transfer reactions in photosynthesis, the application of fast EPR in the nanosecond time scale has the advantage of identifying unambiguously paramagnetic intermediates which either participate in the main photochemical pathway (doublets), or bear direct relation to structure and dynamics (triplets). Using FFT spectroscopy to monitor electron transfer reaction between the photoexcited triplet of ZnTPP and duroquinone, the time-evolved spectra of the anion radical, $DQ^{\bullet -}$, has shown dramatic changes in the line shape, reflecting the mechanism under which the anion radical is being produced. The very early spectrum in time, is produced via the triplet mechanism (TM); in later times the radical pair (P) prevails. It implies that the electron transfer occurs before spin relaxation within the triplet spin states sets in, i.e., $k_{et} > 10^9 s^{-1}$. In a recent study, employing FFT spectroscopy, it has been shown that the production of the anion radical is dictated by the mechanism under which the triplet state of the precursor is formed, namely the selectivity in the intersystem crossing mechanism. These observations allow elucidation of the exact mechanism for the electron transfer process regarding both the donor and the acceptor. With two different chromophores, ZnTPP and MgTPP, acting on a very common quinone, DQ, the production of the $DQ^{\overline{\bullet}}$ strongly depends on selectivity in production of the triplet state of the particular porphyrin. The covalently bound model compounds discussed by this panel should be good candidates for mechanistic studies of electron transfer and charge separation by EPR in the nanosecond time range.

Covalently Bound Model Systems

Significant efforts have been made to mimic photodriven charge separation processes in model molecular systems which link a chromophore and electron acceptor (or acceptors) in various geometries. Interest in this type of model has been enhanced by the growing availability of equipment for monitoring photophysical behavior in such compounds, even into the femptosecond region. While initial efforts did not achieve significantly stable charge separation, more recent development of triad and tetrad molecules have yielded ionic species with long lifetimes. Carotenoid-porphyrin-quinone (C-P-Q) triads have been studied photochemically and can produce a charge separated state of the form C^{+}-P-Q^{-} via the pathway [C-P*-Q $\rightarrow$ C-P^{+}-Q^{-} $\rightarrow$ C^{+}-P-Q^{-}] with half-lives of several hundred nanoseconds. In such molecules, the C-P and P-Q linkages may be systematically altered for investigation of mechanistic details of charge separation and back reaction. A marked distance dependence has been observed for P-Q linkages. Not only may one monitor the fate of the charge separated state, one may use fluorescence lifetimes to measure the initial forward reaction. Further sophistication has followed with substitution of two quinones, -Q_a-Q_b to yield the tetrad C-P-Q_a-Q_b. The photophysics of such species provide a multiplicity of states, intermediate to the ground state and C^{+}-P-Q_a-Q_b^{-}, which may be investigated. The final product may have lifetimes in the microsecond region. Additionally, tetrads of the form C-P_a-P_b-Q and the various transient species intermediate to C^{+}-P_a-P_b-Q^{-} have been investigated. It may be seen that such elegant synthetic efforts and related photophysical characterizations are providing significant insight into the mechanisms by which multistep charge separation and stabilization may occur.

Intramolecular model studies are also found to be of particular value in the elucidation of distal and orientation effects on energy transfer rates. For example, distal dependence for singlet energy transfer in bis-bis porphyrins (H2P-H2P) covalently bound by ether linkages reveal a critical distance of ~12 Å to exhibit 100% energy transfer efficiency. One might extend these studies by introducing a π acceptor to promote CT interaction. Initial efforts with such species again show a critical distance of 12 Å to exhibit optimal folding of the molecule.

Organized Assemblies

Significant attention is being paid to the influence of molecular assemblies on electron transfer and the stabilization of charge separated pairs. For example, zeolites, which are microporous aluminosilicates, can be used as templates for molecules involved in electron transfer. The resulting self-assembly of redox components can act to restrict molecular motions normally leading to neutralization of charge pairs. Covalently linked $Ru(bpy)_3$-diquat cations immobilized on zeolite L exhibit photoinduced charge separated states of about 1 µsec lifetime, as compared to 5 nsec in solution. Addition of a secondary acceptor, benzyl-viologen (BV^{2+}), which can be further removed in the zeolite from the donor, gives a BV^+ species with a lifetime of 35 µsec. In another example, electron transfer between a cyanine dye and methyl-viologen (MV^{2+}) has been studied in lipid bilayer films; donor and acceptor have been separated by a layer of either arachidic acid or a quinthiophene. The thiophene acts as a molecular wire by which MV^+ can be formed and stabilized against back reaction. Efforts to use lipid vesicles to separate and stabilize charges formed in photochemical reaction continue along several lines. Isolation of a photosensitizer in the vesicular inner water pool with a primary acceptor located in the bilayer and a secondary acceptor in the bulk phase provides the basis for attempts to build a thermodynamic barrier to the back reaction.

PHOTOELECTROCHEMICAL PROPERTIES OF III-V ISOTYPE HETEROJUNCTION ELECTRODES

ARTHUR B. ELLIS,* LEE R. SHARPE, AND STEVEN P. ZUHOSKI
Department of Chemistry, University of Wisconsin-Madison,
Madison, WI 53706

R. M. BIEFELD* AND D. S. GINLEY*
Sandia National Laboratories, Albuquerque, NM 87185

INTRODUCTION

The development of metalorganic chemical vapor deposition (MOCVD) [1] and molecular beam epitaxy (MBE) [2] as III-V semiconductor growth techniques has created new opportunities for photoelectrochemical investigations. These techniques have been used to prepare nearly atomically abrupt junctions with fascinating electro-optical properties [1,2].

Recently, we and others have begun to investigate photoelectrochemical cells (PEC's) based on III-V isotype heterojunction (IH) electrodes [3-7]. Our studies, which are summarized in this paper, have focused on electrodes containing either a single internal junction or a periodic arrangement of internal junctions known as strained-layer superlattice (SLS) structures. We demonstrate herein that stable PEC's can be constructed from these materials; that photoluminescence (PL) and electroluminescence (EL) can be used to characterize electric fields in the solids; that matched pairs of SLS electrodes, differing only in their terminating layers, may provide a means for decoupling bulk and surface contributions to photoelectrochemical properties; and that PL can be used to characterize SLS quality. Experimental details supporting our studies have been published elsewhere [3-5].

SINGLE-JUNCTION ELECTRODES

Our initial studies were conducted with MOCVD-prepared IH electrodes, consisting of one n-$GaAs_xP_{1-x}$ composition (x ~0.70) atop another n-$GaAs_xP_{1-x}$ composition (x ~0.59) [3]. The topmost layer, possessing a smaller band gap, had a thickness of only ~0.2 µm, permitting both layers to be photoexcited simultaneously. This material served as the basis for a stable PEC, employing aqueous ditelluride electrolyte, that converted 458-nm light to electricity with ~13% efficiency.

The PL spectral distribution consisted of contributions from both layers, with the bulk contribution dominating as the penetration depth of the exciting light increased. Both PL bands suffered field-induced

Photochemical Energy Conversion
James R. Norris, Jr. and Dan Meisel, Editors

quenching consistent with a dead-layer model, with the surface emission being the more susceptible to quenching. The essence of the dead-layer model is that electron-hole pairs created within a distance on the order of the depletion width are separated by the electric field and do not contribute to PL. The mathematical form of the PL quenching for the bulk layer is given by eq. 1 and reflects its infinitely thick nature (it is actually ~2 μm thick and sits atop several other n-$GaAs_xP_{1-x}$ layers that in turn sit upon a n-GaAs substrate):

$$\Delta D^b = -(1/\alpha'^b)\ln(PL_1{}^b/PL_2{}^b) \qquad (1)$$

Here, ΔD^b is the change in dead-layer thickness (equated with the change in depletion width) in passing between the first and second potentials; $PL_1{}^b$ and $PL_2{}^b$ are the bulk PL intensities at those potentials, wherein the semiconductor is in depletion; and $\alpha'^b = (\alpha^b + \beta^b)$, where α^b and β^b are the bulk absorptivities for the exciting and emitted light, respectively. When the surface layer has a finite thickness X, the analogous expression is given by eq. 2, where the superscript "s" refers to values appropriate to the surface layer and D_2 is the dead-layer thickness at the second potential, often the open-circuit potential in our studies. We have generally assumed the open-circuit potential to be the flat-band potential, thereby making $D_2 = 0$, but we will show below a case where nonfulfillment of this condition can be used to estimate the quality of a SLS electrode.

$$\Delta D^s = (-1/\alpha'^s)\ln\{(PL_1{}^s/PL_2{}^s)[1-\exp(-\alpha'^s(X-D_2))] + \exp(-\alpha'^s(X-D_2))\} \qquad (2)$$

An assumption made in the dead-layer model is that the surface recombination velocity is either independent of applied potential or relatively large [8]. Operationally, the model was successfully tested by interrogating with several excitation wavelengths to demonstrate that ΔD is constant.

The ability to map the electric field in these two-zone electrodes using PL can be extended using EL techniques [3]. When the IH electrodes were used as dark cathodes in 0.5 M $Na_2S_2O_8$ formamide electrolyte, the EL spectrum obtained was dominated by the surface-based emission, indicating that injected holes recombine primarily in the surface layer.

More recently, we have uncovered a dependence of the bulk PL quenching on the magnitude of photocurrent passing through the electrode [9]. Figure 1 presents quenching data for a typical IH electrode, V-018, having a surface composition of x ~0.76 (and thickness of $\leq$0.2 μm) and a bulk composition of x ~0.59; the sample is Se-doped with n ~(4-5) x 10^{17} cm^{-3}. The spectrum consists of two overlapping bands originating from the surface (λ_{max} ~715 nm) and bulk (λ_{max} ~670 nm) layers; the contribution of the bulk layer to the overall spectrum increases with optical penetration depth. As shown in Fig. 1, the surface PL band is substantially quenched in passing from open circuit (-1.8 V vs. SCE) to a potential near short circuit, but the bulk PL band is only slightly quenched. The PL quenching resulting from use of several Ar^+ laser lines could be used to show that the dead-layer thicknesses in the surface and bulk zones increased by about 520 Å and 150 Å, respectively, in passing from -1.8 to -1.0 V vs. SCE.

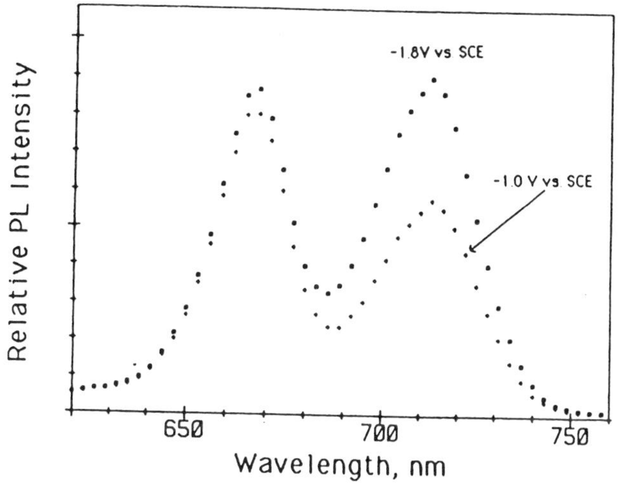

FIG. 1. Uncorrected, low-resolution PL spectra of sample V-018, measured at open circuit, ~-1.8 V vs. SCE, and -1.0 V vs. SCE in ditelluride electrolyte at 22°C. Both spectra were taken in an identical geometry using ultraband gap 514.5-nm excitation from an Ar^+ laser.

When photocurrent and bulk and surface PL intensity are measured as a function of applied potential, Fig. 2, the bulk PL exhibits an unusual effect. Unlike the surface PL intensity, which varies smoothly with potential as is typical for homogeneous electrodes [10], the bulk PL intensity shows almost no potential dependence except in the narrow voltage window wherein photocurrent is changing. This observation suggests that the electric field thickness in the bulk is dependent on the current flow through the internal junction.

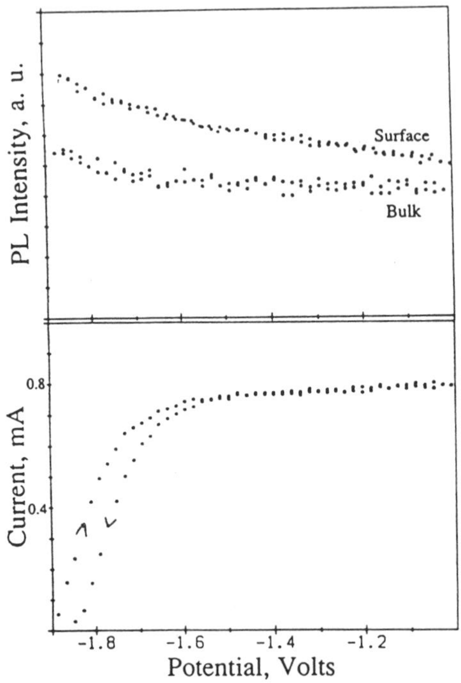

FIG. 2. Photocurrent (bottom panel) and PL intensity for both the surface and bulk regions (top panel) for sample V-018 as a function of applied potential. The data were obtained in ditelluride electrolyte at 4°C. Each PL intensity-voltage curve was taken simultaneously with the current-voltage curve using 514.5-nm light from an Ar^+ laser at a scan rate of 10 mV/s. The PL intensities were measured at 715 and 670 nm for the surface and bulk compositions, respectively.

To probe this effect in more detail, ΔD^b was calculated for different excitation intensities (and corresponding photocurrent densities) as the potential was varied from open circuit (-1.8 to -1.9 V vs. SCE) to -1.0 V vs. SCE. Table I summarizes these data for sample V-018. The key observation is that a five-fold increase in photocurrent density causes little change in ΔD^s but nearly quadruples ΔD^b. A possible explanation for the effect on ΔD^b is provided by Fig. 3, which shows that the flow of electrons into the bulk is impeded by a conduction band offset, a potential barrier, at the junction of the surface and bulk compositions. Electrons can collect at this internal junction, their number being proportional to the current, and in so doing affect the magnitude of the bulk electric

TABLE I. Dependence of Dead-Layer Thickness on Excitation Intensity for Sample V-018-E.[a]

Photocurrent, mA	ΔD^s, Å[b]	ΔD^b, Å[c]
0.20	540	90
0.40	580	120
0.50	600	170
0.65	630	220
0.80	640	280
1.00	650	320

a)Experiments were run at 22°C under N_2 in stirred solutions of 7.5 M KOH/ 0.20 M (di)telluride electrolyte; the redox potential was ~-1.2 V vs. SCE. A 0.05-cm^2 surface area of the electrode was irradiated with 514.5 nm light from an Ar^+ laser. Photocurrents are short-circuit values.
b)The change in dead-layer thickness for the surface zone, calculated using eq. 2, in passing from -1.0 V to open circuit (-1.8 V to -1.9 V vs. SCE).
c)The change in dead-layer thickness for the bulk zone, calculated using eq. 1, in passing from -1.0 V to open circuit (-1.8 V to -1.9 V vs. SCE).

field. If the pileup of electrons can be reduced, ΔD^b should be diminished. Temperature provides a way to test this notion, since the number of electrons surmounting the potential barrier by thermionic emission is an activated process.

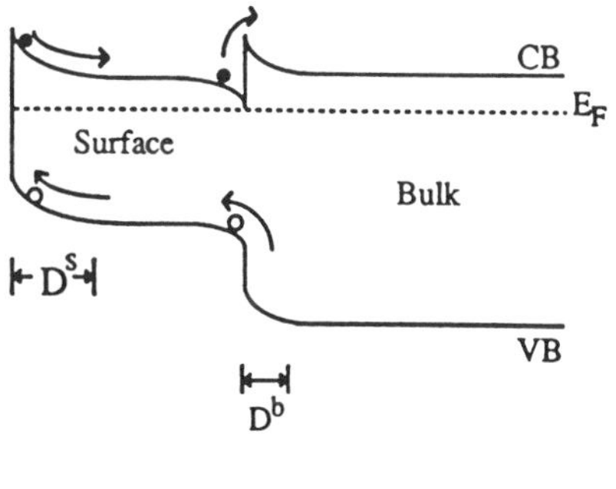

FIG. 3. Band structure of a representative IH electrode having a small band gap $GaAs_xP_{1-x}$ alloy atop a larger band gap alloy. The symbols CB, VB, E_F, D^s, and D^b represent the conduction band edge, valence band edge, Fermi level, and surface and bulk dead-layer thicknesses.

Temperature studies support this notion. Table II reveals that ΔD^b decreases with increasing temperature. By 55°C, bulk PL intensity-voltage profiles show little dependence on applied potential. We should note that in calculating dead-layer values over the range of indicated temperatures, we have used 22°C values, i.e., we assume that ultraband gap absorptivities employed in eq. 1 and 2 are not substantially affected by the temperature.

TABLE II. Temperature Dependence of the Dead-Layer Thickness for Sample V-018-H.[a]

Temperature, °C	ΔD^s, Å[b]	ΔD^b, Å[c]
5	610	630
15	550	430
25	490	250
35	500	220
45	485	190
55	410	130

a)With the exception of temperature, all conditions are as described in footnote a of Table I. The intensity was varied so that a short-circuit current density of 16 mA/cm^2 was maintained at all temperatures.

b)See footnote b, Table I.

c)See footnote c, Table I.

MULTIPLE-JUNCTION ELECTRODES

The SLS electrodes of our studies are composed of thin, strained semiconducting layers that alternate between two different compositions. The layers are generally equal in thickness, being on the order of 120 to 250 Å for all of the samples that we have studied. When the layers are sufficiently thin, the lattice mismatch between the two III-V compositions comprising the SLS is accommodated by uniform elastic strain of the layers, thereby minimizing the formation of misfit dislocations and the accompanying degradation of electronic properties. A several-micron-thick buffer layer, graded from the substrate composition (GaP or GaAs) to the average composition of the two materials that comprise the SLS, further reduces the strain in the SLS. Misfit dislocations that develop in the buffer layer can propagate into the SLS, but they are deflected to the sides of the SLS, leading ultimately to SLS material of good quality.[11] The periodic modulation of composition in the SLS leads to novel electronic properties, including quantization of the continuum energy states that exist in bulk materials.

Our initial studies were conducted with n-GaP/n-$GaAs_{0.4}P_{0.6}$ SLS electrodes.[4] We demonstrated that these electrodes could be stabilized in PEC's employing aqueous ditelluride electrolyte at modest current densities; monochromatic energy conversion efficiencies of a few percent were measured. Near-band-gap PL of these solids exhibited PL quenching in accord with a dead-layer model. A comparison of EL and PL spectral distributions showed them to be similar.

Since the bulk electro-optical properties of SLS electrodes are governed by the alternation in composition, these materials afforded the opportunity to decouple bulk and surface properties: matched pairs of SLS's, consisting of 25 periods differing only in their terminating layer, were grown under identical conditions [5]. One member of each pair terminated with a n-GaP layer, its mate with a n-$GaAs_{0.4}P_{0.6}$ layer. The band structures of these pairs are illustrated in Fig. 4. From the positions of the band edges, holes will tend to be localized in the alloy layers and electrons in the GaP layers.

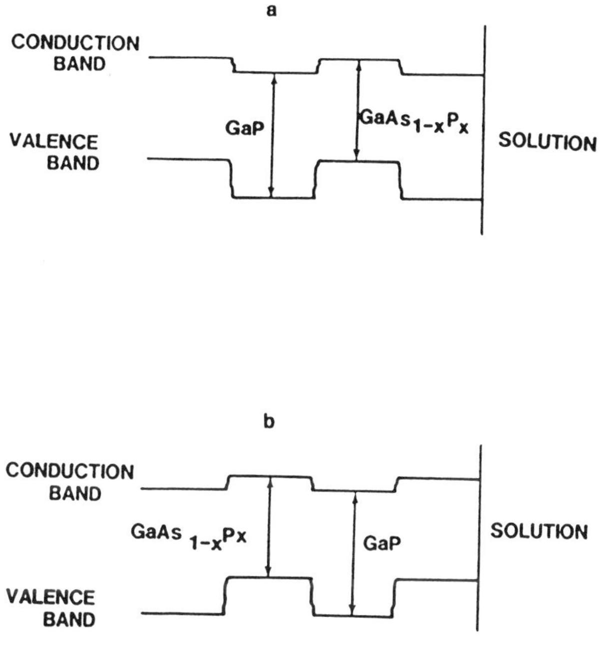

FIG. 4. Band edges for a matched pair of n-GaP/n-$GaAs_{0.4}P_{0.6}$ SLS electrodes. The sketches correspond to termination of the structures in (a) GaP and (b) alloy layers.

As might be expected based on the stability of homogeneous n-$GaAs_xP_{1-x}$ electrodes in aqueous dichalcogenide electrolytes, [12,13] stable PEC's result from use of either member of an SLS pair in ditelluride electrolyte at modest current densities (~1 mA/cm^2). Shown in Table III are current-voltage data, obtained using various excitation wavelengths, for a representative pair of SLS electrodes. Our general observation, typified by the data in Table III, is that the alloy-terminated electrodes are consistently superior to their GaP-terminated counterparts with respect to energy conversion efficiency at ultraband gap wavelengths. This

TABLE III. Current-Voltage Properties for a Representative Pair of SLS Electrodes.[a]

Sample[b]	λ_{ex}, nm[c]	ϕ_x (η_{max})[d]	V (η_{max})[e]	η_{max} (%)[f]
987/alloy	458	0.35	0.72	11.0
	514	0.17	0.68	5.7
	580	0.0077	0.65	0.30
988/GaP	458	0.27	0.46	4.6
	514	0.12	0.49	2.4
	580	0.057	0.51	1.4

a)With the exception of excitation wavelength, all conditions are as described in footnote a of Table I.

b)Sample number and terminating layer.

c)Excitation wavelength.

d)Photocurrent quantum yield, measured at the point of maximum efficiency.

e)Output voltage at the point of maximum efficiency.

f)Maximum output efficiency.

difference may to some extent reflect the superior energy conversion properties of homogeneous n-$GaAs_{0.4}P_{0.6}$ electrodes relative to n-GaP electrodes in chalcogenide electrolytes [12]. But it also reflects the location of barriers to hole migration (each GaP layer) relative to the semiconductor surface, Fig. 4: the measured quantum yields indicate that holes are collected from several superlattice periods, and the superiority of the alloy-terminated samples could arise from the presence, on average, of fewer and more remote barriers to hole capture. In addition, recombination via surface states may be more facile on the GaP-terminated electrodes.

An interesting property of the matched pairs is that all of the GaP-terminated samples exhibit a weak band in their photoaction spectrum near the band edge. As illustrated in Table III, this leads to better energy conversion properties in the GaP-terminated samples compared to their alloy-terminated counterparts. The insensitivity of the band to applied potential indicates that it is surface-localized.

In a recent study, we have found that PL properties of SLS electrodes can be used to nondestructively assess sample quality. Our work stems from the notion that the region of the SLS nearest the buffer layer/SLS junction may possess misfit dislocations that could quench SLS PL intensity. To investigate this possibility we examined two SLS electrodes, each consisting of alternating n-GaAs and n-$GaAs_{0.85}P_{0.15}$ layers with each layer having a thickness of ~220 Å: R907 had a total of 20 SLS periods (each period includes a layer of each composition) for a total thickness of

~8800 Å, and R960 had a total of 65 periods for a total thickness of ~30,000 Å.

The thicker R960 sample displayed a PL band at 860 nm, due to the SLS, that exhibited field-induced quenching when the material was used as a photoanode in aqueous ditelluride electrolyte. Substitution of the quenching data obtained from three exciting wavelengths into eq. 1 gave a consistent value for ΔD of 1800 Å between -1.0 V and open circuit in ditelluride electrolyte.

In contrast, Fig. 5 reveals that the PL spectrum of the thinner SLS sample, R907, is dominated by two bands having maxima at 860 and 825 nm. The former is assigned to the SLS based on its similarity to the R960 band and the fact that this is the only band observed in an EL experiment: EL is expected to be more surface-localized, owing to the initiation of the experiment at the semiconductor-electrolyte interface. Subsequent etching experiments confirmed that the 825-nm band arises from the underlying buffer zone.

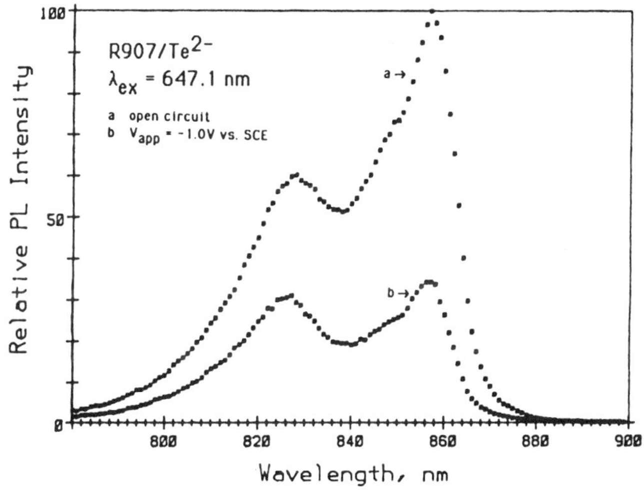

FIG. 5. Uncorrected, low-resolution PL spectra of sample R907, measured at open circuit, ~-1.7 V vs. SCE (a), and at -1.0 V vs SCE (b) in ditelluride electrolyte at 22°C. Both spectra were taken in an identical geometry using ultraband gap 647.1-nm excitation from a Kr^+ laser.

The observation of the buffer PL band in Fig. 5 indicates that the finite-layer form of the dead-layer model, eq. 2, should be used. If we assume that the open-circuit potential is the flatband potential (D_2 = 0), we are unable to obtain a consistent value for ΔD^s with PL quenching data from different exciting wavelengths. However, the value of the exponent, $(X-D_2)$, in eq. 2 can be adjusted until a consistent value for ΔD^s is obtained, which for sample R907 is ΔD^s = 1600 Å, corresponding to $(X-D_2)$ = 3000 Å. But since the total thickness of the SLS is X = 8800 Å, this implies that D_2 = 5800 Å. An electric field thickness this large at open circuit is physically unreasonable, even for the nominally undoped material employed in this study. In reality, it reflects our treatment of the entire 8800 Å of the SLS as being emissive.

The actual value of D_2, the depletion width at open circuit, was estimated from Mott-Schottky plots, which yielded a flatband potential of -2.0 V vs. SCE and a carrier concentration of 1 x 10^{16} cm^{-3}. Knowing the flatband potential, the depletion width can be calculated in the usual way, [8] leading to depletion widths of 2100 Å at open circuit, -1.7 V, and 3800 Å at -1.0 V vs. SCE; the difference, 1700 Å, is in good agreement with the 1600 Å determined for ΔD^s. Thus, D_2 = 2100 Å, making X = 5100 Å to satisfy the PL-derived fit that led to $(X-D_2)$ = 3000 Å. Within the approximations made, our data thus indicate that the 5100 Å of the SLS nearest the electrolyte can be considered to be emissive; the remaining 3700 Å of the SLS nearest the buffer layer (about 17 layers) can be treated as nonemissive, Fig. 6. A study by Gourley et al. using a direct PL imaging technique supports our conclusion that a defect-ridden, poorly emissive zone exists in the SLS immediately adjacent to the buffer region [14]. We should emphasize, however, that our two-zone emissive/nonemissive model is an approximation based on an averaging of properties; the change in PL with distance need not be abrupt.

To further test this model, sample R960 was slowly photoetched in nonstabilizing aqueous hydroxide electrolyte. Good accord with the dead-layer model appropriate for infintely thick samples, eq. 1, was found until about half of the SLS was removed. At this point the PL quenching of the SLS band no longer fit eq. 1.

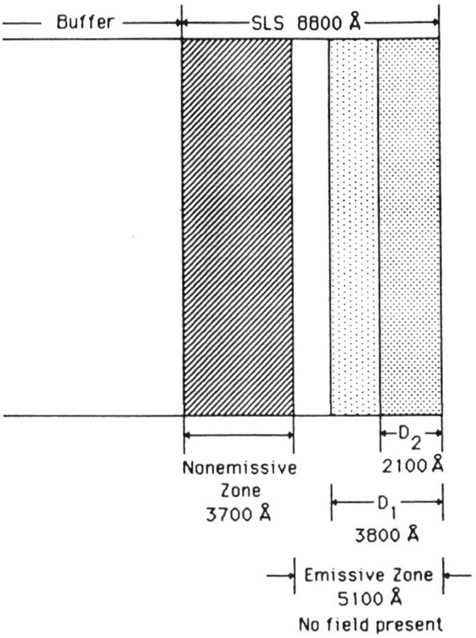

FIG. 6. Schematic of the division of the SLS region of electrode R907 into emissive and nonemissive regions. See text for the derivation of the indicated distances.

REFERENCES

1. R.D. Dupuis, Science 226, 623 (1984).
2. A.C. Gossard in "Treatise on Materials Science and Technology," Vol. 24, K.N. Tu and R. Rosenberg, Eds.; Academic Press: New York, 1982, p. 13.
3. W.S. Hobson, P.B. Johnson, A.B. Ellis, and R.M. Biefeld, Appl. Phys. Lett. 45, 150 (1984).
4. P.B. Johnson, A.B. Ellis, R.M. Biefeld, and D.S. Ginley, Appl. Phys. Lett. 47, 877 (1985).
5. S.P. Zuhoski, P.B. Johnson, A.B. Ellis, R.M. Biefeld, and D.S. Ginley, J. Phys. Chem. in press.
6. A.J. Nozik, B.R. Thacker, and J.M. Olson, Nature 316, 51 (1985).
7. A.J. Nozik, B.R. Thacker, J.A. Turner, J. Klem, and H. Morkoç, Appl. Phys. Lett. 50, 34 (1987).
8. A.A. Burk, Jr., P.B. Johnson, W.S. Hobson, and A.B. Ellis, J. Appl. Phys. 59, 1621 (1986).

9. L.R. Sharpe, Ph.D. Thesis, University of Wisconsin-Madison, 1987.
10. A.B. Ellis in "Chemistry and Structure at Interfaces," R.B. Hall and A.B. Ellis, Eds.; VCH Publishers: Deerfield Beach, Florida; 1986, Chap. 6.
11. P.L. Gourley, T.J. Drummond, and B.L. Doyle, Appl. Phys. Lett. 49, 1101 (1986).
12. C.M. Gronet and N.S. Lewis, J. Phys. Chem. 88, 1310 (1984).
13. A.B. Ellis, J.M. Bolts, S.W. Kaiser, and M.S. Wrighton, J. Am. Chem. Soc. 99, 2848 (1977).
14. P.L. Gourley, R.M. Biefeld, and L.R. Dawson, Appl. Phys. Lett. 47, 482 (1985).

ACKNOWLEDGMENT

We are grateful to the Office of Naval Research (ABE, LRS, SPZ) for supporting this research. Part of this work was supported by the U.S. Department of Energy under contract No. DE-AC04-76DP00789.

PHOTOELECTROCHEMICALLY MODIFIED SEMICONDUCTOR/ELECTROLYTE INTERFACES

Claude LEVY-CLEMENT

LABORATOIRE DE PHYSIQUE DES SOLIDES
CNRS, 1 PLACE ARISTIDE BRIAND
92195 MEUDON CEDEX PRINCIPAL, FRANCE

ABSTRACT

Photoelectrochemical processes can modify the surface of n and p type semiconductors, leading to beneficial effects on the performance of photoelectrochemical cells. Photoelectrochemical etching (photoetching), photoelectrosynthesis of an interphase and photoelectrodeposition of noble metal are some examples. Studies of mechanisms underlying these photomodifications are reported in this paper for several II-VI alloys and lamellar compounds.

INTRODUCTION

Over the last ten years photoelectrochemistry has provided researchers many opportunities to perform fundamental and applied investigations of semiconductor/electrolyte junctions. As a result of a deep understanding of the interfacial processes, photoelectrochemistry has been applied to the areas of the paint industry, microelectronics (dicing of chips) and optical communications (microlenses, microprisms or microgratings) (1,2). Work in the field of efficient solar energy conversion has provided new research opportunities especially in the area of surface modification. Numerous techniques such as chemical derivatization (3), ion adsorption (4), dye sensitisation (5), polymer and metal coatings (6,7) and a number of types of chemical modification(8) have been successfully used to protect the surface of the semiconductor against photocorrosion or to increase the rate of charge transfer at the junction.

Photoelectrochemistry can also be used as a technique to modify the surface of the photoelectrode, leading to profound and positive changes in interfacial properties. In some cases one can take advantage of the photocorrosion of the semiconductor. One example is photoelectrochemical etching (photoetching) which leads to selective attack of surface defects not accessible to usual chemical etchants (9). Another example is illustrated by the beneficial reaction between photocorrosion products and chemical species in the electrolyte solution. Consequently an in-situ thin conductive film is formed on the surface of the semiconductor and a new junction is created (10). On the other hand, when the semiconductor is stable a different type of photomodification, photoelectrodeposition of a noble metal, which has been widely studied by many research groups(11), can be done, leading to a modified solid/liquid junction with catalytic activity and thus to technological applications (microelectronics contacts). It is the purpose of this paper to describe under which conditions these types of

Photochemical Energy Conversion
James R. Norris, Jr. and Dan Meisel, Editors

photomodifications take place, to show their beneficial effects on the junction properties and to investigate different types of photomodified semiconductor/electrolyte interfaces, using conventional and less conventional interface analysis techniques in order to understand the mechanisms underlying these photomodifications. Our work has been centered on II-VI alloys (CdTe, CdSe, $CdSe_{1-x}Te_x$) and lamellar compounds (n and p type InSe); comparison of their behavior to the same type of photoelectrochemical modification provides an opportunity to study the effects of dimensionality on interfacial chemistry.

PHOTOETCHING

Photoelectrochemical etching results from a decomposition reaction induced by photogenerated minority carriers at the interface between the semiconductor and an electrolyte. For the reaction to proceed, the n-type semiconductor must be biased positively with respect to the electrolyte to form a depletion layer in the semiconductor. For p-type semiconductors in contact with an electrolyte a forward bias (anodic) leads to h^+ accumulation at the surface and etching with characteristics similar to those obtained by photoetching is obtained. Photoetched surfaces are fully active after removal of the photocorrosion products, which is generaly obtained by their selective chemical dissolution in a suitable medium. Photoetching of various materials has been shown to bring about considerable improvement in solid state properties of semiconductors (2e), particularly in the performance of photo-electrochemical cells (short circuit current and fill factor) made of photoelectrodes including II-VI compounds and lamellar chalcogenides (12,13). For the III-V compounds this effect has been studied for technological purposes (etching integral lenses on light-emitting diodes (LED's), waveguiding effects, unpinning of GaAs... (2)).

Upon photoetching the morphology of the surface of semiconductors is strongly modified. The surface of photoetched II-VI compounds is corrugated and contains a pattern of small and highly dense ($>10^9/cm^2$) etch pits whose density varies with the doping density of the material (14a) (figure 1). In the case of lamellar compounds the variation of the morphology depends on the crystallographic plane exposed. Neither of the

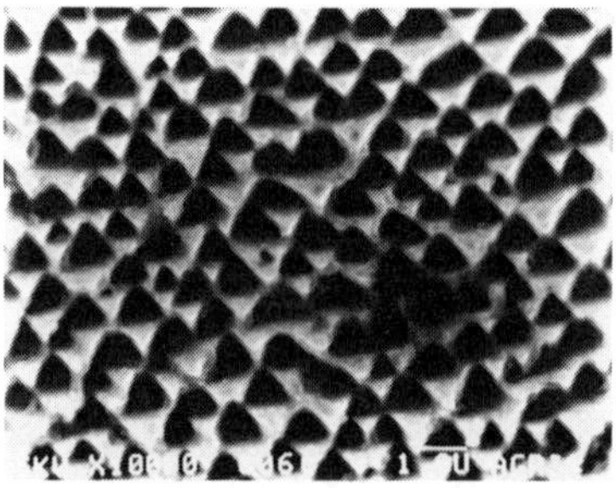

a)

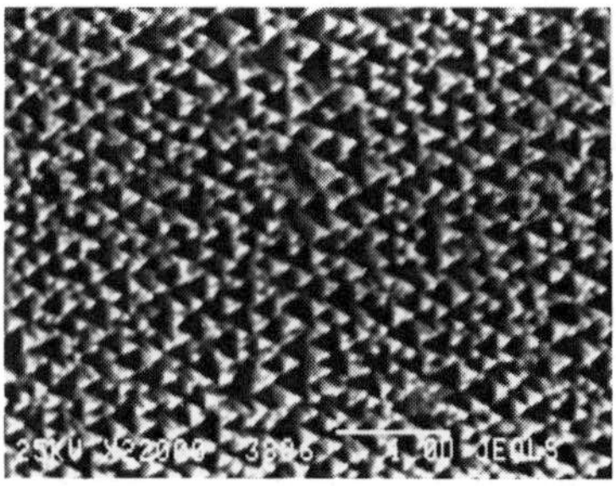

b)

FIG. 1. Scanning electron microscope pictures of photoetched CdSe with different doping density. a) $10^{16}/cm^3$ carriers. b) $10^{17}/cm^3$ carriers.

two different surfaces (cleavage surface and surface perpendicular to the cleavage plane) exhibits a high density of etch pits after photoetching. In the case of InSe no modifications were observed on the cleavage surface, while the perpendicular surface was totally restructured showing a mixture of cleavage planes (001) and perpendicular planes (110) (13a) (figure 2). For WSe_2, another lamellar compound, photoetching takes place

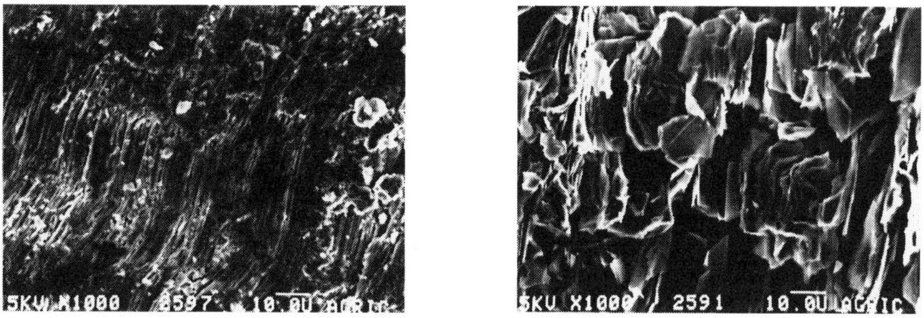

a) b)

FIG. 2. SEM pictures of n-InSe a) before any treatment.
b) after photoetching.

on the cleavage surface specifically where mechanical defects are present(13b-d). The difference in the morphological modification upon photoetching of the surface of tridimensional compounds and bidimensional compounds may be an indication that the mechanism of the photoetching is different in these two types of materials. Investigations of the properties of the surface of several II-VI compounds (CdTe, CdSe, $CdSe_{1-x}Te_x$) and of the lamellar compound WSe_2 before and after photoetching have been carried out to determine the origins of the improvement of the properties of the surface of the photoetched semiconductors.

1 - II-VI SEMICONDUCTORS

Previous studies on the mechanism of the photoetching of II-VI compounds using electrical (i.e. current voltage) and spectral photoresponse measurements (14b), supported by electroluminescence and electrolyte electroreflectance (14c) indicated that surface defects are selectively removed after photoetching. Electron-Beam-Induced current (EBIC) of Au Schottky barriers on CdS shows that the electron-hole recombination is greatly reduced at the Au/CdS interface after photoetching (14d). Cathodoluminescence imaging applied to the (1121) face of CdTe to discriminate between material defects and the uniform etch pit distribution created by photoetching shows that the etch pit pattern is not correlated with gross materials defects (14e). Consequently the source of the etch pits is a result of another property of the semiconductor surface. It was previously shown that the etch pit density depends on the donor concentration of the compound and decreases with increasing wavelength (14a,f), and it was suggested that the etch pit pattern is a manifestation of nonuniformities in the light-induced charge flow, which is the driving force for photo-etching in semiconductor junctions due to the ionized impurities (14a,g). In order to verify this hypothesis photocurrent spectrum and photoluminescence studies on n-CdSe and n-$CdSe_xTe_{1-x}$ have been undertaken.

n-CdSe single crystals with resistivities ranging from 0.3 to 11 Ohm.Cm and with free carriers densities between 10^{16} and $10^{17}/cm^3$ were obtained from Cleveland Crystal. n-$CdSe_{0.62}Te_{0.38}$ single crystals were grown at CNRS-Meudon, their resistivity and doping density were equal to 0.42 Ohm.cm and $2.6\times10^{16}/cm^3$. Photoetching was carried out following a routine procedure (14h).

Photocurrent spectra of the II-VI single crystals were recorded prior to and after photoetching, and after excessive photoetching. Results are reported here for $CdSe_{0.62}Te_{0.38}$ (figure 3a). Using the Gärtner model the bandgap of the compound was determined during different steps of the photoetching (figure 3b). A red shift of the photocurrent spectra is observable after photoetching. The red shift observed in the photocurrent spectra was interpreted using a BURSTEIN-MOSS model for heavily doped semiconductors which predicts a red shift due to a decrease of the doping density (15a).

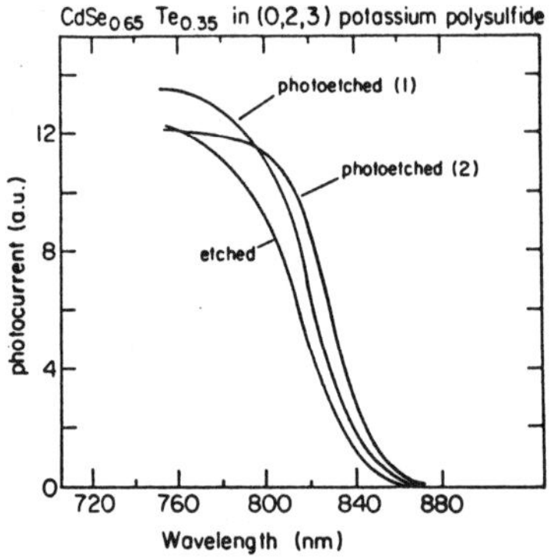

FIG. 3a. Photocurrent spectra of $CdSe_{0.65}Te_{0.35}$ (0001) face prior to and after excessive photoetching.

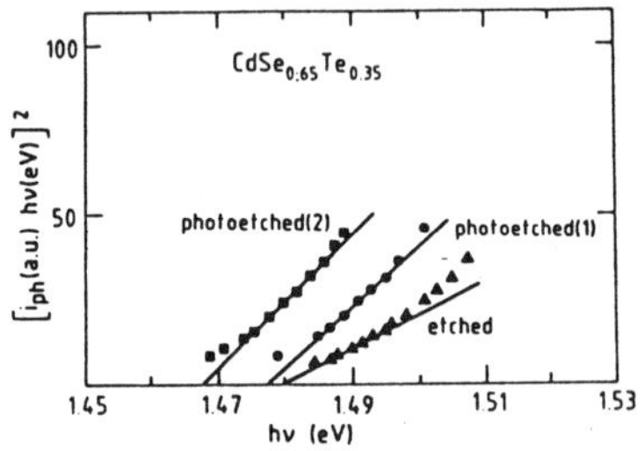

FIG. 3b. Determination of the band gap using the data of FIG. 1.

An alternative, or perhaps additional, explanation is obtained through a model developed by Rothwarf (15b) which shows the improved performance of materials with low minority carrier diffusion length to be due to surface corrugation. In addition, the increase in the slope of the curve $I_{ph}Xh\nu^2$ vs. hν after photoetching can be ascribed to a lower doping density near the semiconductor surface. These results are to be compared with capacitance measurements showing that chemical etching of the II-VI semiconductor does not modify the doping density at the surface (14i). From these results a strong assumption can be made that the photoetching selectively removes the doping impurities (donor states) at the surface.

High resolution photoluminescence of n-CdSe crystals at

1.8K and above has been used to investigate the effect of photoetching on the impurity distribution near the semiconductor surface. Signals of the photoluminescence spectra of CdSe are associated with free and bound excitons and with donor-acceptor bands with their respective phonon replica. We were able to study the changes induced by photoetching in the signal associated with donor states. Photoluminescence spectra of etched and photoetched CdSe were done under the following conditions :

- the intensity of the excitating light was varied from 5.7 to 10^4 mW/cm^2
- the excitation wavelength varied from 5145 to 4579 Å
- the temperature varied from 1.8 to 50K

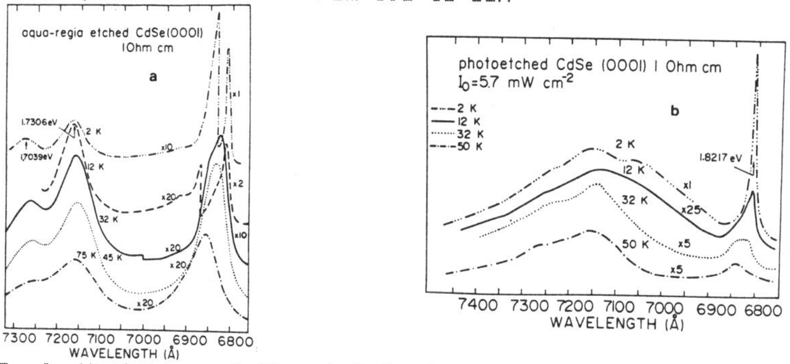

FIG. 4. Variation of the photoluminescence with temperature for a chemically etched (a) and photoetched (b) CdSe. Light intensity is 5.7 mW/cm^2.

Photoluminescence spectra of etched and photoetched crystals have been compared in detail elsewhere (14j). The main features are that upon photoetching the donor-acceptor peak is shifted to the blue (figure 4) and the intensity of the bound (to donor) exciton decreases. These results suggest that shallow donor states are removed from the surface preferentially and hence the surface becomes relatively intrinsic following that surface treatment.

2 - LAMELLAR COMPOUNDS

Interest in the photoetching of lamellar compounds came more recently as it was believed that the best photoelectrochemical performance with these kinds of materials could be obtained only with perfectly cleaved surfaces (16). Work of TENNE et al. demonstrated that photoetching done on the cleavage surface of WSe_2 modifies it and exposes new crystallographic planes which have proved to be as active as the Van der Waals planes (13d). Upon photoetching, a few concentric etch pits of at least 100Å diameter are randomly formed. They nucleate where mechanical defects (dislocations,crossing of steps) are

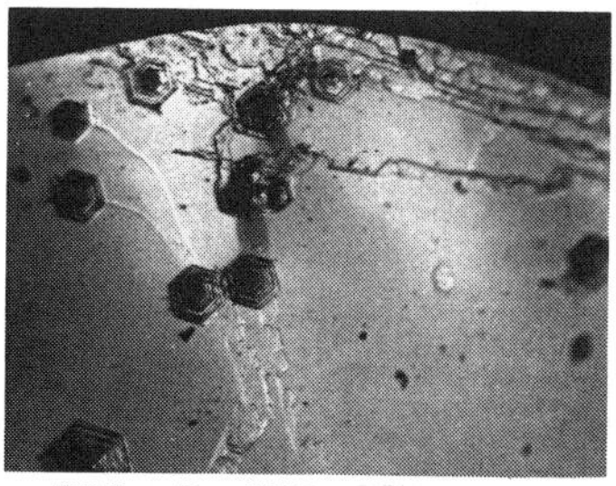

FIG. 5. SEM picture of WSe_2 after photoetching

present (figure 5). With the increase of the duration of the photoetching the etch pits coalesce, losing their observable hexagonal regular shape. EBIC studies demonstrate that mechanical defects still observable with SEM (Scanning Electron Microscope) are no longer electrically active after photoetching. EBIC also shows that a cleaved region which appears morphologically smooth reveals recombination centers, each of them inducing losses of about 25% from the maximum collected signal. The width of these damaged zones extends to a few microns and is determined by the minority carrier diffusion length.

CONCLUSION

Although more work need to be done in order to understand the mechanism of the photoetching of lamellar compounds, it is clear that the mechanisms of photoetching seem to vary appreciably between II-VI compounds and lamellar compounds. In the case of tridimensional compounds it is most probable that the charge flow is non-uniform due to the presence of ionized dopants which create non-uniform microscopic electric fields and consequently photoetching takes place at the atomic level. In lamellar compounds it was suggested (17) that mechanical defects at the surface funnel the photogenerated charge carriers by the non-uniform electric fields that they exert in their vicinity, and photogenerated holes can be viewed as rather delocalized entities with randomized trajectories, steps and dislocations being then preferentially photoetched.

IN SITU PHOTOELECTROCHEMICAL GROWTH OF A SEMICONDUCTOR/ELECTROLYTE INTERPHASE

Photocorrosion of the semiconductor in a photoelectrochemical cell depends not only on the semiconductor/electrolyte interface energetics, but also on the degree of solubility of the photocorrosion compounds produced. If the products are soluble, the semiconductor will photodecompose. However if one of the products is insoluble it may be utilized to form in situ an interfacial film which would avert further decomposition while permitting efficient charge transfer at the newly created interface. Literature in this field shows that such in situ formation of interfacial thin films requires a judicious selection of the cell components. If the thin film is electrically insulating, with a thickness compatible for tunneling of minority carriers, highly stable efficient photoelectrochemical cells may be built. This mechanism has been found in the following cells: CdX(X=S, Se)/polychalcogenide electrolyte, p-InP/V^{2+},V^{3+} and n-$CuInSe_2$/neutral polyiodide (18). In another case, the interfacial film may be viewed as a "surface state" communicating between the valence band and the solution redox species, allowing mediated charge transfer. This situation has been invoked in n-type cadmium chalcogenide $Fe(CN)_6^{3-/4-}$ cells (19). A different situation from those cited above is created when insoluble photocorrosion products chemically react with redox species in solution, generating a thick overlayer of a new insoluble compound on the semiconductor. This situation may be advantageously exploited in a photoelectrochemical cell when the interphase exhibits good electronic transport properties, suitable electronic affinity, and transparency to visible light. The formation of the stable $CuInSe_2/Cu^+,I^-,I_2$,HI solar cell showing a 12% efficiency is a direct consequence of the photoelectrochemical growth of this kind of interphase (20). It has been suggested that a solid state junction is formed between n-$CuInSe_2$ and the interphase which, depending on the growth conditions, is either p-$CuISe_3$-Se^o or $CuIIn_2Se_3$, the electrolyte providing a front ohmic contact. Our interest in this type of photomodified interface relates to the fact that it can be extended to other systems.

We have recently reported that in acidic copper polyiodide solution, n-InSe reacts under illumination in a similar way to $CuInSe_2$ (21). Since InSe is a lamellar compound, the comparison of the behavior of $CuInSe_2$ and InSe to the same kind of surface chemistry provides a unique opportunity to study the effects of dimensionality on the interfacial chemistry.

The surface chemistry under illumination of InSe in contact with a solution of acidic copper polyiodide (1M KI + 4M HI + 0.3M CuI + 0.005M I_2) depends strongly on the nature of the crystallographic plane used. The cleavage surface ((001) plane) of InSe consists of selenium atoms bonded together with covalent forces and the surface perpendicular to the cleavage surface ((110) plane) is made of selenium atoms bonded together by Van der Waals bonds and of indium atoms bonded by metallic bonds. An electrode with the (110) plane in contact with the acidic copper polyiodide photocorrodes very rapidly showing a large decrease of the photoresponse. Analysis of its surface with SEM-EDAX (Energy Dispersive X-ray analysis) shows large particles (>10μm) of an indium compound, probably In_2O_3. On the other hand, the cleaved (001) face of InSe reacts with the acidic copper polyiodide similarly to $CuInSe_2$. An interfacial layer grows on the cleavage surface if InSe is left in contact with the solution for several weeks or by operation as a photoanode for half an hour in the same solution. The interphase generated on InSe was characterized by SEM-EDAX and X-ray diffraction in grazing incidence. It is made of hexagonal platelets of $CuISe_3$ (22) (rhombohedric with a = 14.083 Å and c = 14.187 Å). SEM and EDAX analysis done at different stages of the generation of $CuISe_3$ show that the $CuISe_3$ is generated from the layers of Se, probably Se^o, formed during the photocorrosion process (figure 6). The thickness of the $CuISe_3$ interphase is of the order of several μm.

As observed for $CuInSe_2$, short circuit current (>10 mA/cm² under AM1 illumination), photovoltage (0.5V) and fill factor (0.5V) of a photoelectrochemical cell made of InSe increase after photomodification. There is a further increase after annealing of photomodified InSe at 150°C for 20 minutes. However the stability of illuminated photomodified InSe in acidic copper polyiodide under short circuit condition is not improved by annealing, in contrast to $CuInSe_2$ under the same condition, where stability is extended over months.

Experimental data were collected on InSe and photomodified InSe using spectroscopic techniques (spectral photoresponse, photoreflectance, photoluminescence) in order to determine the mechanism of charge transfer across the n-InSe/$CuISe_3$/copper polyiodide interface. The photocurrent action spectra of n-InSe/$CuISe_3$ done in (2M HI + 1M HI) and (1M KI + 2M HI + 0.05M I_2) electrolytes, are consistent with generation of electron hole pairs in the n-InSe, the long wavelength cut-off at 1.27eV

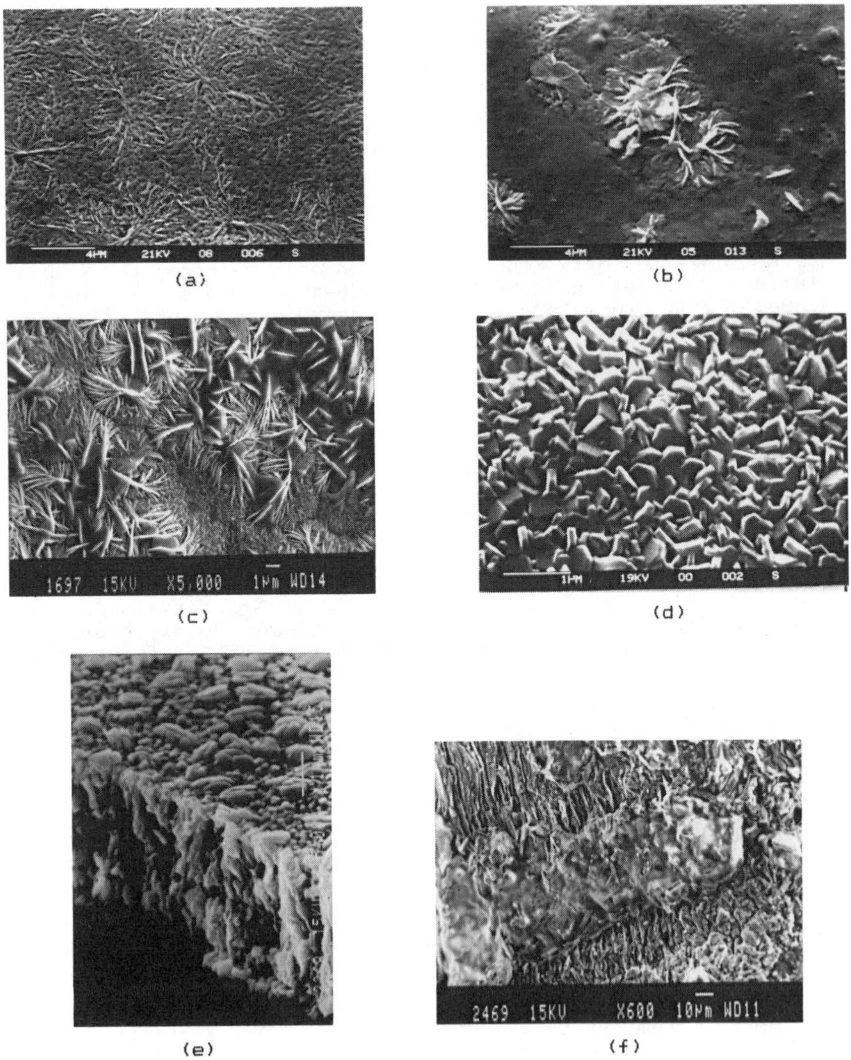

FIG. 6. SEM of films grown on InSe surfaces in I^-,I_3^-,HI,Cu^+, showing: – on the cleavage surface of InSe, (a) the initial formation of a layer of Se^0, (b,c) the first and second stages of the $CuISe_3$ film photoelectrochemically generated, (d,e,) the $CuISe_3$ film after 30 minutes illumination in copper polyiodide; – on the (110) face of InSe, (f) the formation of an In compound (In_2O_3) photogenerated in the same electrolyte.

corresponding to the InSe band gap (figures 7a,b). Comparison between these two figures clearly shows that the fall off at high energy of the photomodified electrode is not due to electrolyte absorption but either to an increase of recombination centers on InSe or to the absorption of ligth with λ<600 nm by the $CuISe_3$ film.

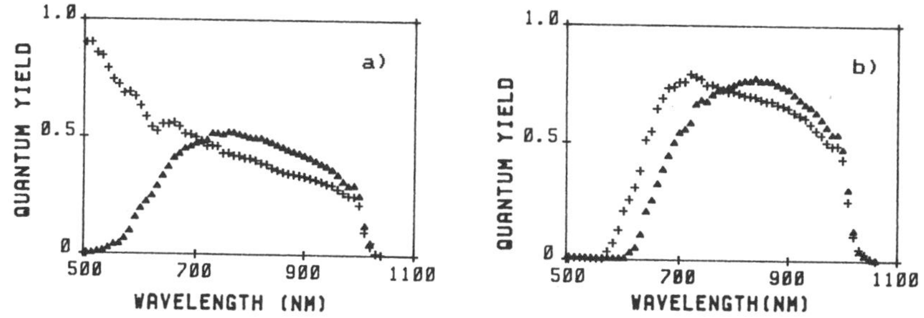

FIG. 7. Spectral photoresponse of +++ InSe and ▲▲▲ InSe/$CuISe_3$ in the two following electrolytes: a) 2M HI + 1M KI, b) 2M HI + 1M KI + 0.005M I_2.

Photoreflectance data obtained at energies between 1.1 to 2.2eV, discussed elsewhere (21), also give direct evidence of the possible role in charge transport of the interphase; a strong assignment was made that the photogenerated $CuISe_3$ is semiconducting with an energy gap of 1.98eV. These results are consistent with an interpretation in which the interphase acts as a window for n-InSe and cleans the surface of trapping centers. These data agree with those obtained with surface microwave conductivity studies of $CuInSe_2$/$CuISe_3$ (20d). Using a value of 5.38eV, for the Fermi level (assigned from UPS study) of $CuISe_3$, which has been found to be p type (20b,d,e), and energies of band edges published elsewhere for n-InSe (23), a schematic energy-band diagram of the three-phase junction n-InSe/p-$CuISe_3$/I^-,I_2,HI,Cu^+ is proposed in figure 8. This scheme suggests that, similarly to $CuInSe_2$/$CuISe_3$, the photovoltaic active junction is a p-n junction. The photoelectric response of a solid state device InSe/$CuISe_3$/Au, represented in figure 9

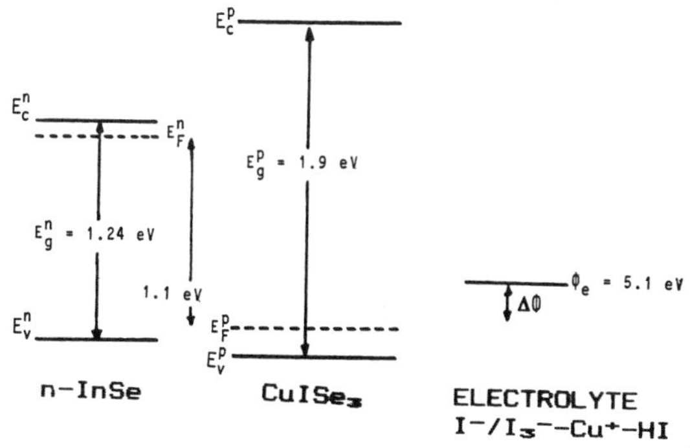

FIG. 8. Schematic of energy levels of n-InSe, p-$CuISe_3$, and redox couple

shows that the active junction is solid state. The fall off in the high energy range corresponds to the absorption of the light by $CuISe_3$ and the energy is a function of its thickness. The low quantum yield observed for the solid state devices may be attributed to a poor InSe/Au Schottky diode junction; on the other hand, the $CuISe_3$/Au contact may not be ohmic. No attempt to optimize the photovoltage and photocurrent efficiency of this device has been made.

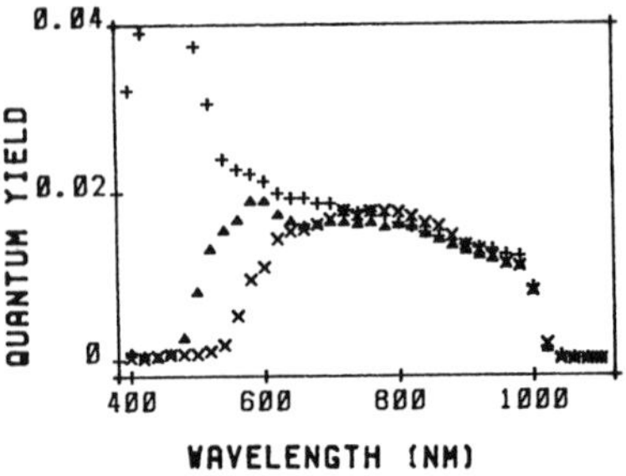

FIG. 9. Spectral photoresponse of solid state devices: +++ n-InSe/Au; ▲▲▲ Thin, xxx thick film n-InSe/$CuISe_3$/Au

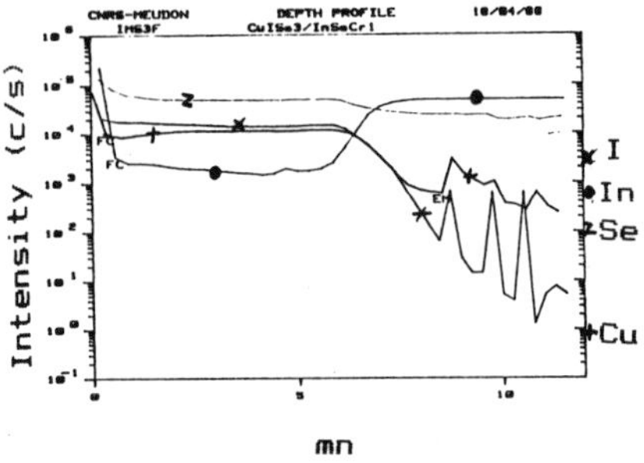

FIG. 10. SIMS depth profile of Se, I, Cu, and In in the n-InSe/$CuISe_3$ junction.

CONCLUSION

Our results show that there is very little difference between the chemistry of the surface modification of $CuInSe_2$ and that of the cleavage surface of n-InSe when modified in acidic copper polyiodide under similar conditions. The data suggest that the interphase act as a window that cleans the surface of recombination centers. The nature of the conduction mechanism through the interphase in both systems is not totally elucidated. Though we have direct proof that the photovoltaic active junction for the InSe system is solid state, we do not know if the interphase is porous and we cannot, therefore, exclude the hypothesis that an ionic conductivity is superposed on the electronic conductivity. The high quantum yield of the InSe/$CuISe_3$/I^-,I_2,HI,Cu^+ junction argues for such an hypothesis.

One puzzling point is the difference in stability of the two systems. We have speculated (21) that the partial stabilisation of the InSe/copper polyiodide may originate from the layer structure of InSe, and on the possibility of "bulk corrosion" due to possible intercalation of the electrolyte through the layers, the interphase being localized on the surface. As a result, only the top layer is stabilized and it will eventually collapse due to continuing corrosion of the internal layers. Indeed, the SIMS (secondary ion mass spectroscopy) profile of a photomodified InSe electrode shows that the InSe/$CuISe_3$ junction is rather abrupt (figure 10). If this hypothesis is supported by direct experimental verification it will have important consequences on the feasibility of stabilizing lamellar compounds by surface modification.

STRUCTURAL STUDY OF PLATNUM ISLANDS PHOTOELECTROCHEMICALLY DEPOSITED ON THE CLEAVAGE SURFACE OF p-InSe

Photoelectrochemical cells which use a photoactive electrode (n or p type semiconductor) and a dark counter electrode require an external bias to accomplish photoelectrolysis of water. The external bias required may be reduced by depositing a thin film of electrocatalysts on the surface of the semiconductor electrode, this effect being known for some time. Literature in the field of catalyzed photoelectrodes is extensive, showing the importance to fundamental and applied research of this area of study (24). Different models have been proposed to explain changes in photoelectrochemical reaction rates of several orders of magnitude occuring after modification of an electrode. There are differences among these models. The model of Heller et al. invokes adsorption of hydrogen on the deposited noble metal, changing its work function in such a way that the semiconductor/noble metal/electrolyte interface should be seen as a semiconductor/H_2 junction (24,25). The model of Bockris et al. suggests that the photocurrent is controlled by the metal-solution interface (26), while recently Tsubomura et al. proposed a theory of quantization of the energy of the carrier near the surface of the semiconductor when the metal layer deposited has a discontinuous island structure, the width of the islands being of the order of 50 Å (27).

Use of a photoelectrochemical process has led to the deposition of noble metals as small islands on the surface of several p-type semiconductors of some interest for solar energy conversion. Enhancement of photoevolution of hydrogen has been observed after photoelectrodeposition of noble metals (Pt, Ru, Rh, Pd) on p-CdTe, p-WSe_2, p-GaP, p-Si, the most spectacular increase being observed up to now, on p-InP (29). The properties of the photomodified semiconductor have been investigated using various techniques including electrochemical and phototoelectrochemical measurements, SEM, TEM (Transmission Electron Microscopy), electric properties and photovoltage of Schottky diodes. TEM studies of platinum islands photoelectrodeposited onto InP have shown that these islands are made of clusters of platinum of 50 Å diameter, with no structural long range order (30). In some particular cases, a prolonged deposition of several hours, under mass transport control, may lead to the evolution of the islands in a porous metallic film transparent to the visible light. The catalytic activity versus hydrogen evolution of this type of film is very high. Until now transparent metallic films made of Pt, Rh or Pd have been photogenerated only on p-InP(30)

Our interest in this area has been to investigate whether some correlation may exist among the conditions of the photoelectrodeposition of the noble metal islands, their catalytic activity and their structural properties. We did this study on p-type InSe on which noble metal had been photodeposited. p-InSe was selected because the redox couple H^+/H_2 is situated between its valence and conduction bands, its lamellar structure makes it possible to obtain a smooth surface favorable for good photoelectrodeposition of a metal, and it shows stability in acidic media. Stuctural investigation was done using surface extended X-ray absorption fine structure

(SEXAFS) spectroscopy, which is a well adapted method to study compounds with structural short range order, allowing the determination of the oxidation state and the local environment of a particular atom (31).

Pt and Rh were photoelectrodeposited from solution of their respective salts, until the optimum photoelectrochemical performance of InSe was achieved. Perfectly cleaved surfaces were used without any pretreatment. When photodeposition was carried out under mass-transport limited conditions (10^{-5} to 10^{-4}M $PtCl_6^{2-}$ or $RhCl_3$, potentiel fixed at 0.1V vs.SCE or continuous variation of the potential between +0.3 and -0.3V), only a low increase of the cathodic photocurrent was noticed while a cathodic dark current appeared. Prolonged photodeposition led to a decrease in the performance of the photocathode. SEM studies showed that the arrangement of the islands was heterogeneous, and formation of a continuous film of metal was not observed. When the concentration of Pt and Rh salts was increased to 10^{-3} or 10^{-2}M, and InSe was potentiostated between +0.3 and -0.3V, a dramatic improvement in efficiency for H_2 photogeneration was observed, Rh deposition leading to a better performance of the photoelectrode (figures 11a,b).

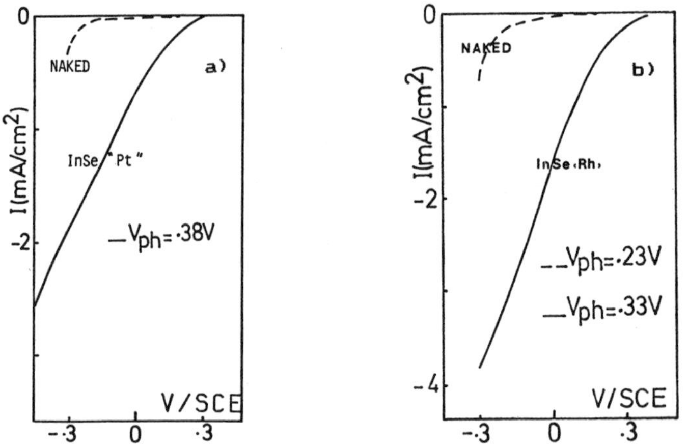

FIG. 11. Photocurrent-voltage curves of ---- naked InSe, (a) —— platinized InSe and (b) —— rhodinized InSe, when around 10^{-7} mol/cm² of Pt or Rh are photodeposited.

SEM analysis showed that islands of 1000 Å average size were homogeneously plated onto InSe, the amount of photodeposited Pt or Rh being of the order of 10^{-8} mol/cm² (figure 12a). Two to five more voltage sweeps increased the amount of Pt or Rh photodeposited without improving the cathodic photocurrent. More than five sweeps under these conditions were counterproductive and led to a rapid decrease of the photocurrent and to the appearance of a cathodic dark current, especially in the case of Pt (amount of photodeposited Pt or Rh $\simeq 10^{-6}$ mol/cm²). Small and large islands coexisted on the same surface (figure 12b).

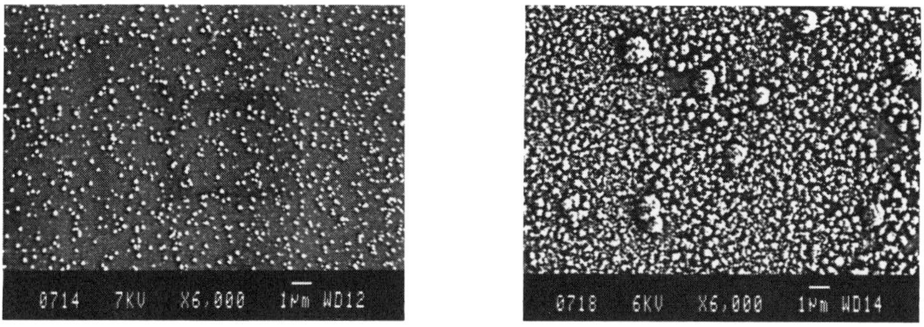

a) b)

FIG. 12. SEM of platinized p-InSe cleavage surfaces. Bright "islands" are Pt and dark are the naked InSe. a) 3.7×10^{-7} Mol/cm^2 Pt. b) 1.8×10^{-6} mol/cm^2 Pt.

In EXAFS studies X-rays, which are absorbed by matter through the photoelectric effect, are used to excite the atoms. The term EXAFS refers to oscillations of the X-ray absorption coefficient on the high energy side of an absorption edge. Each element has its own absorption threshold energy. Differences between EXAFS and SEXAFS studies come mainly from the type of particles that are detected after the absorption. X-rays are detected in EXAFS while only electrons that escape from the surface are detected in SEXAFS. SEXAFS spectra are recorded in reflection mode. Analysis of the data is similar in the two kinds of experiments (31). The number and kind of atoms surrounding the central atom and their distances from the absorber are determined from the EXAFS oscillations.

SEXAFS studies were carried out only on InSe(Pt) electrodes because the intensity of the X-Ray beam of the L.U.R.E. (Laboratoire pour l'utilisation du rayonnement electromagnetique) synchrotron (France) was too low at the energy of the K edge of Rh for the experiments to be performed. Analysis was done on two samples on which different amounts of platinum were deposited leading to different catalytic behavior. One sample (electrode 1), which showed an optimum photoevolution of H_2 had 3.7×10^{-7} mol/cm^2 Pt on the surface. The second sample (electrode 2), with 1.8×10^{-6} mol/cm^2 Pt

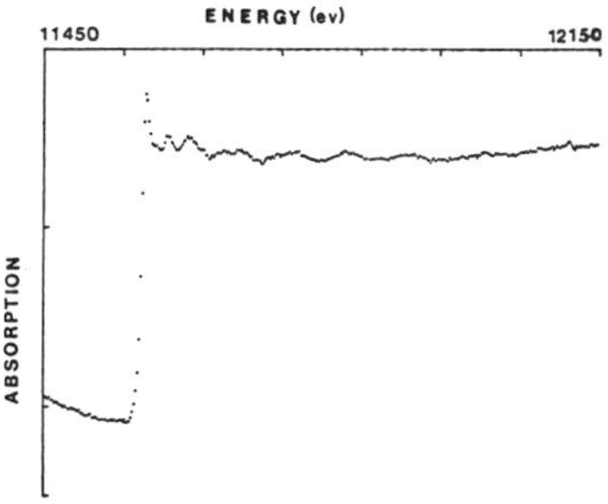

FIG. 13. SEXAFS spectra at L_3 edge of Pt photodeposited on a smooth cleavage surface of p-InSe.

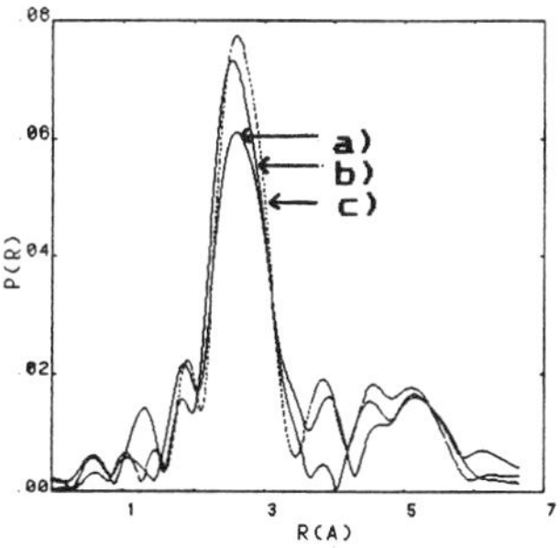

FIG. 14. Fourier transforms of the Pt SEXAFS spectra of: a) 3.7×10^{-7} mol/cm^2, b) 1.8×10^{-6} mol/cm^2 Pt. c) EXAFS spectrum of bulk platinum as a standard.

photodeposited, showed a decreased photocurrent and appearance of a cathodic dark current. SEXAFS spectra were measured at the L_3 edge of platinum. Total data accumulation for the platinum island typically consisted of several hours of summed twenty minute scans (figure 13). Analysis of our results show that a relationship exists between the conditions of the photodeposition of Pt and its local structure. We found for electrode 1 that the local structure of Pt islands is similar to that of bulk platinum (12 atoms of Pt at 2.73_6 A). For electrode 2, where the catalytic effect is strongly reduced, the best agreement between calculated and experimental curves is obtained for 11.5 atoms of Pt at 2.72_5 A and 1 atom of Cl at 2.32 A (figure 15), indicating that some absorption or reaction of chlorine on/with platinum occurs.

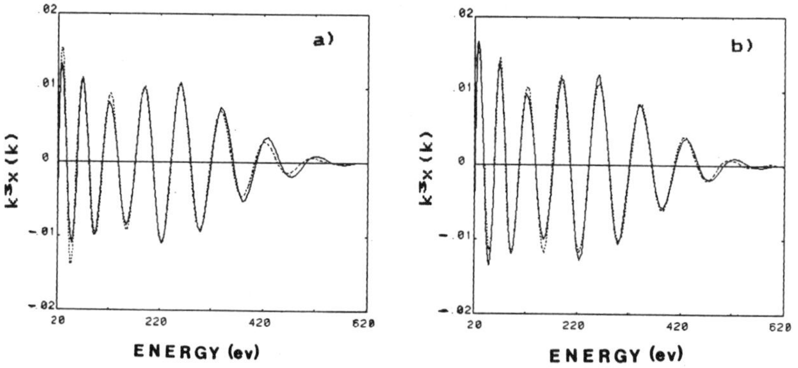

FIG. 15. EXAFS signal (inverse Fourier transform) of the first neighbouring shell of photodeposited Pt: --- experimental curves and —— calculated curves of a) 3.7×10^{-7} mol/cm^2 Pt and, b) 1.8×10^{-6} mol/cm^2 Pt.

CONCLUSION

SEXAFS studies show that platinum photoelectrodeposited on the two electrodes is in both cases present in the form of Pt^o. Above a certain amount of Pt deposited, some chloride is adsorbed on or reacts with it. Our results can be related to the competition between two electrochemical processes: electrochemical growth and corrosion of platinum in chlorine solutions. When small and large Pt islands coexist on the same surface, the electrochemical potential of the small islands is more positive than for large islands. Consequently small islands are more active and will react more rapidly with the chlorine to form $PtCl_6^{2-}$ which dissolves. This mechanism may explain the presence of Cl atoms observed on electrode 2. As the adsorption of Cl^- anion is known to have a negative effect on hydrogen adsorption (32), it may to some extent affect the formation of the platinum hydride alloys on which H_2 is photoevolved (24,25). The InSe/Pt(hydride) Schottky barrier expected to be formed may be a poor barrier, and the electrical contact with Pt become an ohmic one, as the Fermi level of p-InSe has an energy close to the work function of Pt. This may lead to the appearance of a cathodic dark current and to the decrease of the photocurrent. However, one should not eliminate the possibility that above a certain amount of Pt deposited, the absorption of the light will also decrease the photocurrent.

CONCLUDING REMARKS

This paper has presented some examples of modifications of the surface of the semiconductor by different photoelectrochemical processes. Photoetching, photosynthesis of an overlayer and photodeposition of noble metal are processes which lead to beneficial effects of the photoelectrochemical behavior of solid/liquid junctions made of two different types of solids: II-VI and lamellar compounds. Comparison of their behavior to the same type of photomodification has opened the opportunity to study the effects of dimensionality on interfacial chemistry. The photoprocesses discussed here have already proved useful for industrial applications. Further understanding of chemistry of the semiconductor/electrolyte interface may lead to the development of new types of photomodification and to application of photomodification to previously unstudied semiconductor/electrolyte junctions. Studies in these area will find new applications and perhaps even serendipitous effects.

ACKNOWLEDGMENT

I warmly thank all my collaborators for their invaluable contribution to this work: Didier Sedaries, Michael Neumann-Spallart, Jacques Rioux, Robert Triboulet, Henri Mariette, of my laboratory; Reshef Tenne and Diana Mahalu of the Weizmann Institute of Israel; Micha Tomkiewicz and Wu-Mian Shen of the Brooklyn College of New York. I would like to thank Adam Heller for valuable discussions on noble metal photodeposition. Extensive assistance of Claude Godart, Dominique Chandesris and Jean Lecante in the SEXAFS experiments and of Jean François Rommelure in photoluminescence is appreciated. Amy Ryan is gratefully acknowledged for reading the manuscript.

This work is supported by the CNRS (PIRSEM) and AFME through a GRECO 130061 research programm. Funds for the collaborations with R. Tenne are from the French Foreign Office through the DSCI, and with M. Tomkiewicz have been from a NATO grant 86/0647 for international collaboration in research.

REFERENCES

1. A. Heller, Y. Degani, D.W. Johnson Jr., and P.K. Gallagher, J. Phys. Chem. 91, 5987 (1987).
2. (a) F.W. Ostermayer Jr., P.A. Kohl, and R.H. Burton, Appl. Phys. Lett. 43, 642 (1983); (b) R.D. Rauh and R.A. LeLievre, J. Electrochem. Soc. 131, 2811 (1985); (c) J.E. Bowers, B.R. Hemeney and D.P. Wilt, Appl. Phys. Lett. 46, 453 (1985); (d) D.V. Podlesnik, H.H. Gilsen,and R.M. Osgood Jr., ibid. 48, 496 (1986); (e) S.D. Offsey,

J.M. Woodall, A.C. Warren, P.D. Kirchner, T.I. Chappell and G.D. Petit, ibid. 48, 475 (1986); (f) M.M Carrabba, N.M. Nguyen, and R.D. Rauh, J. Electrochem. Soc. 134, 260 (1987); (g) C.W. Lin, Fu-Ren. F. Fan and A.J. Bard, ibid. 134, 1038 (1987).

3. M.S. Wrighton, in Chemically modified surfaces in catalysis and electrocatalysis, Ed. J.S. Miller, ACS Symposium Series 192, 99 (1982).
4. (a) R.J. Nelson, J.S. Williams, J.H. Leamy, B. Miller,H.C. Casey Jr., B.A. Parkinson, and A. Heller, Appl. Phy. Lett. 36, 76 (1980); (b) D.W. Johnston, H.J. Leamy, B.A. Parkinson, and B. Miller, J. Electrochem. Soc. 127, 90 (1980).
5. (a) H Gerischer and F. Willig, Topics in current chemistry 61, 31 (1976); (b) M.T. Spitler and M. Calvin, J. Chem. Phys. 66, 4294 (1977); (c) C. Kavassalis, M.T. Spitler, ibid. 87, 3166 (1983) and references therein.
6. (a) R. Noufi, D. Trench, and L.F. Warren, J. Electrochem.-Soc; 128, 2596 (1981); (b) T. Skotheim, I. Lundstrum, and J. Prejza, ibid. 128, 1625 (1981); (c) G. Horowitz and F. Garnier, ibid. 132, 634 (1985).
7 (a) Y. Nakato, Y. Iwakabe, M. Hiramoto, and H. Tsubomura,J. Electrochem. Soc. 133, 900 (1986) and references therein; (b) L.A. Harris, M.E. Gerstener, and R.H. Wilson, ibid. 124, 1511 (1977); (c) S. Menezes, A. Heller, and B. Miller, ibid. 127, 1268 (1980)
8 (a) Y. Mirovsky and D Cahen, Appl. Phys. Lett. 40 (1982); (b) G.W. Cogan, C.M. Gronet, J.F. Gibbons, N.S. Lewis, ibid. 44, 539 (1984); (c) N. Chazalviel, J. Electroanl. Chem. 233, 37 (1987).
9. G. Hodes, Nature, 285, 29 (1980).
10. S. Menezes, H.J. Lewerenz and K.J. Bachmann, Nature 305, 615 (1983).
11. H. Heller, E. Aharon-Shalom, W.A. Bonner, and B. Miller, J. Amer. Chem. Soc. 104, 6942 (1982) and references therein.
12. R. Tenne and G. Hodes, Surface Science 135, 453 (1983) and references therein.
13. (a) R. Tenne, B. Theys, J. Rioux, and C. Levy-Clement, J. Appl. Phys. 57, 141 (1985); (b) R. Tenne, A. Wold, Appl. Phys. Lett. 47, 707 (1985); (c) R. Tenne, W. Spahni, G. Calzaferri, and A. Wold, J. Electroanal. Chem. 189, 247 (1985); (d) D. Mahalu, A. Jakubowicz, A. Wold, and R. Tenne, Phys. Rev. B to be published.
14. (a) R. Tenne, V. Marcu, and N. Yellin, Appl. Phys. Lett. 45, 1219 (1984); (b) H.H. Streckert, J. Tong,, and A.B. Ellis, J. Am. Chem. Soc. 104, 581 (1982); (c) M. Tomkiewicz, W. Siripala, and R. Tenne, J. Electrochem. Soc. 131, 736 (1984); (d) G. Hodes, D. Cahen, and H. Leamy, J. Appl. Phys. 54, 4676 (1983). (e) R. Tenne and A.K. Chin, Mater. Lett. 2, 143 (1983). (f) R. Tenne, V. Marcu, and Y. Prior, Appl. Phys. A37, 205 (1985); (g) R. Tenne, H. Flaisher, and R. Triboulet, Phys. Rev. B 29, 5799 (1984); (h) C. Levy-Clement, R. Triboulet, J. Rioux, A. Etcheberry, S. Litch, and R. Tenne, J. Appl. Phys. 58, 4703 (1985); (i) R. Garuthara, M. Tomkiewicz, and R. Tenne, Phys. Rev. B 31, 7844 (1985); (j) R. Tenne, H. Mariette, C. Levy-Clement, and R. Jager-Waldau, ibid. 36, 1204 (1987).
15. (a) H. Jager and E. Seipp, J. Appl. Phys. 52, 425 (1981); (b) A. Rothwarf, Proc. 18th Photovol. Spec. Conf., 809-912 (1985).
16. (a) J. Gobrecht, H. Gerischer, and H. Tributsch, Ber. Bunsenges. Phys. Chem; 82, 1331 (1978); (b) H.J. Lewerenz,

A. Heller, H.J. Leamy, and S.D. Ferris, ACS Symp. Ser., 146, 17 (1981); (c) G. Kline, K.K. Kam, D. Canfield and B.A. Parkinson, Solar Energy Mater. 4, 301 (1981).
17. (a) H.J. Lewerenz, A. Heller, and F.J. Disalvo, J. Am. Chem. Soc. 102, 1877 (1980); (b) H. Gerischer, J. Electroanal. Chem. 150, 553 (1983).
18. (a) D. Cahen, G. Hodes, and J. Manassen, J. Electrochem. Soc. 125, 1623 (1978); (b) A. Heller, B. Miller, H.J. Lewerenz and K.J. Bachmann, J. Am. Chem. Soc. 102, 6555 (1980); (c) D. Cahen and Y.W. Chen, Appl. Phys. Lett. 45, 746 (1984).
19. H.D. Rubin, D.J. Arent, B.D. Humphrey, and A.B. Bocarsly, J. Electrochem. Soc. 134, 93, (1987).
20. (a) S. Menezes, Appl. Phys. Lett. 45, 148 (1984); (b) S. Menezes, H.J. Lewerenz, G. Betz, J. Bachmann, and R. Kotz, J. Electrochem. Soc. 131, 3030 (1984); (c) K.J. Bachmann, S. Menezes, R. Kotz, M. Fearheily, and J.H. Lewerenz, Surface Science,138, 475 (1984); (d) S. Menezes, Solar Cells,16, 255 (1986); (e) H.J. Lewerenz and E.R. Kotz, J. Appl. Phys. 60, 1430 (1986); (f) S. Menezes, J. Electrochem. Soc. 134, 2771 (1987).
21. C. Levy-Clement, D. Sedaries, W.M. Shen and M. Tomkiewicz, JCS Symp. Ser., Photoelectrochemistry and electrosynthesis on semiconductors materials. G. Ginley and A. Nozik editors. 1988 to be published
22. W. Milius and A. Rabenau, Mat. Res. Bull. 22, 1493 (1987).
23. C. Levy-Clement, N. Le Nagard, O. Gorochov, and A. Chevy, J. Electrochemical Soc. 131, 790 (1984).
24. D.E. Aspnes and A. Heller, J. Phys. Chem. 87, 4919 (1983) and references theirin.
25. A. Heller, E. Aharon-Shalom, W.A. Bonner, and B. Miller, J.Am. Chem. Soc. 104, 6942 (1982).
26. (a) M. Szklarczyk and J.O'M. Bockris, Appl. Phys. Lett. 42,1035 (1983); (b) M. Szklarczyk and J. O'M. Bockris, J. Phys. Chem. 88, 1808 (1984).
27. Y. Nakato and H. Tsubomura, J. Photochemistry 29, 257 (1985).
29. (a) J.A. Baglio, G.S. Galabrese, D.J. Harrison, E. Kamieniecki, A.J. Ricco, M.S. Wrighton, and G.D. Zoski, J. Am. Chem. Soc. 105, 2246 (1983) and references theirin; (b) A. Heller, Science 223, 1141 (1984).
30. (a) A. Heller, D.E. Aspnes, J.D. Porter, T.T. Sheng, andR.G Vadimsky, J. Phys. Chem. 89, 4444 (1985); (b) J.D.Porter, A. Heller, and D.E. Aspnes, Nature 313, 664 (1985); Y. Degani, T.T. Sheng, A. Heller, D.E. Aspnes, A.A. Studna, and J.D. Porter, J. Electroanal. Chem. 228, 167 (1987).
31. (a) J. Stohr, in X-ray Absorption: Principles Aplications Techniques of EXAFS, SEXAFS and XANES. R. Prins and D. Koningsberger editors. (J. Wiley, New York) (1986); (b) P.H. Citrin: An overview of SEXAFS during the past decade. Journal de Physique 47, C8-437 (1987).
32. S. Gilman, in Electroanalytical Chemistry Vol.2, A.J. Bard editor (M. Dekker, New York).

ARTIFICIAL PHOTOSYNTHESIS VERY EFFICIENT VISIBLE LIGHT ENERGY HARVESTING AND CONVERSION BY SPECTRAL SENSITIZATION OF FRACTAL OXIDE SEMICONDUCTOR FILMS

MICHAEL GRAETZEL,
Institut de chimie physique, Ecole Polytechnique Fédérale, CH-1015 Lausanne, Switzerland

ABSTRACT

The lecture describes a recent and surprising discovery concerning spectral sensitization of wide band oxide semiconductors, such as titanium dioxide. Due to their high stability these materials have attracted wide attention for use as light harvesting units in solar energy conversion devices. However, in view of their wide band gaps their response to sunlight is feeble. Their sensitivity to visible light may be enhanced by spectral sensitization. The problem with this approach has been that the efficiencies so far obtained were disappointingly small. By using charge transfer dyes molecularly engineered to be chemisorbed at the semiconductor/solution interface in conjunction with a novel structured semiconductor surface, a breakthrough in the field of spectral sensitization has been achieved. This could have far reaching implications for the design of efficient, durable and low cost solar energy conversion devices.

INTRODUCTION

The response of the semiconductor/electrolyte interface to band gap irradiation is analogous to that of the semiconductor/metal Schottky barrier, and it has therefore been widely investigated as an alternative to the photovoltaic solid-state cell. [1,2] It must, however, be recognised that an electrolyte under solar irradiation is a very hostile environment for long-term operation of a semiconductor device. Photocorrosion is therefore a serious problem. It particularly affects those narrow gap semiconductors, such as silicon, gallium arsenide or cadmium selenide which would otherwise have an acceptable solar energy conversion efficiency due to their good match to the solar spectrum. Stable semiconductor electrodes tend to be wide band gap oxides whose sensitivity is limited to the short wavelength visible and ultraviolet regions of the spectrum.

One approach to resolving this dilemma is to separate the functions of light absorption as well as charge carried separation and transport, which in a standard photoelectrochemical system are both carried out within the depletion region of the semiconductor [3]. Our recent studies concern devices where optical absorption is carried out by a dye in intimate association with the semiconductor surface. Light induced charge separation takes place at the dye/semiconductor interface. The principle has been recognised for many years [4] and refinements of the technique appeared in successive reviews. [5,6] However, until recently the efficiency of these dye-sensitized semiconductor devices under sub-band gap irradiation has been extremely low, and practical applications seemed remote. In the present work a new semiconductor surface texture and derivatisation procedure are applied, with striking improvement of quantum efficiency.

Theory of Dye Sensitization of Semiconductor Electrodes

Any reactive species presented to an electrolyte adopts a redox state which may be specified with reference to the vacuum level [7] of the electron or to a known electrochemical equilibrium, e.g. the standard calomel electrode (SCE). This principle is applied both to the semiconductor electrode and

Published 1989 by Elsevier Science Publishing Co., Inc.
Photochemical Energy Conversion
James R. Norris, Jr. and Dan Meisel, Editors

to the dye. According to the standard model for photoelectrochemistry the semiconductor band edge energy levels at the interface with a given electrolyte are fixed. Oxide semiconductors are usually n-type and their conduction band edge position i.e. the flat band potential, E_{cb}, depends on the pH value of the electrolyte according to the Nernst equation:

$$E_{cb} = E_{cb}(\text{pHO}) - 0.059 \text{ pH} \qquad (1)$$

where the factor 0.059 applies to $T = 298^{o}K$. The flat band potential is also a function of the type of oxide used. [8] Thus, the energetic position of the conduction band edge can be adjusted over a wide range by judicious selection of the oxide materials and solution pH.

Figure 1 shows the principles of operation for a dye sensitized photoelectrochemical cell. The sensitizer must be chosen such that its ground state potential lies within the band gap of the semiconductor while that of the excited state lies above the conduction band edge. Incident visible light is absorbed by the sensitizer raising it to the excited state from which a thermodynamically favorable decay mode is electron injection into the conduction band of the semiconductor. The local electrostatic field present in the depletion layer of the semiconductor prevents the recombination of the injected electron with the oxidized state of the parent dye molecule (D^{+}). The latter is reduced to the initial state (D) by reaction with an electron donor agent. At the counter electrodes a cathodic reaction takes place, carried out by the photo-injected electrons which have passed through the external circuit performing electrical work. If the cathodic reaction is the reverse of the oxidation carried out by the dye cation, then the cell is a current producing system, analogous to a solid state photovoltaic cell. If this is not the case, for example if the two reactions are respectively the oxidation and reduction of water, then the cell is photosynthetic and in this example produces oxygen and hydrogen gases.

A system which meets these energy level conditions may nonetheless remain inefficient on mechanistic grounds. For example, if the dye is merely dissolved in the electrolyte, the excited state is so rapidly quenched, typically within 10^{-8} s, that diffusion to the electrode surface and charge transfer may be ruled out. [9] Surface-attached species alone can contribute to the sub-band gap photoresponse of the device. On such a modified semiconductor surface only the first adsorbed monolayer can transfer charge, thicker dye layers tending to be insulating. The light harvesting efficiency (LHE) of such a device is given by the expression:

$$\text{LHE} = (I - (I/I^{o})) = (1 - 10^{-\varepsilon\Gamma}) \qquad (2)$$

where ε is the decadic extinction coefficient expressed in units of cm^{2}/mol and Γ is the surface concentration in mol/cm^{2} of the sensitizer. Suppose the dye has an ε value of 2×10^{4} M^{-1} cm^{-1} corresponding to 2×10^{7} mol/cm^{2} and that it occupies an area of 100 $\AA^{2}$/molecule at the surface. On a flat surface Γ would be 1.67×10^{-10} mol/cm^{2} at monolayer coverage, yielding LHE = 7.65×10^{-3}. Such a value is unattractively low for a light energy conversion device. In the present system, the semiconductor suface is highly structured yielding a roughness factor of ca. 200. Therefore Γ is increased 200 times, i.e. $\Gamma = 3.34 \times 10^{-8}$ $mols/cm^{2}$ and the LHE value is 0.785. This is only a lower bound since multiple scattering of the light within the structured surface further increases the LHE. Thus, practically quantitative light harvesting is feasible with such a system.

Efficient light harvesting is only the first requirement for a dye sensitized photo-electrochemical cell to be suitable in solar energy conversion. A second condition is that the dye has a high quantum yield for photocurrent

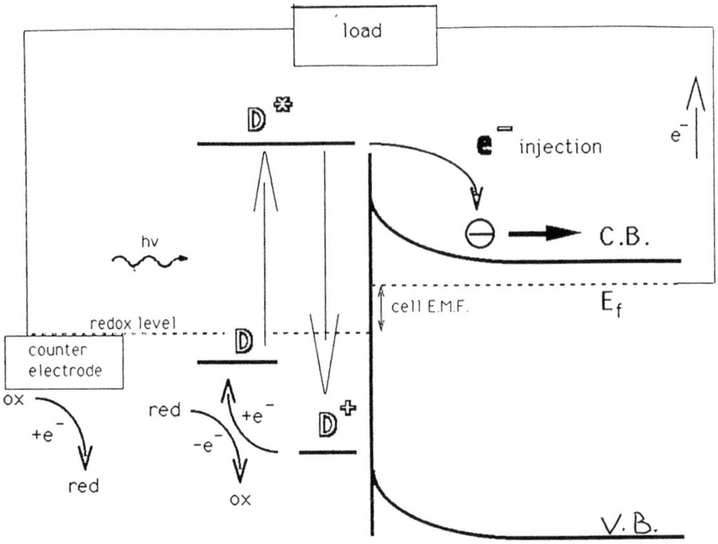

Fig. 1.

generation (ϕ_{pc}). A very important parameter characterising the performance of the device under short circuit conditions is the incident monochromatic photon to current conversion efficiency (IPCE)

$$IPCE = LHE \times \phi_{pc} \tag{3}$$

The parameter ϕ_{pc} is the product of the quantum yield for charge injection and the probability of charge separation, P_{cs}. The introduction of the latter is necessary since in general not all electrons injected by the dye are drawn off as a current.

$$\phi_{pc} = \phi_{inj} \times P_{cs} \tag{4}$$

$(1 - P_{cs})$ designates the probability for an injected electron to be recaptured by the oxidized sensitizer or other acceptors in solution. From Eq. (2) to (4) one obtains for the incident monochromatic photon to current conversion efficiency:

$$IPCE = (1 - 10^{-\varepsilon'\Gamma}) \times \phi_{inj} \times P_{cs} \tag{5}$$

The elementary reactions involved in dye sensitization are comprised of:

i) Dye excitation by light producing excited states:

$$D \xrightarrow{h\nu} D^* \tag{6}$$

ii) Charge injection:

$$D \xrightarrow{k_{inj}} D^{+} + e^{-}_{cb}\ \text{(surface)} \qquad (7)$$

iii) Electron escape from the surface of the semiconductor to the bulk:

$$e^{-}_{cb}\text{(surface)} \xrightarrow{k_{esc}} e^{-}_{cb}\ \text{(bulk)} \qquad (8)$$

iv) Electron recapture by the oxidized dye molecule and electron scavenging by other acceptors (A) in the electrolyte:

$$e^{-}_{cb}\text{(surface)} + D^{+} \xrightarrow{k_{rec}} D \qquad (9)$$

$$e^{-}_{cb}\text{(surface)} + A \xrightarrow{k_{scav}} \text{products} \qquad (10)$$

and

v) Reduction of the oxidized dye by the donor present in solution:

$$D^{+} + R \xrightarrow{k_{r}} D + R^{+} \qquad (11)$$

Spitler [10] has applied a one-dimensional Onsager model for a kinetic description of these processes. This was used to quantitatively interpret the photocurrent-potential curves observed in the sensitization of WSe_2 crystals. The present studies are concerned with charge injection into oxide semiconductors. These materials have, typically, a high dielectric constant rendering the thermalization distance of the electron so short (< 2 Å) that the simpler conventional model [11] for electron transfer at the semiconductor/electrolyte interface becomes applicable.

Preparation of Structured TiO_2 Layers

The TiO_2 (anatase) films are prepared by thermal decomposition of titanium alkoxide. The solution was spread over a 4 cm^2 Ti sheet and the layer was hydrolyzed at room temperature and 48% relative humidity.[12] Pyrolysis at 450°C yields a layer of TiO_2, of polycrystalline anatase structure. The procedure was repeated several times until a sufficient thickness of anatase, typically 20 μm, was accumulated. n-type conductivity in TiO_2 is generally a consequence of substoichiometry in oxygen: the electrodes were therefore partially reduced by heating for 35 min. at 550°C in flowing high purity Ar. A scanning electron micrograph of the TiO_2 layer produced in this way is shown in Fig. 2. The specific texture and high surface roughness is apparent from the presence of numerous pores and crevices in the film.

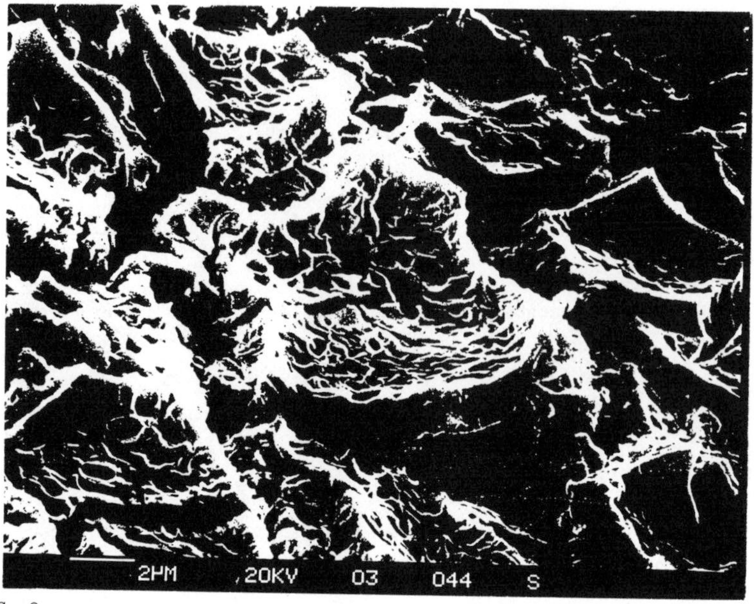

FIG. 2

Charge Transfer Sensitizers

A charge transfer dye must fulfill several criteria in order to be suitable for application in a sensitized photo-electrochemical cell. It must be stable and should have a strong and broad absorption band in the visible. Furthermore, the charge injection for the excited state into the semiconductor has to occur with a quantum yield of close to unity. A condition for efficient electron transfer, alluded to earlier, is that there is intimate contact between the sensitizer and the oxide surface. Only in such a case is the electronic coupling of the excited state wave function with the conduction band states of the oxides large enough to ensure that charge injection can compete with other excited state deactivation pathways. Scheme I shows molecular structures for several charge transfer dyes that have been employed so far. Ruthenium complexes with two [13] or three carboxylated bipyridyl ligands [14], alizarine-S, coumarin 343 as well as zinc-tetra (4-carboxyphenyl) porphyrin [15] exhibit very promising features. The highest photon to current conversion efficiency, i.e. 82% has been obtained with coumarin 343. The role of the carboxylate function in these molecules is to enhance adsorption to the titanium dioxide surface on contact with their solution. As for the ferrocyanide, this system is well-known for its facility in forming complexes with other metallic species. In an independent experiment when TiO^{2+} ions were added to a $Fe(CN)_6^{4-}$ solution, or Ti^{3+} to $Fe(CN)_6^{3-}$ an orange colored precipitate was immediately formed, presumably as titanium analogue of "Prussian blue". Its reflectance spectrum is similar to that of $Fe(CN)_6^{4}$ treated TiO_2, so it is therefore proposed as the photoelectrochemically active dye in this system [16].

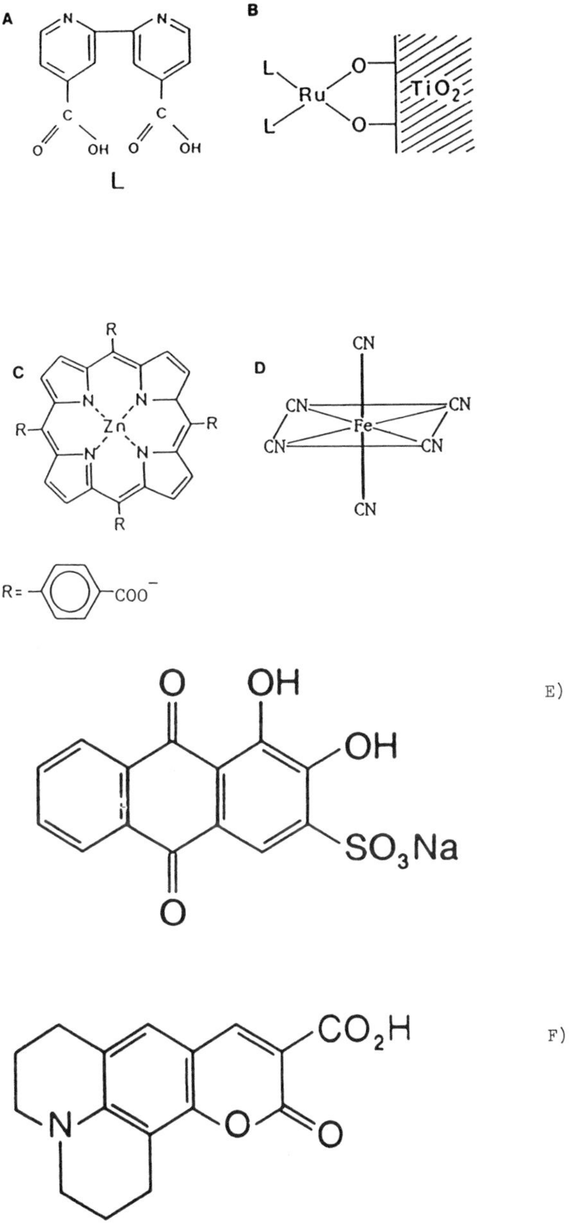

Charge transfer dyes: (A) 2,2'-bipyridyl-4,4'dicarboxylate ligand (L) as in tris (RuL_3) ruthenium complexes; (B) oxygen-bridged surface attachment of RuL_2 to TiO_2; (C) zinc tetra (4-carboxaphenyl) porphyrin; (D) iron hexacyanide complex; (E) 3-alizarinsulfonic acid; (F) coumarin 343.

The characterisation of the derivatized electrodes by conventional solid state techniques presents some difficulties. In ESCA, for example, the ruthenium 3d lines coincide with carbon, while one of the much weaker 3p lines is closely adjacent to a titanium emission. Given that the atomic fraction of ruthenium, even in a well-derivatized surface layer is so much smaller than those of carbon or titanium, it is not surprising that its detection is difficult and the results ambiguous. Rutherford-particle backscattering (RBS) is in this case much more informative and establishes the presence of ruthenium up to 0.7 μm below the geometric surface of the sample. This is taken as evidence of the porous structure of the anatase layer used in the present work, which permits the penetration in depth of the dye while maintaining an intimate dye-semiconductor contact for charge transfer.

Dye Coating of the Electrode, Surface Roughness Factor and Fractal-Like Character of the TiO_2 Films

The chromophores are coated onto the TiO_2 layer by dipping the electrode in an aqueous solution of the dye. For example, in the case of RuL_3, the TiO_2 film assumed an orange/red color due to adsorption of the RuL_3. In acidic aqueous medium and in the absence of complexing agents which compete for adsorption sites on the surface of TiO_2, the RuL_3 adhesion is irreversible and no leaching of the dye into the solution could be detected at pH < 4 and moderately high electrolyte concentrations.

RuL_3 desorption from the TiO_2 surface occurs in alkaline medium and this can be used to determine the surface roughness factor of the electrode. The RuL_3 coated TiO_2 sheet was immersed in 10^{-2} M aqueous NaOH for 24 h resulting in the complete desorption of the dye. The solution spectrum agreed with that for RuL_3 and an extinction coefficient of 2.1×10^4 M^{-1} cm^{-1} at 470 nm was used in order to derive its concentration. According to Furlong et al [17], one RuL_3 molecule occupies 100 Å TiO_2 surface at saturation monolayer coverage. Based on this area, the surface roughness factor is 181. This should be regarded only as a lower limit since the surface may contain micropores which are inaccessible [18] to an adsorbent with the size of RuL_3.

We have performed preliminary studies aiming at characterizing the TiO_2 films in terms of a fractal-type dimension. It has been suggested that various porous materials have fractal structures or fractal pore structures over certain length scales [19]. A number of oxides have been analyzed as fractal structures [20]. In our investigation, we have employed impedance analysis of the TiO_2 films in contact with aqueous electrolyte following the method of Pajkossy [21]. The logarithm of the impedance was found to be a linear function of the logarithm of the frequency over several decades at constant phase angle. From the slope of the straight line the fractal dimension was estimated as 2.7. However, the model of Pajkossy et al has recently been criticized [22]. Therefore more detailed measurements are required to confirm the fractal dimension of our TiO_2 electrode. Nevertheless, this result is instructive in as much as it confirms the highly textured nature of the oxide film.

Performance Characteristics of Regenerative Photovoltaic Cells Based on Dye Sensitization

The effect of sensitized charge injection in fractal oxide semiconductors can be exploited for the conversion of light into electrical power. The performance characteristics of such a photovoltaic device are established in the following manner. First, the monochromatic photon to current conversion efficiency (IPCE) is measured as a function of incident wavelength. As an example, we show in Fig. 3 results obtained with coumarin 343 as a sensitizer. The photocurrent action spectrum was obtained in 10^{-3} M $HClO_4$ in the presence of 0. 1 M NaI as electron donor. For comparison, results obtained with the same electrode prior to dye coating are also presented[11].

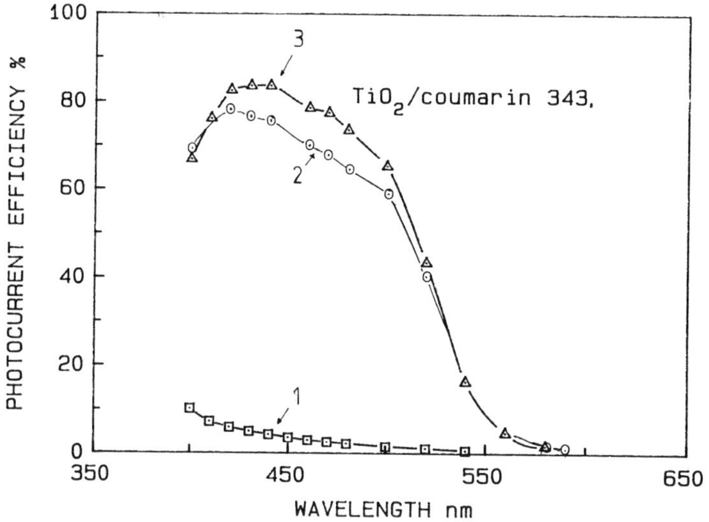

FIG. 3. Photocurrent action spectrum for polycrystalline TiO_2 films coated with coumarin 343 as sensitizer. Electrolyte is 0.1 M NaI in aqueous $HClO_4$ (10^{-3}M). The photoresponse of a bare TiO_2 film is also shown (curve 1).

Whereas TiO_2 alone shows only a weak response in the visible, the Coumarin coated electrodes exhibited very high photocurrents. Thus, in the presence of 1 M KI, the photocurrent density at 440 nm at an incident light flux of 1.14 W/m^2 was 34 A/cm^2 which corresponds to an IPCE value of 84% The achievement of such high conversion efficiencies in spectral sensitization is unprecedented. Matsumura et al [23] have obtained an IPCE of 22% using Al-doped sintered ZnO electrodes in conjunction with rose bengal as a sensitizer and iodide as a donor. Alonso et al [24] achieved an IPCE of 1.5% with a similar electrode and alkyl-substituted $Ru(bipy)_3^{2+}$ as a sensitizer.

Once the spectral response of the dye coated semiconductor film is established the influence of the electrode potential on the photocurrent needs to be determined. From the onset of the photocurrent and the steepness of its increase under anodic bias of the semiconductor, one derives information on the strength of the electrical field necessary to effect charge separation. It is important to assess the magnitude of the voltage drop in the depletion layer of the semiconductor required to prevent recombination of the injected electron with the oxidized dye.

Fig. 4 shows photocurrent-potential plots obtained with RuL_3-coated electrodes and solutions containing iodide, hydroquinone or bromide as an electron donor [12]. The photocurrent onset for the iodide containing solutions is at -0.3 V (SEC), which is close to the flatband potential of the TiO_2 electrode which is -0.4 V at pH 3 in the absence of electron donor. Thereafter, the IPCE rises steeply, the plateau value of 73% being already attained at -0.1 V. The implication of this result is that relatively small band banding (∿200 mV) within the depletion layer suffices to afford practically complete charge separation. Under these conditions P_{cs} in Eq. (5)

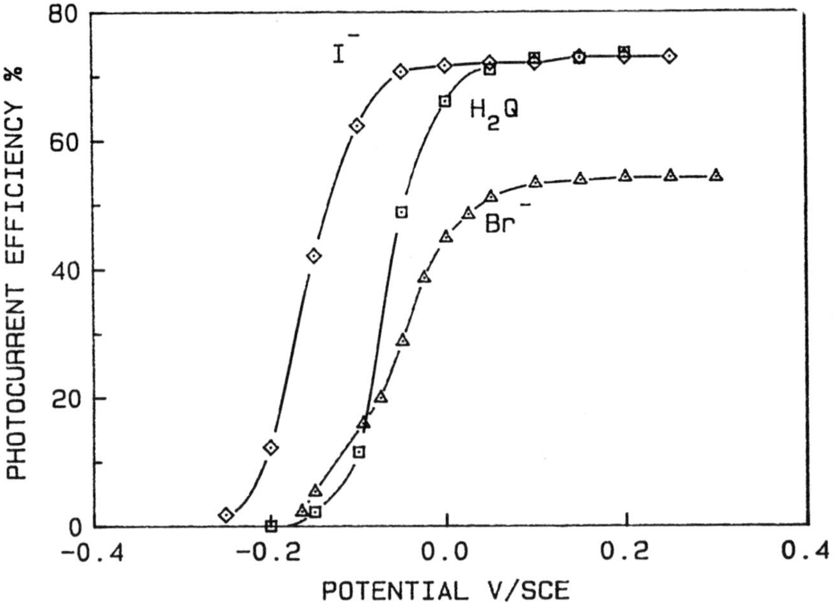

FIG. 4. Photocurrent efficiency (IPCE) - potential curve recorded during 470 nm irradiation of a RuL_3-coated TiO_2 film in aqueous 10^{-3} M $HClO_4$ containing:

0.1M Kl
0.1M hydroquinone + 0.1M $LiClO_4$
0.1M NaBr

is unity. This result is surprising in view of the polycrystalline nature of the TiO_2 layers. The defects and surface states present at the semiconductor solution interface are expected to act as recombination centers. That this is not the case is evident from the steep edge of the photocurrent-potential characteristic and rapid attainment of saturation. This indicates that recombination, i.e. recapture of conduction band electrons by the oxidized sensitizer following charge injection, is slow even at small band bending within the TiO_2 film. These results contrast with the findings [25] obtained with Ru(II) bis (2,2'-bipyridyl)$_2$(2,2'-bipyridyl-dicarboxylate) chemically attached to TiO_2 where recapture of the photo-injected electrons was efficient even at large band bending resulting in a very poor quantum yield of sensitization, i.e. 0.25%. The current action spectrum observed with the attached latter dye is structureless, indicating that the chemical derivatization of the TiO_2 surface produces semiconductor

(t_{2g})/sensitizer surface states acting as recombination centers. This sharply increases the rate of reaction (9) reducing drastically the efficiency of sensitization. The photocurrent action spectrum for RuL_3 matches its absorption characteristics indicating that this type of interaction is absent. Employing colloidal TiO_2 particles, the values of k_{inj} and k_{rec} have been determined to be $3.2 \times 10^7 s^{-1}$ and $4 \times 10^5 s^{-1}$, respectively [26]. Therefore, in the case of RuL_3, the rate of electron injection in the TiO_2 conduction band is 80 times faster than that for recombination explaining the very effective light induced charge separation.

The photocurrent onset observed with hydroquinone (HQ) as an electron donor is shifted by ca. 100 mV in the positive direction with respect to that of iodide, while for Br^- the displacement is ca. 150 mV. Apparently, the oxidation of these donors by RuL_3 at the TiO_2 solution interface (reactions (12) and (13) occur at a smaller rate than that of iodide.

$$2RuL_3^+ + H_2Q \longrightarrow 2RuL_3 + 2H^+ + Q \qquad (12)$$

$$2RuL_3^+ + 2Br^- \longrightarrow 2RuL_3 + Br_2 \qquad (13)$$

At electrode potentials close to flat band conditions, reactions (12) and (13) are not fast enough to compete with charge recombination. Therefore no, or very small, photocurrents are observed. As the band bending increases, the rate constant for the back reaction (9) decreases resulting in a steep augmentation of the photocurrent. In the case of hydroquinone, the ICPE reaches a similar plateau value as for iodide. On the other hand, with bromide as a donor the maximum ICPE observed was lower, i.e. between 56 and 61%. This is likely to arise from partial bleaching of the dye in the photo-stationary state. If the incident light flux is high enough to render charge injection faster than reaction (6) a significant fraction of RuL_3 is converted to the oxidized form. This reduces the light energy harvesting efficiency of the device and hence its IPCE.

For practical applications it is important to check the stability of the dye under long term irradiation. A RuL_3-coated $4 cm^2$ area TiO_2 film was exposed to polychromatic irradiation (Xe lamp, 420 nm cut-off filter) in 0.1 M NaI solution. The photocurrent was 1.2 m A and remained practically constant (fluctuations $< 8\%$) during 100 hours of illumination corresponding to 430 Coulombs of photoinduced charge and a turnover number of RuL_3 of 3.7×10^4. This result confirms the unusually high stability of this charge transfer sensitizer.

We shall finally explore the behavior of the RuL_3 sensitized fractal TiO_2 layers in regenerative photocells. The principal features of such a device have already been discussed in Fig. 1. Light excitation of the sensitizer is followed by charge injection in the semiconductor producing the oxidized form of the dye and conduction band electrons. The original state of the sensitizer is reformed by reaction with an electron donor in the electrolyte. Regeneration of the oxidized form of the donor occurs by reaction with the conduction band electrons at the cathode connected to the semiconductor through the external circuit. The electromotive force is given by the difference of the quasi-Fermi level of the conduction band electrons and solution redox potential. For optimal conversion, the E^o value of the donor should be as positive as possible. The Br_2/Br^- couple in aqueous solution has an E^o - value of 1.06V (NHE) while that of the I_2/I^- redox system is only 0.53 V. The open circuit voltage of a cell with bromide as a donor is expected to be 500 mV larger than that obtained with iodide. Therefore, the following studies were conducted with the Br_2/Br^- redox xouple.

Fig. 5 shows the current voltage characteristics of a cell consisting of the RuL_3-coated 2 x 2 cm sized TiO_2 film as photoanode and a Pt counterelectrode. The aqueous solution contained 0.1 M LiBr, 10^{-3} M Br_2 and 10^{-3} M $HClO_4$. Monochromatic light of 470 nm wavelength and 1.58 W/m^2 intensity produced a short circuit current of 135 μA corresponding to an ICPE of 56.4%. As the load resistance or voltage is increased, the current at first stays fairly constant and then falls to zero at an open circuit voltage of 0.75 V. The maximum power delivered is represented by the area of the largest rectangle that can be fitted under the curve; in this case 0.75 mW at 0.62 V. A measure of squareness of the characteristic is the fill factor defined as:

fill factor = maximum output power/(short curcuit current x open circuit voltage)

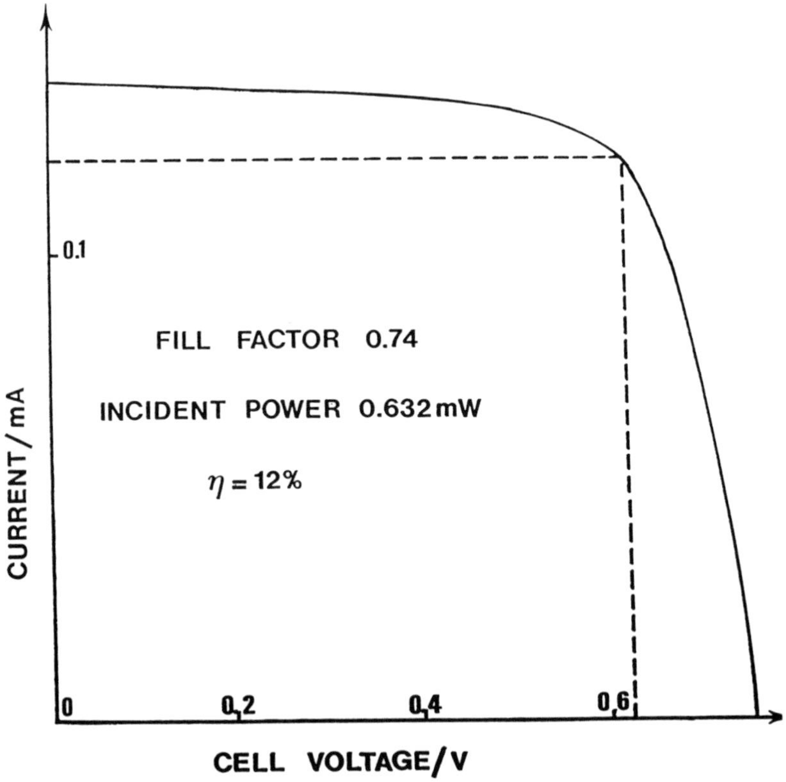

FIG. 5. Photocurrent-voltage characteristics of a regenerative photoelectrochemical cell consisting of the RuL_3-coated TiO_2 film (area 4 cm^2) as photoanode and a Pt counterelectrode. Electrolyte contains 10^{-3} M $HClO_4$, 0.1M LiBr and 10^{-3} M Br_2. Excitation wavelength 470 nm.

From Fig. 5 the fill factor is determined as 0.74 which is astonishingly high for a device based on a polycrystalline semiconductor film. This indicates that the loss mechanism, such as recombination, normally encountered in semiconductor photoelectrochemistry has been minimized. The performance characteristics achieved here are comparable to those of high quality single crystal photovoltaic cells. The conversion efficiency is the maximum power output expressed as a percentage of the input light power:

$$\eta\ (\%) = \text{max. output power} \times 100 / (\text{irradiance} \times \text{area})$$

The initial studies gave η = 12% for 480 nm light, a figure which has been improved in the meanwhile to 14%. This is by far the highest monochromatic conversion yield achieved until now in dye sensitized regenerative photoelectrochemical cells. The maximum monochromatic efficiency reported previously by Matsumara et al [22] is 2.5% for 540 nm light and sintered ZnO pellets coated with rose bengal as a sensitizer.

The cell characteristics remain excellent even under polychromatic excitation at light intensities comparable to the solar flux. Open circuit voltages close to 1V and fill factors as high as 77% have been obtained under these conditions. When the entire λ > 420 nm light output of the high pressure Xe lamp was used for excitation of the RuL_3-coated TiO_2 film the power conversion yield was ca. 4 times smaller than the monochromatic efficiency at 480 nm. These performance parameters may be further improved by optimizing the experimental conditions. Also, it should be noted that polychromatic efficiencies depend on the emission features of the light source. Sunlight experiments providing the solar conversion characteristics are under way.

CONCLUSIONS

The findings presented here confirm the very promising properties of high surface area anatase films employed in conjunction with suitable sensitizers, as visible light energy harvesting and conversion devices. Dye coated films display incident monochromatic photon to current conversion efficiencies exceeding 80%. For RuL_3 turnover numbers are at least several ten thousands showing the extremely rugged character of this chromophore. The quantum yield for destructive side reactions is below 10^{-5}. Another attractive feature of the latter electron transfer dyes is the high redox potential in the ground state allowing the sensitized electron injection in the semiconductor to be coupled to thermodynamically demanding oxidation reactions. For the first time the sensitized generation of Br_2 from Br^- was achieved at a large band gap semiconductor electrode. This reaction is exploited to generate electrical power in a photodriven electrochemical cell. Even without optimization, a monochromatic conversion efficiency as high as 14% and a fill factor of 0.77 have been obtained. Due to their durability, low cost and high efficiency these systems appear to have the potential to serve in practical solar energy conversion devices.

ACKNOWLEDGMENT

This work was supported by the Gas Research Institute, Chicago, Ill., USA (subcontract with the Solar Energy Research Institute (SERI), Golden, Col., USA), the Swiss National Energy Foundation (NEFF) and the Swiss Office for Energy (OFEN). We are grateful to Dr Ersi Vrachnou for assistance with the chemical actionmetry. We also acknowledge discussions with Prof. Martin Fleischmann, University of Southampton, U.K., concerning the fractal character of the TiO_2 film.

REFERENCES

1. H. Gerischer, Angew. Chem. 100, 63C (1988).
2. "Energy Resources Through Photochemistry and Catalysis" M. Grätzel, ed. Academic Press (1983).
3. A.S. Feiner, A.J. McEvoy and P.P. Infelta, Surface Sci. 189/190, 411 (1987)
4. H. Meier, J. Phys. Chem. 69, 724 (1965)
J. Bourdon, J. Phys. Chem. 69, 705 (1965)
K. Heuffe and J. Range, Z. Naturforsch. B. 238, 736(1968).
H. Tributsch and H. Gerischer, Ber. Bunsenges. Phys. Chem. 72, 437(1968).
5. T. Watanabe, A. Fujishima and K. Honda: in Energy Resources Through Photochemistry and Catalysis, M. Grätzel, ed. (Academic Press, New York 1983).
6. R. Memming, Progr. Surface Sci. 17, 7 (1984)
7. IUPAC Recommendations: J. Electroanal. Chem. 209, 417.
8. H.H. Kung, A.S. Jarrett, A.W. Sleight and A. Ferretti, J. Appl. Phys. 48, 2466 (1977).
9. N. Vlachopoulos, P. Liska, A.J. McEvoy and M. Grätzel, Surface Sci. 189, 823 (1987)
10. M. Spitler, J. Electroan. Chem. 228, 69 (1987); for related work c.f.: D.F. Blossey, Phys. Rev. 139, 5183 (1974) and F. Willig, Chem. Phys. Lett. 40, 331 (1976).
11. O. Enea, N. Vlachopoulos, G. Rothenberger and M. Grätzel to be submitted to J. Electroanal. Chem.
12. For details see Ref. 14 and C. Stalder and J. Augustynski, J. Electrochem. Soc. 126, 2007 (1979).
13. P. Liska, N. Vlachopoulos, M.K. Nazeeruddin, P. Comte and M. Grätzel, J. Am. Chem. Soc., in print.
14. N. Vlachopoulos, P. Liska, J. Augustynski and M. Grätzel, J. Am. Soc., 110, 1216 (1988).
15. K. Kalyanasundaram, N. Vlachopoulos, V. Krishnam, A. Monnier and M. Grätzel, J. Phys. Chem. 91, 2342 (1987).
16. E. Vrachnou, N. Vlachopoulos and M. Grätzel, J. Chem. Soc. Chem. Comm. 868 (1987).
17. D.N. Furlong, D. Wells and W.H.F. Sasse, J. Phys. Chem. 90, 1107 (1986).
18. D. Avnir, J. Am. Chem. Soc. 109, 2931 (1987).
19a) U. Even, K. Rademann, J. Jortner, N. Manor and R. Reisfeld, Phys. Rev. Lett. 52, 2164 (1984).
b) D. Avnir, D. Farin and P. Pfeifer, Nature (London) 308, 261 (1984).
20a) P. Pfeifer, D. Avnir and D. Farin, J. Stat. Phys. 36, 699 (1984).
b) D.W. Schaefer and K.D. Keefer, Phys. Rev. Lett. 56, 2199 (1986).
c) D. Rojanski, D. Huppert, H.D. Bale, X. Dacal, P.W. Schmidt, D. Farin, A. Seri-Levy and D. Avnir, Phys. Rev. Lett. 56, 2505 (1986).
21) L. Nyikos and T. Pajkossy, Electrochim. Acta 30, 1533 (1985).
22. M. Keddam and H. Takenouti, Electrochim. Acta 33, 445 (1988).
23a) M. Matsumura, Y. Nomura and H. Tsubomura, Bull. Chem. Soc. Japan 50, 2533 (1977)
b) M. Matsumura, S. Matsudaira, H. Tsubomura, M. Takata and H. Yanagida, I & E Prod. Res. Devel, 19, 415 (19807.
24. N. Alonso, V.M. Beley, P. Chartier and V. Ern, Revue Phys. Appl. 16, 5 (1981).
25. S. Anderson, E.C. Constable, M.P. Dare-Edwards, J.B. Goodenough, A. Hammet, K.R. Seddon and R.O. Wright, Nature (London) 280, 6139 (1979).
26. J. Desilvestro, M. Grätzel, L. Kavan, J. Moser and J. Augustynski, J. Am. Chem. Soc. 107, 2988 (1985).

SOLAR PHOTOCHEMISTRY AND HETEROGENEOUS PHOTOCATALYSIS: A CONVENIENT AND PRACTICAL UTILIZATION OF SUNLIGHT PHOTONS

NICK SERPONE
Department of Chemistry, Concordia University
1455 deMaisonneuve Blvd. West, Montréal, Qué.,
Canada H3G 1M8

ABSTRACT

The search for alternative energy supplies continues since the oil crisis of 1973. One energy vector is dihydrogen, H_2. Of the group VI hydrides, water has been the focus of most studies in harnessing solar energy and generating H_2. Two basic photochemical strategies have been employed: (i) molecular photocatalytic systems, and (ii) semiconductor based photocatalytic systems. The results have not met with the euphoric expectations of the mid-1970's because of the difficulties encountered in H_2O splitting (E° H_2O/O_2 = -1.23 eV, NHE); however, the spin-offs from these studies have been rewarding. Hydrogen sulfide (E° S^{2-}/ S = +0.51 eV, NHE) is another vehicle tapped as a potential source of H_2. Heterogeneous photocatalysis utilizing semiconductor particulates and sunlight as the photon source has been successful with interesting quantum efficiencies. To this end, novel photocatalytic devices have been developed; one of these uses two coupled semiconductors to achieve vectorial displacement of the photogenerated reducing and oxidizing equivalents. An important area in which sunlight photons will see practical application is the photodegradation of environmental contaminants (e.g., chloroaromatics, chloroaliphatics, surfactants, cyanide) and the recovery of trace metals from the aqueous environment (e.g., mercury, lead, gold, platinum, palladium, silver, and others).

INTRODUCTION

The search [1-3] for alternative (non-fossil fuel) energy supplies continues to dominate the fascination of photochemists, particularly the prospects [4-6] in a sunlight-driven water splitting process. Conversion of solar energy (eqn. 1) and storage of chemical energy as a fuel (e.g., H_2) from H_2O

$$H_2O \xrightarrow{hv} H_2 + 1/2\ O_2 \qquad (1)$$

is one of the few processes by which energy can be generated without the simultaneous production of chemical and/or thermal pollution. The euphoria of the mid-1970's had no parallel as witnessed by the number of early reports [7] and by the staging of these Conferences dealing specifically with conversion and storage of solar energy [8]. Fig. 1A depicts the relative number of yearly publications from 1973; it shows a near-exponential rise peaking in 1982 [9]. While the results have not met with the early expectations, the spin-offs have been phenomenal particularly in the area of excited state electron transfer [10]. The decrease seen after 1982 is no reflection of decreased interest. Rather, it is a "moratorium" reflecting our collective attempts to understand the various factors underlying potentially viable systems based on reaction 1 or others (cf. Table 1).

Several reactions of Table 1 occur [11,12] either <u>in vivo</u> in plant

Photochemical Energy Conversion
James R. Norris, Jr. and Dan Meisel, Editors

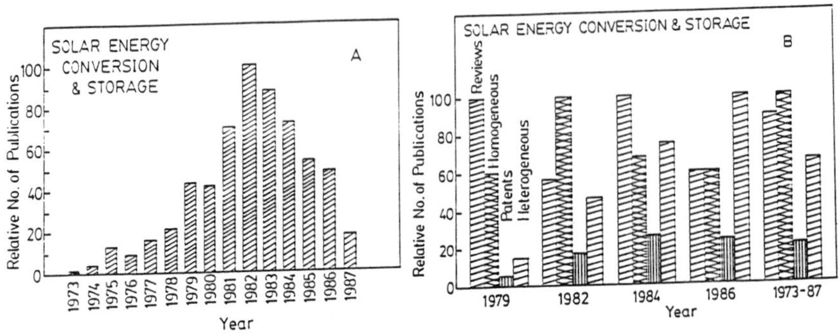

FIG. 1. Relative number of publications for the period indicated.

TABLE I.- Thermodynamic characteristics for some reactions involving water and atmospheric gases employable in the photocatalyzed conversion of solar energy to chemical energy [11,12].

Reactions[a]	Number of Electrons Transferred	ΔG^{o}_{298} kcal mol^{-1}	ΔH^{o}_{298} kcal mol^{-1}
H_2O -----> H_2 + $1/2O_2$	2	56.7	68.3
H_2O + CO_2 ---> HCOOH + $1/2O_2$	2	68.4	65.4
H_2O + CO_2 ---> H_2 + CO + O_2	4	118.2	135.9
H_2O + CO_2 ---> CH_2O + O_2	4	124.8	134.6
$2H_2O$ + CO_2 ---> CH_3OH + $3/2O_2$	6	167.9	173.6
$2H_2O$ + CO_2 ---> CH_4 + 2 O_2	8	195.5	212.8
$3H_2O$ + $2CO_2$ ---> C_2H_5OH + $3O_2$	12	318.3	336.8
H_2O + CO_2 ----> 1/6 $C_6H_{12}O_6$	4	114.7	111.6
$3/2H_2O$ + $1/2N_2$ ---> NH_3 + $3/4O_2$	3	81.1	91.4
$2H_2O$ + N_2 ---> N_2H_4 + O_2	4	181.3	148.7

[a] Data given for liquid H_2O, HCOOH, CH_3OH, C_2H_5OH, N_2H_4; solid $C_6H_{12}O_6$; the remaining compounds are considered in the gas phase.

photosynthesis or in atmospheric nitrogen fixation. Unfortunately, the processes do not occur on illumination with sunlight. The production of high energy fuels through these reactions has necessitated development of photocatalytic systems sometimes complex, implicating photocatalysts in the photon-assisted sequence and catalysts involved in the subsequent dark reactions that ultimately generate fuels. Two major avenues have been explored: (i) molecular photocatalytic systems and (ii) semiconductor based photocatalytic systems. Since the pioneering work of Fujishima and Honda [13], systems have been developed that achieve solar energy conversion with efficiencies that reach several percent [see refs. 8c,11,12,14-18]. The field of molecular photocatalytic systems has not been neglected [see refs. 11,12,15, 16,18-23].

Fig. 1B depicts literature reports classified as (a) reviews, (b) research on a pure homogeneous phase, (c) patents, and (d) research in heterogeneous media. The number of reviews in 1979 and 1984 outweighs the other publications. The number of studies in heterogeneous media has continually grown. This recognizes the necessity for a heterogeneous catalyst surface (colloidal Pt to produce H_2 in eqn. 1) in an otherwise homogeneous system (see below).

Polypyridyl metal complexes, organic dyes, and metalloporphyrins have been used as "photosensitizers" to effect reaction 1 [3,7,8,24-26]. A disadvantage with homogeneous solutions is the energy-wasting back electron transfer. Results from studies in organized molecular assemblies (micelles, vesicles, micro-emulsions) are encouraging [27-29]. The near diffusion controlled rates of back electron transfer, following an endergonic photoredox process in homogeneous systems, bring focus to heterogeneous semiconductor (SC) particulates and colloidal systems as the light harvesting units.

STRATEGIES USED IN THE CONVERSION OF LIGHT ENERGY TO CHEMICAL ENERGY

Electron transfer processes play a key role in practical energy conversion devices. The charge separation event is of primordial importance. In natural photosynthesis, charge separation is achieved through vectorial electron transfer across the photosynthetic membrane [30]. The distinct environment present in molecular assemblies provides the kinetic control on the charge transfer events.

Two major strategies have been followed [31] in designing devices that photogenerate H_2 and O_2 from H_2O. One approach is the molecular photocatalytic system in which a photosensitizer S promotes electron transfer between suitable electron relay species, R. The reduced and oxidized forms of R, in the presence of some appropriate catalyst(s), can liberate H_2 and O_2, respectively. Fig. 2 shows a photosystem for H_2 generation implicating R and a catalyst; the sacrificial electron donor D regenerates S. An analogous photo system can be constructed for O_2 generation. Fig. 3 depicts the coupling of

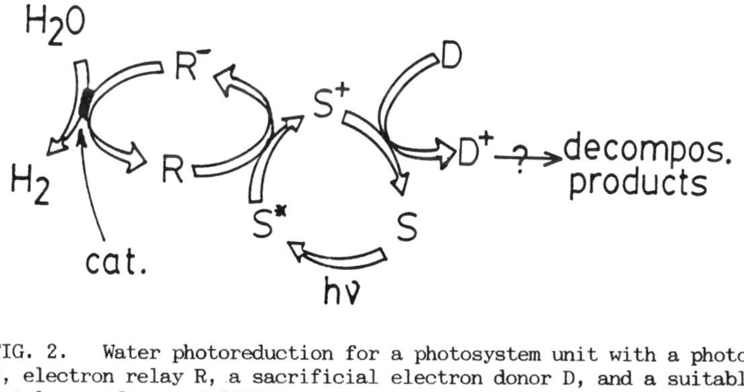

FIG. 2. Water photoreduction for a photosystem unit with a photosensitizer S, electron relay R, a sacrificial electron donor D, and a suitable redox catalyst. After ref.[32].

these two photosystem units: in photosystem 1, the excited $*S_1$ (e.g., a water-soluble metalloporphyrin) reduces an electron acceptor A (e.g., methyl viologen, MV^{2+}). The oxidized form of the sensitizer, S_1^+, is subsequently reduced by the electron donor D, recycling S_1; A^- generates H_2 from H_2O in the presence of Pt. In photosystem 2, D^+ (e.g., EDTA or triethanolamine), oxidizes S_2 (e.g., $Ru(bpy)_3^{2+}$; bpy is 2,2'-bipyridine) to S_2^+ which oxidizes water to O_2 in the presence of RuO_2 (or IrO_2 or PtO_2). While both photosystem units have been extensively investigated [31], the coupling of Fig. 3 has not materialized. An alternative but analogous molecular system is portrayed in

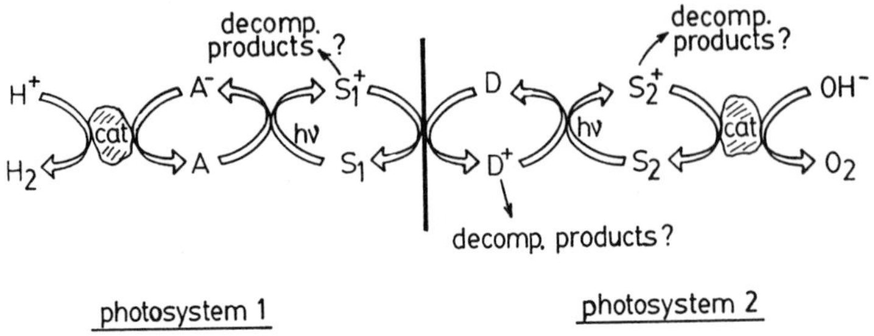

FIG. 3. Coupled photosystems to effect the water cleavage with two photosensitizers and two electron relays. After ref. [32].

Fig. 4; this system utilizes a single photosensitizer (S), one electron relay R and two appropriate redox catalysts to effect water cleavage [33].

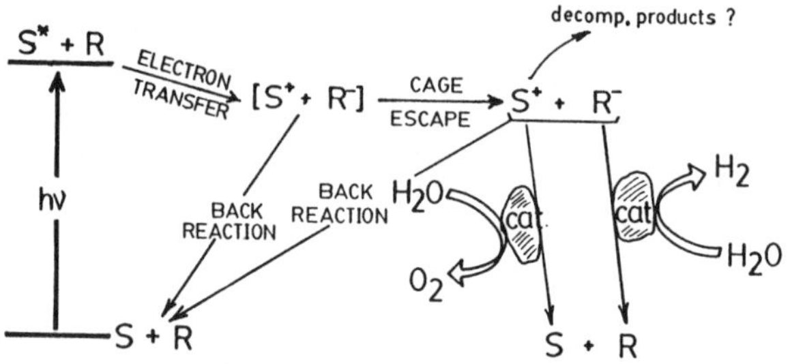

FIG. 4. Alternative molecular system in the photocleavage of water. After ref. [33].

The second approach uses inorganic semiconductor materials (e.g., TiO_2, $SrTiO_3$, CdS). The relay R in Fig. 4 is replaced by a semiconductor particle. The light harvesting unit suitable for water splitting is shown in Fig. 5. S is adsorbed on the SC particle. Charge injection from the excited state S* to the conduction band of the SC occurs, following which the e^- is channeled to a catalytic site for H_2 evolution. The RuO_2 codeposited on the particle mediates O_2 generation from S^+ and H_2O, and regenerates S. In another case, the SC particle is the light harvesting unit (Fig. 6); irradiation with hν greater than the energy gap (E_{bg}) generates electron/hole pairs (e^-/h^+) that separately migrate to the particle surface where water reduction and oxidation occurs mediated by surface-adsorbed redox catalysts. The light harvest ing units depicted in Figs. 5 and 6 are better than the simple S/R pair of Fig.4,

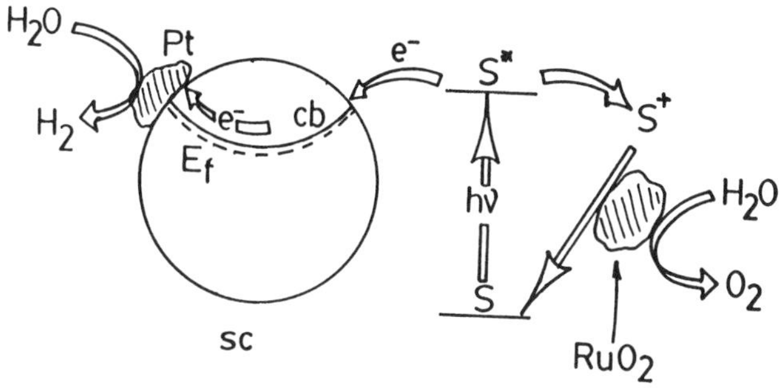

FIG. 5. Water splitting scheme using an adsorbed sensitizer and a semiconductor. After ref. [33].

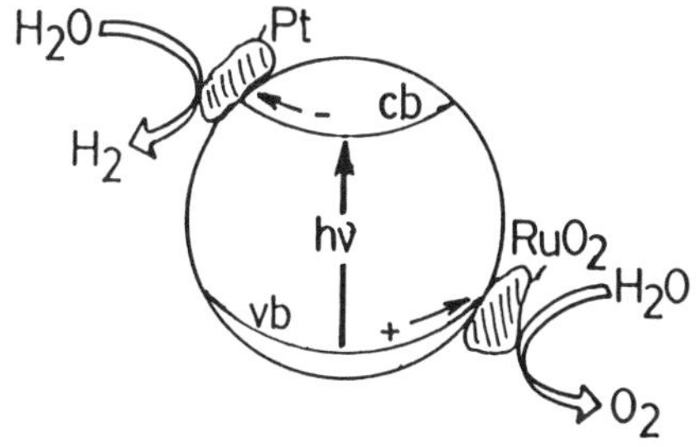

FIG. 6. Illustration of a water cleavage process using a semiconductor particle as the light harvesting unit. After ref. [33].

because light induced charge separation and redox catalysis are concentrated in a very small and confined reaction space [33]. Here, the water splitting events take place on a single SC particle, thus avoiding bulk-phase diffusion of the reactants.

Colloidal SC particulates offer distinct advantages in heterogeneous photocatalysis: (1) high absorption cross sections, (2) fast carrier diffusion to the interface, and (3) suitable redox levels of the valence and conduction band edges that can yield high efficiencies in light energy conversion processes. The rapid movement of the charge carriers in their respective band and the presence of local electrostatic fields at the semiconductor/electrolyte interface leads to charge separation and to subsequent formation of fuels from light. Moreover, the particle surface can be molecularly engineered to improve the absorption and redox characteristics. Unlike homogeneous systems, photoredox reactions that occur at the SC/solution interface between the excited SC* and the redox species in solution occur in one direction and, generally, are not reversible [34].

The common denominator in both strategies is that light is first used to generate reduction equivalents (R^-, A^-, or e^-) and oxidation equivalents (D^+, S^+, or h^+). This light induced process is subsequently coupled to some dark (possibly catalytic) reactions generating H_2 and O_2 from water and regenerating the starting chemicals. The overall fuel generating steps (Fig. 3) are described by reactions 2 and 3. Four oxidation equivalents are needed to pro-

$$4D^+ + 2H_2O \longrightarrow 4D + 4H^+ + O_2 \qquad (2)$$

$$2A^- + 2H_2O \longrightarrow 2A + 2\ OH^- + H_2 \qquad (3)$$

duce O_2, and two reduction equivalents to generate H_2. The formation of O_2 by a single electron oxidant, D^+ or S^+, is a difficult task as it involves four subsequent steps with such high energy intermediates as H_2O_2, OH, or O_2^- radicals. Electron storage catalysts have been developed to avoid these intermediates. Reactions 2 and 3 are multi-electron transfer processes connected with high kinetic barriers; in the absence of suitable redox catalysts, either the reactions are not observed or they are very inefficient.

It is unlikely that simple homogeneous systems will find application in artificial devices that convert sunlight to chemical energy. They have serious disadvantages: (i) the light-driven electron transfer processes are diffusion-controlled; and (ii) there is no kinetic barrier to inhibit the back electron transfer event which degrades photons to heat. The electron donor is often sacrificed and the photosensitizer itself {e.g., $Ru(bpy)_3^{2+}$} is often not photochemically stable for extended periods. This notwithstanding, investigations continue to prepare and characterize other sensitizers [35] and electron relays [32,33,36] having greater stability. Balzani and coworkers have summarized the requirements for ideal electron relays and photosensitizers [37], and for an assembly of molecular components to obtain a photochemical molecular device possessing suitable photochemical and photophysical characteristics [38].

PHOTOCLEAVAGE OF HYDROGEN SULFIDE AND INTERPARTICLE ELECTRON TRANSFER (IPET)

Hydrogen sulfide is a group VI hydride, like water a potential source of hydrogen. It is a major constituent of natural gas [39-42] and is the principal product of hydrodesulfurization of petroleum and coal. A recent estimate [40-42] suggests that over 10^6 metric tons of H_2S are produced each year from all sources. Recovery of all the H_2 from H_2S alone could yield ~8×10^9 m^3/year [40-43]. Unlike water, which is difficult to oxidize {E^o H_2O/O_2 = -1.23 V, NHE} requiring four oxidizing equivalents (eqn. 2), oxidation of H_2S is thermodyn-amically a facile process {E^o S^{2-}/S = +0.51 V, NHE}: $S^{2-} + 2h^+ \longrightarrow S$.

Dihydrogen is photogenerated when H_2S is bubbled through an alkaline aqueous CdS dispersions [44-47]. Reaction 4 has been studied in some detail with most studies concerned with the preparation of efficient catalytic sys-

$$H_2S \xrightarrow[\text{CdS}]{h\nu} H_2 + S \qquad (4)$$

tems [44-46]. Fig. 7 summarizes the process for H_2 generation on irradiation of CdS with visible light (E_{bg} = 2.4 eV; 520 nm). While CdS is not a particularly good electrode for water reduction, photoevolution of H_2 proceeds at reasonable rates (driving force ~250 mV) without the intervention of a redox catalyst (Pt or RuO_2). Table II summarizes the rates of H_2 evolution {$r(H_2)$} along with quantum yields and energy conversion efficiencies (where measured).

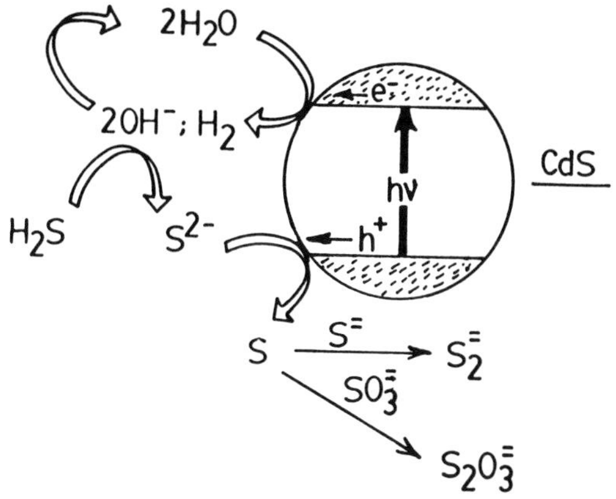

FIG.7. Scheme showing H_2 generation on irradiating a CdS suspension containing S^{2-} and/or SO_3^{2-} in alkaline media (pH 14).

TABLE II- Photogeneration of hydrogen in various systems and for various semiconductor dispersions [48-52].[a]

Dispersion	$r(H_2)$ (lit.H_2/g.cat/hr/lit.soln)	$\Phi(H_2)$[b]	Energy Conv. Efficiency[b]
	<u>0.1 M S^{2-}/1 M OH^-</u>		
CdS	2.6		
CdS/1wt%RuO_2	38.6		
CdS + TiO_2/0.5wt%RuO_2	55.4	-	1%
CdS/0.4wt%Rh	64.4		
	<u>0.1 M S^{2-}/0.1 M SO_3^{2-}/1 M OH^-</u>		
CdS	5.4		
CdS + TiO_2/0.5wt%RuO_2	56.8	-	2.0%
CdS/0.2wt%Rh[c]	146	23%	2.6%
	<u>Pure methanol</u>		
CdS	0.6		
CdS + 20wt%Pt	4.2		
CdS/20wt%Pt	27.4		
CdS + TiO_2/20wt%Pt	56		
CdS + TiO_2/10wt%Pt	45.4		
	65.8 (ethanol)	27%	2.0%
	56.6 (isopropanol)		

[a] Argon-purged solutions; irradiation source, 900-W Hg/Xe lamp; ~245-265 mW cm^{-2}. [b] Lower limits; [c] Turnover number = 2200 for Rh and 15 for CdS [51].

The complete cleavage of H_2S at pH 14 in the presence of SO_2 and irradiated CdS yields H_2 and thiosulfate $S_2O_3^{2-}$ in a 1:1 stoichiometric ratio [45, 46]. The greater $r(H_2)$ with anions present [47] arises from suppression of formation of polysulfides (Fig. 7); these interfere with light absorption by

CdS. In the presence of SO_3^{2-} ions, the net valence band process in Fig. 7 is

$$2h^+ + S^{2-} + SO_3^{2-} \longrightarrow S_2O_3^{2-} \qquad (5)$$

(E^o = -0.72 V, NHE); the overall reaction 6 requires <u>two</u> light quanta for the photocleavage of H_2S. While reaction 6 stores 0.11 eV of free energy per absorbed photon (pH 14, 25°C) compared to 0.36 eV for H_2S cleavage (eqn.4),

$$S^{2-} + SO_3^{2-} + 2H_2O \xrightarrow{2hv} S_2O_3^{2-} + 2OH^- + H_2 \qquad (6)$$

process 6 gives high yields of H_2 without formation of insoluble products (sulfur); also, H_2 evolution is sustained for longer periods.

Coupling of the process in eqn. 6 with one capable of recycling $S_2O_3^{2-}$ back to S^{2-} and SO_3^{2-} has been suggested [53] as a means of sustaining H_2 evolution in a cyclic system. Irradiation of TiO_2 (E_{bg} = 3.2 eV) dispersions containing only $S_2O_3^{2-}$ produces S^{2-} and SO_3^{2-} (eqn. 7). Oxidation by h^+ (eqn. 8) yields $S_4O_6^{2-}$ which disproportionates in alkaline media (eqn. 9). Reaction 10 describes the net valence band process; the process on TiO_2 is that des

$$S_2O_3^{2-} + 2\ e^-(TiO_2) \longrightarrow S^{2-} + SO_3^{2-} \qquad (7)$$

$$2S_2O_3^{2-} + 2\ h^+(TiO_2) \longrightarrow S_4O_6^{2-} \qquad (8)$$

$$1.5H_2O + S_4O_6^{2-} \longrightarrow SO_3^{2-} + 1.5S_2O_3^{2-} + 3H^+ \qquad (9)$$

$$1.5H_2O + 0.5S_2O_3^{2-} + 2h^+(TiO_2) \longrightarrow SO_3^{2-} + 3H^+ \qquad (10)$$

$$1.5H_2O + 1.5S_2O_3^{2-} \longrightarrow 2SO_3^{2-} + S^{2-} + 3H^+ \qquad (11)$$

scribed by reaction 11. This reaction stores 0.38 eV (pH 14, 25°C). Coupling of reaction 6 with reactions 7 and 8 is illustrated in Figure 8; the overall process is summarized by reaction 12; note that <u>three</u> moles of H_2 are produced instead of <u>one</u> mole as in the simple cleavage of H_2S (eqn. 4). The

$$3H_2O + S^{2-} \xrightarrow{hv} SO_3^{2-} + 3H_2 \qquad (12)$$

energy stored by the cyclic photosystem of Fig. 8 is 1.4 eV (G^o_{298} = 32 kcal mol^{-1}). No accumulation of sulfur or $S_2O_3^{2-}$ would take place in this photosystem.

Recently we introduced [48] interparticle electron transfer (IPET) between two different types of semiconductors as a novel strategy to accomplish vectorial displacement of charges and avoid electron/hole recombination. We first successfully applied this strategy to the visible light induced decomposition of H_2S [48-51] and later to the photooxidation of alcohols (Table II) [52]. The process is depicted in Fig.9 and summarized by reactions 13-17:

$$CdS \underset{heat}{\overset{hv}{\rightleftharpoons}} e^-(CdS) + h^+(CdS) \qquad (13)$$

$$2h^+(CdS) + S^{2-} \longrightarrow S \qquad (14)$$

$$2e^-(CdS) + 2H_2O \longrightarrow H_2 + 2OH^- \qquad (15)$$

$$e^-(CdS) + TiO_2 \longrightarrow e^-(TiO_2) + CdS \qquad (16)$$

$$2e^-(TiO_2) + 2H_2O \longrightarrow H_2 + 2OH^- \qquad (17)$$

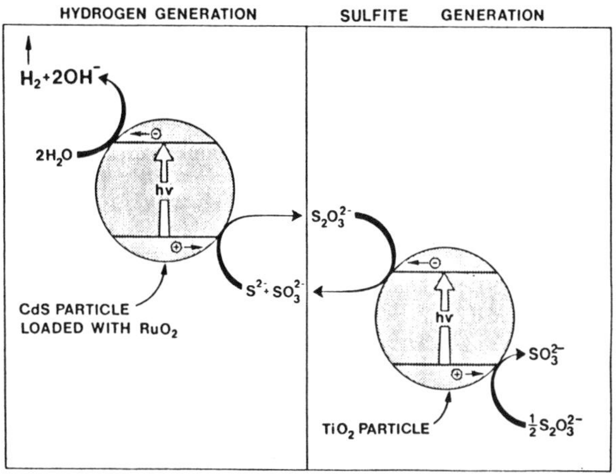

FIG. 8. Coupling of two processes in a cyclic photosystem for hydrogen and sulfite generation. From ref. [53].

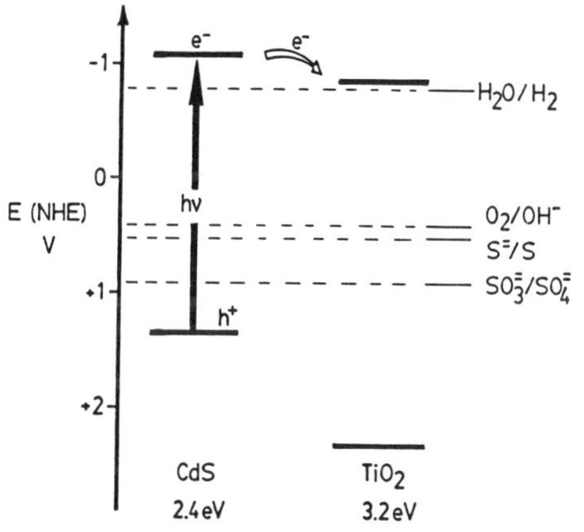

FIG. 9. Flatband potentials of CdS and TiO_2 at pH 14 and 0.1 M Na_2S in aqueous media. The positions of other redox couples are also indicated.

The driving force for reaction 16 is ~180 mV. Once trapped by the TiO_2 particles, the conduction band electrons can still generate H_2 (eqn.17). However, water reduction is slow on naked TiO_2 (driving force ~50 mV) thereby requiring an efficient redox catalyst. Addition of naked TiO_2 to CdS dispersions decreases the efficiency of photoinduced H_2 generation from CdS [48]. Equally important, CdS in combination with TiO_2/RuO_2 performs better than CdS/RuO_2 alone (Table II). The energy conversion efficiency of the former is ~ 1%, but with sulfite present in the dispersion, the efficiency is ~ 2%. Through the combination of these two semiconductor materials, the extent of electron/ hole recombination has diminished. Charge separation is achieved through selective transfer of conduction band electrons from CdS to TiO_2 particles; the transfer of holes between valence bands is thermodynamically inhibited [48]. Further evidence for the IPET process comes from the photooxidation of alcohols (Table 2); the coupled semiconductor system is CdS + TiO_2/Pt. In both cases, quantum yields are high (Table 2). IPET between two semiconductor particles has since been confirmed photoelectrochemically [54-56], by photoconductance [57] and by spectrofluorimetric techniques [58].

Evolution of H_2 occurs with a quantum efficiency of ~23% and an energy conversion efficiency close to ~3% for a Rh-loaded CdS catalyst in alkaline media of H_2S and SO_2 has been achieved (Table 2) [51]. The performance of this catalyst appears superior to that of the CdS-TiO_2/cat. combination. The high quantum yield and the facile preparation of the Rh-loaded CdS catalyst make it practical in photogenerating H_2 with the removal of two pollutants: H_2S and SO_2. Of significance, hydrogen evolution is independent of the presence of air in this system.

Two principal factors must also be considered in heterogeneous photocatalysis: (i) long term stability of the catalyst, and (ii) separation of the products from the reactants and from the catalytic slurry. The former is particularly important for CdS known to anodically and cathodically photocorrode. Redox catalyst loaded CdS particles appear more stable. By contrast TiO_2 is very stable to photo and chemical corrosion [59]. Binding the photocatalyst on a suitable support (polycarbonate matrix [60], glass beads [61]) may have practical applications. For example, binding CdS/RuO_2 onto a polycarbonate matrix yields H_2 from H_2S for several cycles on irradiation without loss of activity or catalyst material; quantum yield $\geq$ 10%, energy conversion efficiency ~0.5% [60].

PHOTODEGRADATION OF ENVIRONMENTAL CONTAMINANTS

An area ripe for exploitation by heterogeneous photocatalysis is the photodegradation (disposal) of environmental contaminants by SC particulates irradiated with AM1 <u>simulated sunlight</u> or <u>natural sunlight</u>. We examine below some of our recent work on four classes of pollutants: (1) cyanide [62-64]; (2) chlorinated aromatic hydrocarbons [61,63,65-72]; (3) surfactants [63,73-76]; and (4) trace noble and/or toxic metals [59,61,63,64,77,78].

Cyanide

Cyanide is a frequent pollutant occurring from such sources as rinse waters following steel surface hardening treatments, electroplating, gold extraction, metal cleaning, and in mining processes. Its removal is usually carried out by chemical oxidation (O_3 or Cl_2)) and by electrochemical means at some considerable cost [79]. Removal of CN^- from waste waters can be coupled to the photocleavage of H_2S in alkaline media. Polysulfide ions, S_n^{2-}, react with CN^- to form thiocyanate SCN^- (reaction 18) [80]: Irradiation of a

$$S_2^{2-} + CN^- \longrightarrow SCN^- \qquad (18)$$

CdS/Rh catalytic slurry containing 0.1 M Na_2S, 0.1 M NaCN and 1 M NaOH evolves H_2 and produces SCN^- in 1:1 stoichiomeric ratio [62]. The quantum efficiency of this process is $\geq$ 40% depending on the CdS preparation [81]. CN^- can be converted to CNO^- on irradiated TiO_2 suspensions [82]. The advantage of CN^-/SCN^- is the high quantum efficiency and SCN^- is less toxic than CNO^- [83]. Because of the high toxicity of CdS (Cd^{2+} on photocorrosion), a better approach to degrade CN^- is UV/H_2O_2 oxidation to CO_2 and NH_3 (or N_2) [64].

Chlorinated Hydrocarbons

Chlorinated hydrocarbons are environmentally hazardous owing to their toxicity and their continued widespread utilization as pesticides and insecticides [84]. Some 57 million metric tons of non-radioactive hazardous wastes were generated (1980; US EPA) by the manufacturing sector in the United States alone. Chlorinated aromatics such as pentachlorophenol (PCP) find use in a variety of industrial, agricultural, domestic, and health-related applications [85].

The total mineralization (eqn. 20; for PCP) of several contaminants (Fig. 10) has been achieved in aqueous media with TiO_2 suspensions under exposure to simulated sunlight [63,67]. Both HCl and CO_2 are observed in

$$HO\text{-}C_6Cl_5 + 4.5O_2 + 2H_2O \xrightarrow[TiO_2]{hv} 6CO_2 + 5HCl \qquad (20)$$

quantitative and stoichiometric quantities. Photodegradation occurs at significant rates under sunlight illumination [69, 86]. Results of our extensive survey (Fig. 10; Table III) suggest that the initial event(s) in the mineralization process is substitution of Cl ring substituents by OH groups [69].

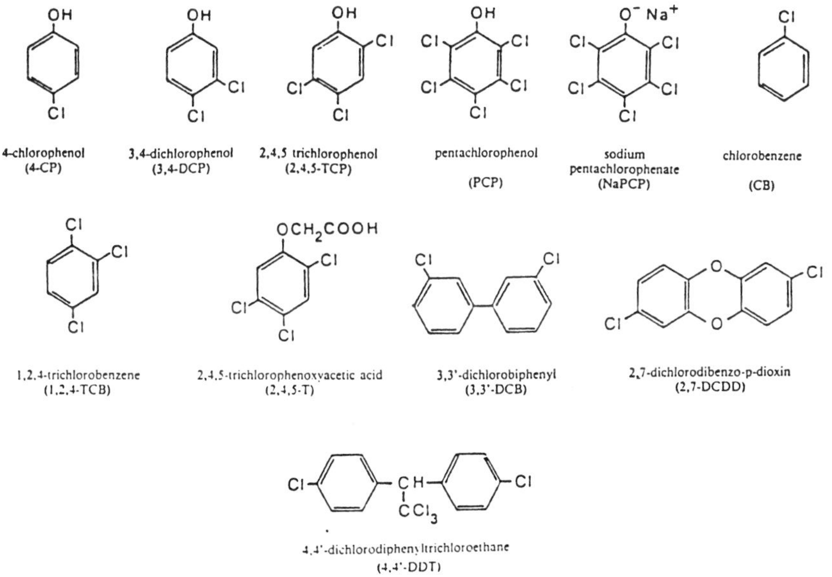

FIG. 10. Chlorinated aromatics investigated using heterogeneous photocatalytic methods [63,67].

A survey of different SC materials for the photodegradation of PCP is shown in Fig. 11. In the presence of particulates, PCP partitions between the aqueous bulk and the surface of the semiconductors. Illuminated TiO_2 and ZnO rapidly photodegrade (~2 hrs) PCP. Slower degradation rates are evident with

TABLE III - Half-lives for the total photodegradation of contaminants assisted by TiO_2 on exposure to simulated sunlight [63,67].

Contaminant	Concentration (ppm)	pH	$t_{1/2}$ (min)
4-CP	6	3.0	14
3,4-DCP	18	3.0	45
2,4,5-TCP	20	3.0	55
PCP	12	3.0	20
NaPCP	12	10.5	15
CB	45	2.5	90
1,2,4-TCB	10	3.0	24
2,4,5-T	32	3.0	40
4,4'-DDT	1	3.0	46
3,3'-DCB	1	3.0	10
2,7-DCDD	0.2	3.0	46

CdS, while the degradative process is drastically slower in WO_3 and SnO_2 slurries. Note that both O_2 and H_2O are essential in the degradation process; no mineralization occurs in the absence of one or both of these.

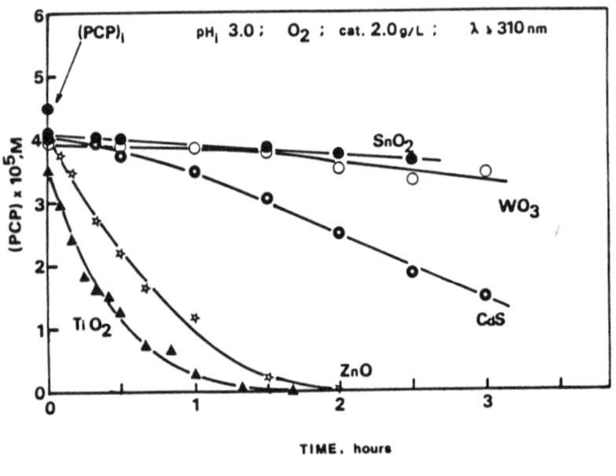

FIG. 11 - Photodegradation of PCP in the presence of various semiconductor dispersions [66].

Fig. 12 shows the results of an important experiment in which the PCP/TiO_2 aqueous slurry was exposed to natural sunlight. Total degradation is evidenced after ~80 min of sunlight exposure; $t_{1/2}$ ~ 8 min. While direct UV photolysis of chlorophenols (250-300 nm) proceeds via homolytic fission of the C-Cl bond [87] (dechlorination), the degradation with semiconductors occurs via oxidation by OH radicals. Also, direct photolysis is slow for PCP (100 hrs at pH 3.3 and 3.5 hrs at pH 7.3 [88]). Direct photolysis does not lead to mineralization to CO_2 and HCl; rather, the photolysis products are tetrachloro-catechol, -resorcinol, -hydroquinone, -phenols, trichlorophenols, and acyclic ketone [88].

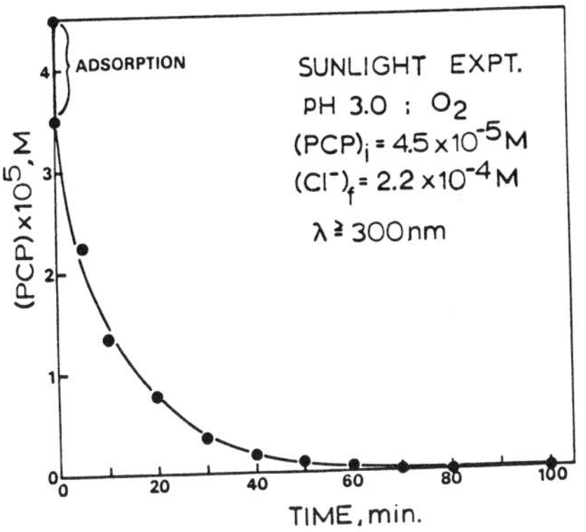

FIG. 12 - Photodegradation of PCP under natural solar light exposure (July 17/84) [66].

Surfactants

Surfactants pose severe ecological problems particularly in Japan, aggravated by the fact that their biodegradation is often too slow and too inefficient. For example, sodium dodecylbenzenesulfonate (DBS) requires 2 days to biodecompose; branched isomers of DBS are not biodegraded even after 1 week of exposure to bacteria [89]. Anionic [73,74], cationic [75], and non-ionic [76], surfactants can be photodegraded rapidly under solar or simulated sunlight exposure in the presence of aerated aqueous TiO_2 suspensions. Fig. 13 depicts a typical case for DBS [74].

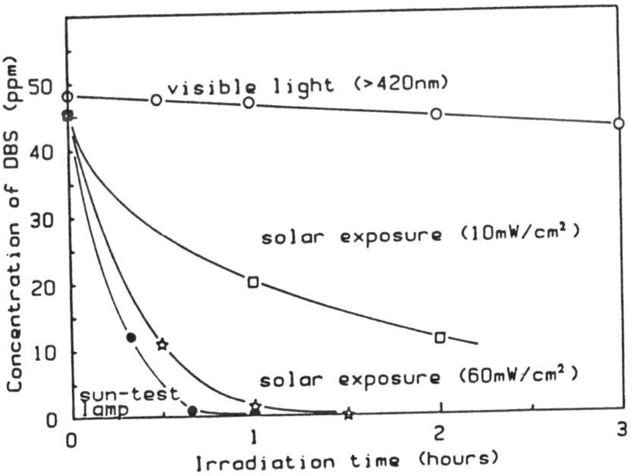

FIG. 13 - Photodegradation of DBS in aqueous TiO_2 dispersions under visible light (> 420 nm) exposure, natural sunlight exposure at two different light intensities, and under simulated sunlight exposure under a suntest lamp (118 mW cm^{-2}) [74].

In the dark (TiO_2 present) or after 7 hrs of light exposure (no TiO_2), the DBS surfactant is not decomposed. Slight degradation occurs upon light exposure for 4 days. By contrast, no traces of DBS are evident in TiO_2 slurries after ~1 hr of simulated sunlight exposure [74]. The OH radicals formed on TiO_2 oxidize the phenyl fragment of DBS rapidly (evidenced by nmr). The aliphatic chain is slower to decompose under these conditions; however, inasmuch as the postirradiated TiO_2 suspension no longer showed surface activity (foaming power), the remaining solution became environmentally acceptable.

Disposal of Trace Toxic Metals

In today's highly industrialized society, the environment is also charged with a number of toxic metals. These possess infinite lifetimes and are not subject to biodegradation, nor are they subject to chemical decontamination.

Noble metals can easily be reduced on irradiated TiO_2 particles (UV or simulated sunlight) and thus removed from solution. Application of this process to the selective recovery and refining of metals in wastes or in mixtures [77] was obvious as this might afford an efficient and facile route to concentrate such noble metals as Pt, Au, Pd, Rh and Ag from very diluted and dirty solutions. An equally important consequence was the potential preparation of single or mixed metal redox catalysts on TiO_2 [90,91], as well as the elimination of trace metal contaminants from waste waters {eg., Hg(II)} [78].

Fig. 14 summarizes the results of a typical experiment in which an air-equilibrated mixture consisting of 30 ppm gold(III), 13 ppm rhodium(III) and 45 ppm platinum(IV) was exposed to simulated sunlight in the presence of 2 g/L TiO_2 (pH 7.39) [77]. Pt(IV) is photoreduced in < 10 min, Au(III) within ~30 min, while Rh(III) remains in solution.

The photoreduction of Hg(II) in air-equilibrated TiO_2 aqueous suspensions (initial pH 4.65) occurs efficiently in < 20 min of irradiation with simulated sunlight (Fig. 15A); driving force ~ 0.73 V (NHE). Photoreduction of 5×10^{-4} M of Hg(II) yields the stoichiometric quantity of H^+ (1×10^{-3}M).

Methylmercury, CH_3HgCl, is not photoreduced under conditions of natural

$$Hg^{2+} + H_2O \xrightarrow[TiO_2]{h\nu} 2H^+ + 1/2\ O_2 + Hg(0) \qquad (21)$$

pH in a TiO_2 aqueous suspension. However, in methanolic aqueous media (Fig. 15B), where methanol acts as a hole scavenger, photoreduction (eqn. 22) and therefore disposal occurs readily: ~ 30 min of simulated sunlight illumination.

$$CH_3HgCl + H_2O \xrightarrow[TiO_2]{h\nu} H^+ + 1/2\ O_2 + Cl^- + Hg^o + CH_4 \qquad (22)$$

CONCLUDING REMARKS

A potential does exist for the exploitation of heterogeneous photocatalysis in the removal of various environmental contaminants such as H_2S, SO_2, cyanides, chlorinated hydrocarbons, and toxic metals. The technique has utilized either simulated sunlight or natural sunlight as the source of photons, along with semiconductor particulates (e.g., TiO_2 anatase). With the exception of studies on the selective recovery of noble metals from industrial mix-

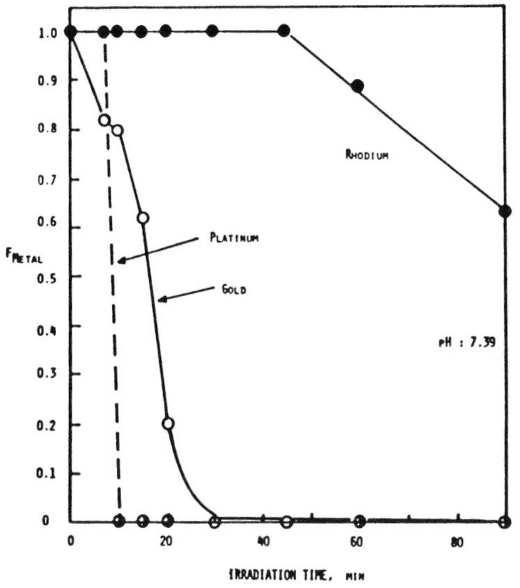

FIG. 14 - Photochemical separation of Rh(III) from Au(III) and Pt(IV) on TiO_2 in an aerated aqueous suspension under AM1 simulated sunlight exposure [77].

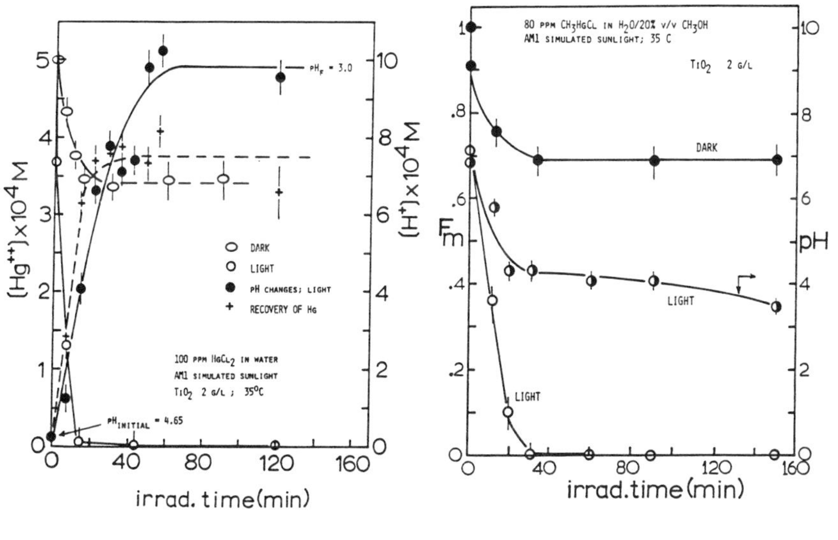

(A) (B)

FIG. 15 - Photoreduction on AM1 simulated sunlight irradiated TiO_2 aqueous suspensions of (A) 100 ppm of Hg(II), and (B) 80 ppm of methylmercury (20% vol.% methanol/water) [77].

tures [64], photodegradation of chlorinated aromatics has thus far been carried out on single compounds. But the real world consists of mixtures of pollutants (organics and inorganics). We continue to investigate the photodegradation processes [92] to understand the factors (kinetic, thermodynamic, mechanistic) underlying the decomposition of components in a multi-component mixture that might mimic the real world situation [93].

The real challenge now will be to demonstrate the economic feasibility as well as the technical feasibility of heterogeneous photocatalysis to the treatment of wastes [86,94,95]. Scaled-up pilot plant operation of the photocleavage of H_2S (SO_2) is being carried out in Spain [96].

ACKNOWLEDGEMENTS

It is a pleasure to acknowledge past and present collaborators, particularly Profs. E. Pelizzetti (Italy), H. Hidaka (Japan), M. Gratzel (Switzerland), M.A. Fox (USA), and Dr. P. Pichat (France) for many useful discussions. Not least, we are indebted to Dr. E. Borgarello (presently at EniRicerche, Milan) and Dr. M. Barbeni (Torino) without whose experimental skills and painstaking work in heterogeneous photocatalysis much of this work would not have materialized. Our research is supported by N.S.E.R.C. of Canada, and by a NATO grant (No. 843/84; Serpone, Pelizzetti, Pichat, Fox).

REFERENCES

1. V. Balzani, L. Moggi, M.F. Manfrin, F. Bolletta, and M. Gleria, Science, 189, 852 (1975).
2. M. Calvin, Photochem.Photobiol., 23, 425 (1976).
3. G. Porter and M.D. Archer, Interdisc.Sci.Rev., 1, 119 (1976).
4. S.N. Paleocrassas, Solar Energy, 16, 45 (1974).
5. J.L. Lambert and J.V. Paukstelis, "Project Completion Report for Period July 1, 1974 to June 30, 1976 on the Photochemical Generation of Hydrogen from Water", Kansas Water Resources Research Institute, Kansas State University, Manhattan, Kansas, November 1977.
6. S. Claesson, Ed., "Photochemical Conversion and Storage of Solar Energy", Swedish Energy Board Report, Stockholm, 1977.
7. See e.g., (a) J.R. Bolton, Ed., "Solar Power and Fuels", Academic Press, New York, 1977; (b) D.O. Hall, Fuel, 57, 322 (1978); (c) J.R. Bolton and D.O. Hall, Annu.Rev.Energy, 4, 353 (1979); (d) M. Kirch, J.M. Lehn, and J.P. Sauvage, Helv.Chim. Acta, 62, 1345 (1979).
8. See e.g., (a) J.S. Connolly, Ed., "Photochemical Conversion and Storage of Solar Energy", Academic Press, New York, 1981; (b) J. Rabani, Ed., Israel J.Chem., 22, No.2 (1982); (c) M. Gratzel, Ed., "Energy Resources through Photochemistry and Catalysis", Academic Press, New York, 1983.
9. This Figure was obtained from a computer search of Chemical Abstracts from 1970 to January 1988 using as keywords: conversion and storage of solar energy.
10. M.A. Fox and M. Chanon, Eds., "Photoinduced Electron Transfer", Elsevier, Amsterdam, The Netherlands, 1988.
11. K.I. Zamaraev and V.N. Parmon, in reference 8c, p. 123.
12. V.N. Parmon and K.I. Zamaraev, in "Photocatalysis", N. Serpone and E. Pelizzetti, Eds., Wiley-Interscience, New York, 1989 (in press).
13. A. Fujishima and K, Honda, Nature (London), 238, 37 (1972).
14. A.J. Bard, Science, 207, 139 (1980).
15. M. Gratzel, Accts.Chem.Res., 14, 376 (1981).
16. M. Gratzel, Discuss.Faraday Soc., No.70, p.359 (1981).
17. A.J. Nozik, Discuss.Faraday Soc., No.70, p.7 (1981).
18. T.C. Dzhabiev and A.E. Shilov, Zh.Khim.Ova., No.5, p.503 (1980).
19. M. Gratzel, Ber.Bunsenges.Phys.Chem., 84, 981 (1980).

20. A.W. Maverick and H.B. Gray, Pure Appl.Chem., 52, 2339 (1980).
21. A. Moradpour, Actual.Chim., No.2, p.7 (1980).
22. Kh.S. Bagdasar'yan, Khim.Fiz., 1, 391 (1982).
23. K.I. Zamaraev and V.N. Parmon, Catal.Rev., 22, 283 (1980).
24. V. Balzani, F. Bolletta, M.T. Gandolfi, and M. Maestri, Topics Curr.Chem., 75, 1 (1978).
25. K. Kalyanasundaram, Coord.Chem.Rev., 46, 159 (1982).
26. K.I. Zamaraev and V.N. Parmon, Catal.Rev., 22, 261 (1980).
27. J. Kiwi, K. Kalyanasundaram, and M. Gratzel, Struct.& Bonding, 49, 37 (1982).
28. K. Kalyanasundaram, Chem.Soc.Rev., 7, 453 (1978).
29. N.J. Turro, A.M. Braun, and M. Gratzel,Angew.Chem.Int.Edn.Engl., 19, 675 (1980).
30. M. Gratzel, Pure Appl.Chem., 54, 2369 (1982).
31. K. Kalyanasundaram, in reference 8c, p. 71.
32. A. Harriman, Platinum Metal Rev., 27, 102 (1983).
33. E. Amouyal and B. Zidler, Israel J.Chem., 22, 117 (1982).
34. M. Gratzel, in "Modern Aspects of Electrochemistry", R.E. White, J.O'M. Bockris, and B.E. Conway, Eds., Plenum Press, New York, 1983, p. 83.
35. (a) A. Juris, V. Balzani, P. Besler, and A. von Zelewsky, Helv.Chim. Acta, 64, 2175 (1981); (b) F. Barigelletti, A. Juris, V. Balzani, P. Besler, and A. von Zelewsky, Inorg.Chem., 22, 3335 (1983); (c) V. Balzani, A. Juris, F. Barigelletti, P. Besler, and A. von Zelewsky, Sci.Papers Inst.Phys.Chem.Res., 78, 78 (1984).
36. (a) F. Pina, M. Ciano, Q.G. Mulazzani, M. Venturi, V. Balzani, and L. Moggi, Sci.Papers Inst.Phys.Chem.Res., 78, 166 (1984); (b) F. Pina, M. Ciano, L. Moggi, and V. Balzani, Inorg.Chem., 24, 844 (1985); (c) F. Pina, Q.G. Mulazzani, M. Venturi, M. Ciano, and V. Balzani, Inorg.Chem., 24, 848 (1985).
37. V. Balzani, A. Juris, and F. Scandola, in "Homogeneous and Heterogeneous Photocatalysis", E. Pelizzetti and N. Serpone, Eds., NATO ASI Series C174, D. Reidel Publ.Co., Dordrecht, The Netherlands, 1986, p. 1.
38. V. Balzani, L. Moggi, and F. Scandola, in "Supramolecular Photochemistry", V. Balzani, Ed., NATO ASI Series C214, D. Reidel Publ.Co., Dordrecht, The Netherlands, 1987, p. 1.
39. J.F. Reber, in "Photoelectrochemistry, Photocatalysis and Photoreactors", M. Schiavello, Ed., NATO ASI Series C146, D.Reidel Publ.Co., Dordrecht, The Netherlands, 1985, p. 321.
40. T.N. Veziroglu, W.D. VanVorst, and J.H. Kelly, Eds., "Hydrogen Energy Progress IV", Pergamon Press, New York, 1982.
41. T.N. Veziroglu and A. Taylor, Eds., "Hydrogen Energy Progress V", Pergamon Press, New York, 1984.
42. E. Pelizzetti, Energia e Materie Prime, 17, 31 (1981).
43. R.W. Bartlet, D. Cubicciotti, D.H. Hildebrand, D.D. McDonald, K. Jemran, and M.E.D. Raymont "Preliminary Evaluation of Processes for Recovering Hydrogen from Hydrogen Sulfide", Final Report, SRI Project No. 8030, J.P.L. Contract No. 955272, SRI International, Menlo Park, California, 1981.
44. E. Borgarello and M. Gratzel, in reference 40, p. 739.
45. E. Borgarello, W. Erbs, M. Gratzel, and E. Pelizzetti, Nouv.J.Chim., 7, 195 (1983).
46. D.H.M.W. Thewissen, K. Timmer, M. Eemwhorst-Reinten, A.H.A. Tinnemans, and A. Mackor, Nouv.J.Chim., 7, 191 (1983).
47. (a) N. Buhler, K. Meier, and J.F. Reber, J.Phys.Chem., 88, 3261 (1984); (b) J.F. Reber and K. Meier, J.Phys.Chem., 88, 5903 (1984).
48. N. Serpone, E. Borgarello, and M. Gratzel, J.Chem.Soc.Chem.Commun., 342 (1984).
49. M. Barbeni, E. Pelizzetti, E. Borgarello, N. Serpone, M. Gratzel, L. Balducci, and M. Visca, Int.J.Hydrogen Energy, 10, 249 (1985).
50. E. Borgarello, N. Serpone, M. Gratzel, and E. Pelizzetti, Inorg.Chim. Acta, 112, 197 (1986).

51. E. Borgarello, N. Serpone, E. Pelizzetti, and M. Barbeni, J.Photochem., 33, 35 (1986).
52. N. Serpone, E. Borgarello, E. Pelizzetti, and M. Barbeni,Chim.Ind. (Milan), 67, 318 (1985).
53. E. Borgarello, J. DeSilvestro, M. Gratzel, and E. Pelizzetti, Helv.Chim. Acta, 66, 1827 (1983).
54. H. Gerischer and M. Lubke, J.Electroanal.Chem., 204, 225 (1986).
55. N. Serpone, P. Pichat, J.-M. Herrmann, and E. Pelizzetti, in "Supramolecular Photochemistry", V. Balzani, Ed., NATO ASI Series C214, D.Reidel Publ.Co., Dordrecht, The Netherlands, 1987, p. 415.
56. N. Serpone, E. Borgarello, and E. Pelizzetti, J.Electrochem.Soc., submitted.
57. P. Pichat, E. Borgarello, J. Disdier, J.-M. Herrmann, E. Pelizzetti, and N. Serpone, J.Chem.Soc.Faraday Trans.I., 84, 261 (1988).
58. L. Spanhel, H. Weller, and A. Henglein, J.Am.Chem.Soc., submitted.
59. E. Borgarello, R. Harris, and N. Serpone, Nouv.J.Chim., 9, 743 (1985).
60. E. Borgarello, N. Serpone, P. Liska, W. Erbs, and M. Gratzel, Gazz.Chim. Ital., 115, 599 (1985).
61. N. Serpone, E. Borgarello, R. Harris, P. Cahill, M. Borgarello, and E. Pelizzetti, Sol.Energy Mater., 14, 121 (1986).
62. E. Borgarello, R. Terzian, N. Serpone, E. Pelizzetti, and M. Barbeni, Inorg.Chem., 25, 2135 (1986).
63. E. Borgarello, N. Serpone, M. Barbeni, C. Minero, E. Pelizzetti, and E. Pramauro, Chim.Ind.(Milan), 68, 53 (1986).
64. N. Serpone, E. Borgarello, M. Barbeni, E. Pelizzetti, P. Pichat, J.-M. Herrmann, and M.A. Fox, J.Photochem., 36, 373 (1987).
65. M. Barbeni, E. Pramauro, E. Pelizzetti, E. Borgarello, M. Gratzel, and N. Serpone, Nouv.J.Chim., 8, 547 (1984).
66. M. Barbeni, E. Pelizzetti, E. Borgarello, and N. Serpone, Chemosphere, 14, 195 (1985).
67. E. Pelizzetti, M. Barbeni, E. Pramauro, N. Serpone, E. Borgarello, M.A. Jamieson, and H. Hidaka, Chim.Ind.(Milan), 67, 623 (1985).
68. M. Barbeni, E. Pramauro, E. Pelizzetti, E. Borgarello, N. Serpone, and M.A. Jamieson, Chemosphere, 15, 1913 (1986).
69. M. Barbeni, E. Pramauro, E. Pelizzetti, M. Vincenti, E. Borgarello, and N. Serpone, Chemosphere, 16, 1165 (1987).
70. M. Barbeni, E. Pelizzetti, E. Borgarello, and N. Serpone, Chemosphere, in press (1988).
71. E. Pelizzetti, M. Borgarello, C. Minero, E. Pramauro, E. Borgarello, and N. Serpone, Chemosphere, in press (1988).
72. R. Borello, C. Minero, E. Pramauro, E. Pelizzetti, N. Serpone, and H. Hidaka, Water Res., submitted.
73. H. Hidaka, H. Kubota, M. Gratzel, N. Serpone, and E. Pelizzetti, Nouv.J. Chim., 9, 67 (1985).
74. H. Hidaka, H. Kubota, M. Gratzel, E. Pelizzetti, and N. Serpone, J.Photochem., 35, 219 (1986).
75. H. Hidaka, Y. Fujita, K. Ihara, S. Yamada, K. Suzuki, N. Serpone, and E. Pelizzetti, J.Jpn.Oil Chem.Soc., 36, 836 (1987).
76. H. Hidaka, K. Ihara, Y. Fujita, S. Yamada, E. Pelizzetti, and N.Serpone, J.Photochem.Photobiol., A, in press (1988).
77. E. Borgarello, N. Serpone, G. Emo, R. Harris, E. Pelizzetti, and C. Minero, Inorg.Chem., 25, 4499 (1986).
78. N. Serpone, Y.K. Ah-You, T.P. Tran, R. Harris, E. Pelizzetti, and H. Hidaka, Sol.Energy, 39, 491 (1987).
79. (a) S.P. Tucker and G.A. Larson, Environ.Sci.Technol., 19, 215 (1985); (b) A.D. Mehrkam, Met.Prog., 108, 103 (1975); (c) K.F. Cherry, "Plating Waste Treatment", Ann Arbor Science, Ann Arbor, Michigan, 1982.
80. E. Schulek, Z.Anal.Chem., 65, 352 (1925).
81. N. Serpone, E. Borgarello, M. Barbeni, and E. Pelizzetti, Inorg.Chim. Acta, 90, 191 (1984).

82. S.N. Frank and A.J. Bard, J.Am.Chem.Soc., 99, 203 (1977); J.Phys.Chem., 81, 1484 (1977).
83. H.E. Christensen and E.J. Fairchild, Eds., "Registry of Toxic Effects of Chemical Substances", US Department of Health, Education and Welfare, Rockville, Maryland, June 1976.
84. (a) D.Y. Lai, J.Environ.Sci.Health, C2, 135 (1984); (b) C.R. Pearson, in "Handbook of Environmental Chemistry", O. Hutzinger, Ed., Springer-Verlag, Berlin, 1982, vol.3, part B, pp. 86-116; (c) Y.R. Harvey, W.Y. Steinhauer, and J.M. Teal, Science, 180, 643 (1973).
85. (a) C. Rappe, in "Handbook of Environmental Chemistry", O. Hutzinger, Ed., Springer-Verlag, Berlin, 1980, vol.3, part A, pp. 157-180; (b) R.A. Bailey, H.M. Clarke, J.P. Ferris, S. Krause, and B.L. Strong, "Chemistry of the Environment", Academic Press, New York, 1978; (c) J. Josephson, Environ. Sci.Technol., 17, 124A (1983); (d) P.A. Jones, "Chlorophenols and their Impurities in the Canadian Environment", Report EPS 3-EC-81-2, Environmental Impact Control Directorate, Environment Canada, March 1981, and see also Supplement, Report EPS 3-EP-84-3, March 1984.
86. D.F. Ollis, Environ.Sci.Technol., 19, 480 (1985).
87. J.H. Carey, J. Lawrence, and H.M. Tosine, Bull.Environ.Contam.Toxicol., 16, 697 (1976).
88. A.S. Wong and D.G. Crosby, J.Agric.Food Chem., 29, 125 (1981).
89. R.D. Swisher, J.Am.Oil Chem.Soc., 40, 648 (1963); J.Water Poll.Control. Fed., 35, 877, 1537 (1963).
90. B. Krautler and A.J. Bard, J.Am.Chem.Soc., 100, 4317 (1978).
91. E. Borgarello, J. Kiwi, E. Pelizzetti, M. Visca, and M. Gratzel, J.Am.Chem.Soc., 103, 6324 (1981).
92. H. Al-Ekabi and N. Serpone, J.Phys.Chem., submitted.
93. N. Serpone, E. Pelizzetti, M.A. Fox, and coworkers, manuscript in preparation.
94. D.F. Ollis, J.Catal., 97, 569 (1986).
95. D.F. Ollis, in "New Trends and Applications of Photocatalysis and Photoelectrochemistry for Environmental Problems", M. Schiavello, Ed., D.Reidel Publ.Co., Dordrecht, The Netherlands, 1988.
96. M. Gratzel, personal communication (1987).

CATALYSIS IN CONVERSION AND STORAGE OF SOLAR ENERGY: PHOTOCATALYSIS VERSUS THERMOCATALYSIS. ADVANCES AND UNSOLVED PROBLEMS

VALENTIN N. PARMON AND KIRILL I. ZAMARAEV
Institute of Catalysis, Prospekt Akad. Lavrentieva 5, Novosibirsk 630090, U.S.S.R.

INTRODUCTION

In 1970-1980 one could witness the rise of the unprecedeur interest towards research aimed at the development of chemical methods of solar energy conversion. This interest resulted in the appearance of quite promising concepts and approaches in this unique and very intriguing area of research, which lies at the junction of photochemistry, photobiology and catalysis. Simultaneously one could also witness the birth and failure of some scientific sensations, which nonetheless, gave each time much of the impetus for the studies in this area. As the result joint efforts of the team of researchers from various countries led to fast accumulation of positive knowledge which can be regarded as a reliable basis for current and future investigations.

By the moment, the world has calmed down after the shock of the energy crisis of the 1970's, which stimulated great interest towards analyzing the potentialities of nontraditional energetics, and especially of a large-scale solar energetics. Thus, time has come for a "cold mind" reasoning of advances and problems left unsolved in the field of direct chemical methods of solar energy conversion. This will make us better prepared to the expected second wave of interest towards solar energetics in the coming future. This time the wave of interest won't be, probably, caused by the threat of energy crisis; rather, it may come from a critical situation in ecology. In this connection two factors should be particularly mentioned: (i) the nowadays widely discussed danger of irreversible global climate changes due to overfilling of the atmosphere with carbon dioxide, that may lead to the necessity to reduce consumption of hydrocarbon fossil fuels, and (ii) an evidently shaken confidence in the reliability and safety of nuclear power plants.

In this article we shall discuss the present state of art in the field of direct conversion of solar energy into the chemical energy of fuels via endoenergetic catalytic processes. Two different types of catalytic processes for direct solar-to-chemical energy conversion will be discussed: (i) <u>photocatalytic</u> ones that are based on photochemical reactions in which solar <u>light quanta</u> serve as energy-rich reagents providing accomplishment of otherwise thermodynamically forbidden catalytic processes, (ii) <u>thermocatalytic</u> ones, that are based on endothermic catalytic reactions proceeding upon <u>heating</u> the reagents with solar light. We shall not discuss here photoelectrochemical and photoelectrocatalytic systems for solar energy conversion, since these are not "pure catalytic" in the strict sence of this term (one of serious distinguishing features of any electrochemical systems is the existence of a visualized electrical circuit for electron transfer).

Photochemical Energy Conversion
James R. Norris, Jr. and Dan Meisel, Editors

THE STATE OF ART IN DEVELOPING PHOTOCATALYTIC SYSTEMS FOR SOLAR ENERGY CONVERSION

At early stages of research the photocatalytic method of solar-to-chemical energy conversion seems to have attracted much more attention than the thermocatalytic one. The main reason for this seems to be the well known fact that photocatalytic method is successfully used by living nature to store solar energy via photosynthesis in plants and bacteria. Indeed, photosynthesis is known to provide a high efficiency of photocatalytic conversion and storage of solar energy in the form of energy-rich organic chemical substances (fuels), usually of carbohydrate type.

In Table I some endoenergetic chemical processes are listed that initially have been considered as most attractive candidates for accomplishing photocatalytic solar-to-chemical energy conversion.

TABLE I. Some redox reactions involving water and atmospheric gases that can be used for photocatalytic conversion of solar energy

Reactions	ΔG^o_{298} [a)], eV	λ_o [b)], nm
$H_2O \rightarrow H_2+1/2O_2$	1.23	1008
$H_2O+CO_2 \rightarrow HCOOH+1/2O_2$	1.48	836
$H_2O+CO_2 \rightarrow H_2+CO+O_2$	1.28	969
$H_2O+CO_2 \rightarrow CH_2O+O_2$	1.35	916
$2H_2O+CO_2 \rightarrow CH_3OH+3/2O_2$	1.21	1025
$2H_2O+CO_2 \rightarrow CH_4+2O_2$	1.05	1176
$3H_2O+2CO_2 \rightarrow C_2H_5OH+3O_2$	1.15	1077
$H_2O+CO_2 \rightarrow 1/6C_6H_{12}O_6+O_2$	1.24	997
$3/2H_2O+1/2N_2 \rightarrow NH_3+3/4O_2$	1.17	1059
$2H_2O+N_2 \rightarrow N_2H_4+O_2$	1.97	629

a) the difference in redox potentials of the pair of products formed. Data are given for liquid H_2O, HCOOH, CH_3OH, C_2H_5OH, N_2H_4, and solid $C_6H_{12}O_6$ (sugar), the other compounds are considered to be gases

b) λ_o is the threshold wavelength for the photocatalytic pathway of the reaction

These processes mimic function of plant photosynthesis in a sense that they allow one to produce energy-rich compounds (chemical fuel) from cheap, easily available raw-materials such as water and atmospheric gases. Note that similarly to photosynthesis in plants, dioxygen is also produced in all these reactions.

Later, it became evident that practical accomplishing via the photocatalytic pathway of even the simplest process among those listed in Table I, namely cleavage of water into dihydrogen and dioxygen upon illumination with visible light was a rather difficult task demanding long enough work.

A somewhat more easy task occurred to be the development of artificial photocatalytic systems mimicing photosynthesis in bacteria. In contrast to plants, bacteria do not evolve dioxygen in the process of photosynthesis. This means that in bacterial photosynthesis reduction of carbon dioxide to carbohydrates is coupled with oxidation of a more easily oxidized substance than water. Thus the most difficult for modeling step of dioxygen evolution (see below) does not need to be present in photocatalytic systems mimicing photosynthesis in bacteria. As a result, by the moment rather promising developments have already been made in this area.

In order to model the energy storing function of photosynthesis both in bacteria and in plants one has to design artificial photocatalytic systems capable of providing three key reaction steps that are intrinsic for natural photosynthesis:

(1) photochemical charge separation assisted by a specific substance -- photocatalyst, PhC, which can absorb visible light and then participate in reversible chemical reactions producing strong reductant and oxidant; the latter compounds serve as starting reagents for further steps of transformations eventually leading to production of a pair of energy-rich substances, i.e. a chemical fuel;

(2) production of the reductive component of the fuel with the aid of the photogenerated reductant;

(3) production of the oxidative component of the fuel with the aid of the photogenerated oxidant. When modeling the function of photosynthesis in plants, the oxidant has to oxidize water to dioxygen, while for mimicing the function of bacterial photosynthesis compounds with stronger electron-donating properties than water are to be oxidized.

It should be noted that light energy is stored only at step (1), while at subsequent steps (2) and (3) the stored energy is partly dissipated.

In case of most successful accomplishing of this three-step scheme for carrying out the processes listed in Table I one can expect to achieve ca. 30% efficiency of conversion of solar energy into Gibbs' energy of the fuels produced. This is the theoretical maximum for any single-step quantum method of solar energy conversion and is twice as high as the theoretical maximum for photosynthesis in plants (see, e.g., [1,2]).

For processes mimicing bacterial photosynthetizing function one can find in the literature two approaches towards calculating the efficiency of solar-to-chemical energy conversion. One of them is the same as for processes mimicing photosynthesis in plants, i.e. the ratio of the stored Gibbs' energy (or enthalpy) to the energy of the absorbed light quanta is taken as the efficiency. This is the efficiency in its traditional meaning.

The second approach consists in calculating the so-called "commercial efficiency" of solar-to-chemical energy conversion [1]. The reason for using "commercial" rather than traditional efficiency is as follows: for processes mimicing bacterial photosynthesis, the oxidative component of the fuel produced with light is not dioxygen. Thus it can be burnt along with the reductive component in the atmosphere of dioxygen. The energy released upon combustion of these two components often exceeds substantially the light energy stored in the photocatalytic process. However the released energy is just the energy in which its consumer may be interested in. Consequently "commercial efficiency" could be defined as the ratio of the Gibbs' energy (or enthalpy) released upon combustion of either both reductive and oxidative components of the fuel or only one of these components to the energy of the absorbed light quanta.

For systems mimicing photosynthesis in plants "commercial efficiency" coincides with the traditional one. The value of "commercial efficiency" for systems mimicing bacterial photosynthesis can in principle exceed 100%.

Thus the systems mimicing bacterial photosynthesis appear to be also of practical interest for solar energetics, despite the fact that, because of smaller values of the Gibbs' energy stored in the photocatalytic process, the traditional efficiencies for such systems are usually expected to be notably smaller than those for systems listed in Table I.

Below the results are presented that have been obtained at the Institute of Catalysis (Novosibirsk) in developing photocatalytic systems that mimic both plant and bacterial photosynthesis. From these data as well as from those obtained by other groups working in this area [3-5], one can see the present state and estimate the perspectives of the photocatalytic method of solar-to-chemical energy conversion.

Cleavage of water in molecular photocatalytic systems. Among the processes listed in Table I, this is the photocatalytic cleavage of water into dihydrogen and dioxygen that attracts the greatest attention today. The development of the photocatalytic systems for water cleavage involves the development of catalysts for three types of processes: photocatalysts, PhC, for photochemical charge separation

$$A + D \xrightarrow[PhC]{h\nu} A^- + D^+ , \qquad (1)$$

which leads to formation of starting reagents for water cleavage, namely strong reductant A^- and oxidant D^+, and catalysts for subsequent multielectron reduction of water to H_2 by the reductant:

$$2A^- + 2H_2O \xrightarrow[cat_{H_2}]{} 2A + H_2 + 2OH^- \qquad (2)$$

and oxidation of water to O_2 by the oxidant:

$$4D^+ + 2H_2O \xrightarrow[cat_{O_2}]{} 4D + O_2 + 4H^+ \qquad (3)$$

This scheme is similar to the sequence of processes leading to light energy storage during photosynthesis in plants (Fig. 1).

Depending on chemical nature of PhC, photocatalytic systems for solar energy conversion are commonly divided into two classes: semiconductor and molecular systems.

Here we discuss only molecular photocatalytic systems for water cleavage. Note that such systems seem to mimic plant photosynthesis more closely than semiconductor systems.

A detailed analysis of the state of art with developing semiconductor photocatalytic systems for water cleavage could be found in other articles of this book, as well as in books [3-5]. It is important to note that the systems based on suspended semiconductors have been already developed, which allowed one to accomplish photocatalytic cleavage of water into dihydrogen and dioxygen though still with rather low quantum yields and, as a rule, only under the action of UV-light. A somewhat higher photocatalytic activity has been achieved for the massive bicomponent (so-called "photodiode") semiconductor structures, but it is also still notably lower than that of the best photoelectrocatalytic systems of water cleavage that demonstrate an efficiency of upto 8.2% for the conversion of solar energy to chemical energy of $H_2+1/2O_2$ mixture (see, e.g., [3-6]).

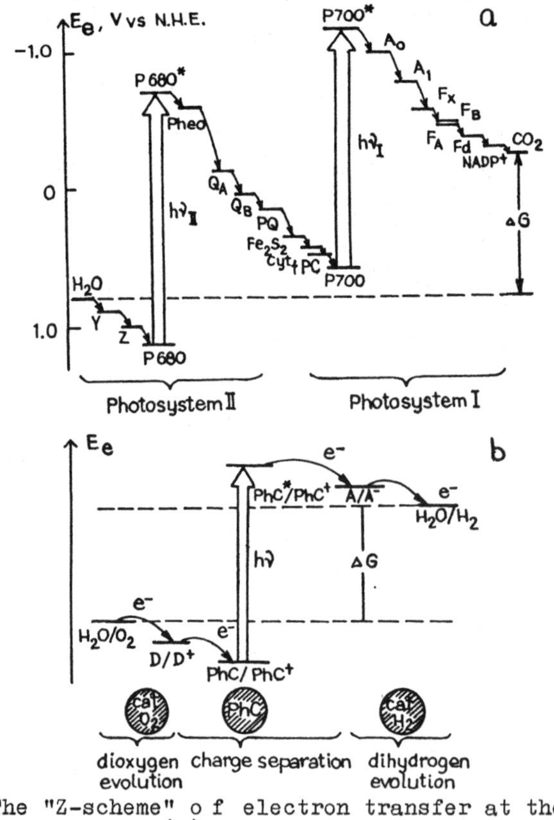

FIG. 1. The "Z-scheme" of electron transfer at the green plant photosynthesis (a) and the scheme of the process of water cleavage in a molecular photocatalytic system (b). Here and below E_e denotes the energy of the transferred electron; hν is the energy of the absorbed photon; ΔG is the net stored chemical energy. The elements of the electron-transfer chain are marked as usually in papers on plant photosynthesis, symbols P680, Pheo, P700, A_o denoting chlorophyll-like molecules. The curved arrows show paths for electron transfer [1].

Step of dihydrogen evolution from water. Numerous catalysts for this step are currently available (see review-articles [1,7-9] and references therein). Good catalysts for reaction (2) are colloidal and finely dispersed metals, both noble (Pt, Rh, Pd, Au, Ag, etc.) and non-noble such as Cu. Finely dispersed Ni, Co and Mo show some catalytic activity as well. Of great interest are heterogeneous catalysts in which small metallic particles are attached to polymers or inorganic supports. Enzymes-hydrogenases, that provide dihydrogen evolution in metabolism of some microorganisms and are most active among available catalysts for reaction (2) (see below), have also long been known. Recently, a broad class of homogeneous catalysts for dihydrogen evolution from water has been found. These are iso- and heteropolycompounds of tungsten and molybdenum. There is also some evidence that homogeneous catalysis of reaction (2) can be provided by complexes of cobalt with macrocyclic ligands.

Methods of preparation of many catalysts for dihydrogen evolution from water are quite simple. For example, the most comprehensively studied heterogeneous catalyst based on finely dispersed metallic rhodium attached to polymers containing amine groups, is obtained by soft reduction of amine type polymeric complexes of Rh(III) in a water medium under the action of V(II) and Cr(II) compounds. The role of the active component of such catalysts is provided by finely dispersed (ca. 20 Å in size) Rh particles that are chemically bound with a polymeric support.

An important advantage of catalysts on polymeric supports is their stability in aqueous solutions over a wide range of pH variations, as well as the possibility of preparing catalysts in the form of granules and films (transparent films including) with controllable geometric parameters.

For most active and stable rhodium catalysts based on some granulated anionites, the activity per unit total catalyst mass can exceed 0.5 mol H_2/g cat. • h and is restricted by diffusion processes within the catalyst granules. For hydrogenases the activity ranges from 0.2 to 200 mol H_2/g enz. • h. While being inferior to the best hydrogenases in activity per unit mass, the Rh-polymer catalysts are far more stable than the hydrogenases, especially in acidic media. Besides, activated rhodium-polymer catalysts can be stored quite long (up to one year or longer) in the air without any loss in activity.

The activity of polymeric catalysts containing non-noble metals (e.g., Cu) can be close to that of catalysts based on noble metals. This latter fact may happen to be important in view of the future possible application of metal-polymeric catalysts in practical photocatalytic systems for dihydrogen evolution from water. The specificity of the metal nature for the catalysts of such type reveals itself mainly in the fact that the pH-region of high catalytic activity is shifted towards more acidic region for non-noble metals compared to noble ones.

Catalysts based on iso- or heteropolycompounds contain several (up to 12) cations of metals of variable valency (W, Mo, etc.) which creates favourable conditions for them to participate in multielectron redox transformations. In this sence these catalysts can be considered as some analogues of such reaction sites of enzymes-hydrogenases as multinuclear iron-nickel-sulfur clusters, that are also capable of multielectron redox transformations. The catalytic activity towards reaction (2) is due to the fact that deeply reduced forms of iso- and heteropolycompounds can spontaneously evolve dihydrogen from acidic water solutions, the oxidized forms being easily reduced back by various reductants A^-.

Step of dioxygen evolution from water. In contrast to catalysts for dihydrogen evolution from water, many of which had been known long before the research on photocatalytic water cleavage was initiated, the first artificial (i.e. non-biological) catalysts for water oxidation to dioxygen were found only in the end of the 1970's. Those were heterogeneous catalysts-oxides of some transition metals, the most interesting being RuO_2, as well as homogeneous and microheterogeneous catalysts based on compounds of some transition metals, iron, cobalt, copper and some other (see review-articles [1, 10,17] and references therein).

One of the most important parameters of the artificial catalysts of reaction (3) is the selectivity of their action, in other words, their ability to direct the strong oxidant D^+ toward reaction (3) but not toward other possible side processes of oxidation, e.g. of organic compounds, which can be present in the solution, or even of the ligands of the complex-oxidant itself.

Thermodynamic and photochemical properties of ruthenium complexes with ligands of bipyridyl (bipy) type, allow one to consider these complexes as promising candidates for the role of water cleavage photocatalysts, whose oxidized form could serve further as particle D^+ in reaction (3) (see below). Therefore, the main part of researches on the development of catalysts for reaction (3) has been performed using these complexes as D^+.

With $Ru^{III}(bipy)_3$ used as oxidant D^+, the activity and selectivity of both homogeneous and heterogeneous catalysts for reaction (3) were found to depend not only upon the nature of catalysts, but also upon their concentration and pH of the medium. A high, nearly 100% selectivity of water oxidation to dioxygen, appeared to be achieved only in exceptional cases; as a rule the selectivity of the catalyst action in using D^+ to obtain O_2 does not exceed 70-90%. The remainder D^+ particles provide an oxidation of organic compounds present in the solution, including bipyridyl ligands of the oxidant itself.

The data obtained so far suggest that at least for cobalt and iron compounds the active forms of catalysts appear to be oligonuclear hydroxide species.

The most active and selective heterogeneous catalysts of water oxidation to dioxygen have been obtained in recent years by anchoring very small (ca. 10 Å in size) colloidal particles of freshly formed Co(III) hydroxide to inert supports (oxides or hydroxides) [12,13].

Note that heterogeneous catalysts, which are almost as highly selective as the best available homogeneous or microheterogeneous catalysts, are, as a rule, far more stable.

The most serious and still existing difficulty in the development of catalysts for water oxidation is their marked specificity with respect to the oxidant D^+. For example, despite that all of the above-mentioned catalysts were found to be quite active with respect to $D^+ = Ru^{III}(bipy)_3$ (which is an oxidized form of a very promising photocatalyst $Ru^{II}(bipy)_3$) as well as to analogs of this complex, there is little evidence for the possibility of applying the same catalysts in systems with other potential classes of photocatalysts, e.g. metal porphyrines.

Another shortcoming of the existing catalytic systems for dioxygen evolution is that even for $D^+ = Ru^{III}(bipy)_3$ the problem of D^+ degradation (via oxidation of the bipy ligand) in the presence of the catalysts for O_2 evolution, has not as yet been completely overcome.

Step of the photocatalytic charge separation. Photoseparation of charges in structurally organized molecular assemblies. The main problem in the development of the photocatalytic systems for water cleavage is how to suppress the rcombination reaction

$$A^- + D^+ \longrightarrow A + D , \qquad (4)$$

which is a simple strongly exothermic process and typically proceeds much more rapidly than complex catalytic reactions (2) and (3). In particular, for this reason it seems rather difficult to accomplish efficiently photocatalytic water cleavage in simple homogeneous systems.

It looks likely, that in order to overcome the mentioned difficulty, molecular photocatalytic systems must contain structurally organized molecular assemblies (SOMA) resembling reactive centers of natural photosynthetizing "apparatus", in which particles D and A (and, consequently, particles D^+ and A^- formed during reaction (1)) are spatially separated. Such "apparatus" may be made sufficiently versatile to be used for water cleavage as well as for other endergonic redox reactions during which light energy can be stored (for example, fixation of carbon dioxide by reducing with water, etc.).

At the present state of art, microheterogeneous systems based on microemulsions, micellar solutions and lipid vesicles are considered to be the most interesting for the development of photocatalytic systems of SOMA type. The results obtained in this field using systems based on lipid vesicles should be particularly mentioned (see other articles of this book and review-articles [1,3,5,9,11,14,15,16]).

Lipid vesicles are microscopic (300-1000 Å in diameter) spherical particles formed by a bilayer lipid membranes, whose inner and outer volumes can contain unmiscible aqueous solutions of different composition. The inner volume can contain, for example, particle D, whereas the outer volume can contain A (see, e.g., Fig. 2). Reaction (1) in such systems has to occur via directed (vectorial) phototransfer of the electron across the membrane.

To achieve a reasonably high quantum yield of charge separation in solutions of vesicles, certain mechanistic studies of electron phototransfer across the membrane occured to be necessary. Considered below are typical examples of SOMA systems whose characteristics have been improved step-by-step on the base of the mechanistic studies. The examples are taken from the studies that have been carried out at the Institute of Catalysis (Novosibirsk).

System I. Among the first systems for which the mechanism of direct electron phototransfer across the membrane of a lipid vesicle was studied, were solutions of lecithin vesicles with a water-insoluble complex, zinc(II)tetraphenylporphin (ZnTPP), embedded as a photocatalyst into the membrane and localized near its inner and outer surfaces. In the presence of an irreversible electron donor (e.g. anions of ethylendiaminetetraacetic acid, $EDTA^{2-}$), which does not penetrate across the membrane, in the inner volume of the vesicles, and of a reversible electron acceptor (e.g., methylviologen bication,

MV^{2+}=CH_3-$^{+}$N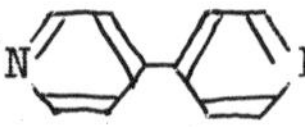N^{+}-CH_3), which also does not penetrate across the membrane, in the outer volume, the photocatalytic electron transfer occurs through the membrane from $EDTA^{2-}$ to MV^{2+} under the action of visible light. However, the quantum yield, ϕ , of this transfer did not exceed 0.2%.

Based on the mechanistic study, the following two-quantum scheme for electron phototransfer across the membrane in system I was suggested:

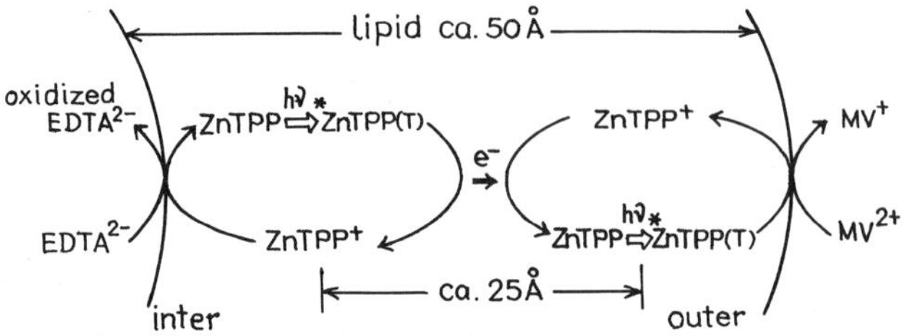

Here *ZnTPP(T) denotes the triplet excited state of ZnTPP.

As the first chemical act this scheme assumes the electron phototransfer at the outer boundary of the vesicle membrane:

$$^{*}ZnTPP(T)_{out} + MV^{2+} \xrightarrow{< 2 \cdot 10^{-5}\ s} ZnTPP^{+}_{out} + MV^{+} ,$$

which is followed by the tunnel electron transfer across the membrane from an excited ZnTPP molecule located near the internal membrane surface into $ZnTPP^{+}_{out}$:

$$^{*}ZnTPP(T)_{int} + ZnTPP^{+}_{out} \longrightarrow ZnTPP^{+}_{int} + ZnTPP_{out} . \quad (5)$$

Finally, irreversible oxidation of $EDTA^{2-}$ takes place:

$$ZnTPP^{+}_{int} + EDTA^{2-} \longrightarrow ZnTPP_{int} + \text{products of } EDTA^{2-} \text{ oxidation.}$$

Process (5) seems to be the rate-determining step for this scheme. It was found also, that the rate of processes (5) is affected markedly by a lateral (along the membrane surface) motion of ZnTPP particles, while a transmembrane motion of the same particles has little effect on this rate.

These data, together with the results of the study of photochemical charge separation in "water in oil" microemulsions [3,9,14], have allowed one to propose several ways to improve the efficiency of electron transfer across the lipid membrane.

First, special efforts were made to suppress the electrostatic field that hinders electron transfer across the membrane. Indeed, in the process of phototransfer the internal volume of the vesicle acquires a positive electric charge and its outer volume -- a negative one. This creates an electric field that may slow down electron transfer. To suppress this field, valinomicine which is a selective carrier of potassium cations across the membrane, was introduced into the membrane, along with the photocatalyst. The simultaneous transport of electrons and K^{+} cations across the membrane prevented accumulation of positive charges in the inner volume or even created a deficiency of these charges. As a result, the quantum yield ϕ of electron phototransfer across the lipid membrane increased by a factor of 2-3.

System II. A further success in increasing the quantum yield was achieved by expoloiting the following idea. Since in system I process (5) seems to be the rate-determining step, one can try to accelerate electron-phototransfer following the example of photosynthesizing organisms. This could be done by additional excitation of the molecule of ZnTPP on the inner surface of the membrane via energy transfer from an "antenna", i.e. a sensitizer which accumulates light and transfers the excitation to "reactive centers" (ZnTPP molecules embedded into the membrane) in the same was as does the pull of molecules of "antenna" chlorophyll in chloroplasts. The water-soluble complexes $Ru^{II}(bipy)_3$ introduced into the inner volume of vesicles was used as a sensitizer-"antenna". The thermodynamic properties of the excited state of $Ru^{II}(bipy)_3$ are such that it is capable of providing not only the energy transfer to ZnTPP, but also the electron transfer onto $ZnTPP^+$ and even onto ZnTPP (which may result in the formation, on the internal membrane surface, of the radical-anion $ZnTPP^-$, which, in principle, can also transfer an electron onto the radical-cation $ZnTPP^+$ formed on the external membrane surface). The lifetime of the electron-excited triplet state of $Ru^{II}(bipy)_3$ (ca. $6 \cdot 10^{-7}$s) is sufficient for excited particles to collide many times with the wall of the internal volume (which are 150-220 Å in diameter) of the vesicle before they are spontaneously deactivated. It occured that the presence of $Ru^{II}(bipy)_3$ leads not only to the spectral sensitization of the process (owing to additional light absorption by the ruthenium complex in the regions where absorption of ZnTPP is negligible) which is accompanied by approximately 6-fold increase of MV^+ accumulation, but also to a 2-3-fold increase of the quantum yield ϕ (which was calculated taking into account the light absorbed both by ZnTPP and by $Ru^{II}(bipy)_3$). The registered value ϕ = (0.6 ± 0.2 per cent) indicates a notable improvement of conditions for the electron phototransfer compared to system I. Note that no valinomicine was used in these experiments.

System III. The next step was modification of system II to develop an asymmetric (from the photochemical point of view) membrane, containing a reversible electron carrier, instead of the photocatalyst, inside the lipid bilayer. The exact composition of the system with such a carrier was selected on the basis of the experiences with systems I and II as well as of the experiments with the photocatalytic electron transfer across the interface in microemulsions of the "water in oil" type. In system III, like in system II, $Ru^{II}(bipy)_3$ was introduced as a photocatalyst into the inner volume of the vesicles. No additional photocatalyst was inserted into the lipid bilayer. As the electron carrier water-insoluble cetylviologen bication, CV^{2+}, was used. The molecules of CV^{2+} are distributed near the lipid-water interface and practically cannot diffuse across the membrane. As found, in the presence of an irreversible donor, $EDTA^{2-}$, in the inner volume, the electron transfer from $EDTA^{2-}$ onto the CV^{2+} particle, located near the inner surface of the membrane, occurs under illumination. The radical-cation CV^+ formed appeared to be capable of transferring its "excess" charge across the membrane via two co-existing mechanisms: by trans-membrane diffusion of CV^+ particle itself and by long-distance electron tunneling onto the CV^{2+} bication located on the external wall of the membrane [16]. In the presence of an appropriate acceptor such as the $Fe(CN)_6^{3-}$ anion in the outer volume, the "excess" electron of the CV^+ particle can easily be transferred into this acceptor. In this case, the quantum yield ϕ of electron phototransfer across the membrane was increased up to (15 ± 5)%. The most important reason for the large value of ϕ is here a high frequency of collision of excited $Ru^{II}(bipy)_3$ particles with the inner surface of the acceptor-containing membrane (or even their adsorption on the membrane) and, consequently, a low probability of useless deactivation of the photocatalyst. In fact, inversion of system III, in other words, exchange of the contents of the inner and outer volume leads to a more than 100-fold decrease in ϕ due to the lowered efficiency of electron transfer from the excited photocatalyst to CV^{2+};

Even higher (up to 50-60%) quantum yields of electron phototransfer across the lipid membrane were achieved in the systems where the charge separation occurred upon interaction of the photocatalyst with a primary electron acceptor not on the boundary layer "water-lipid membrane", but in a homogeneour aqueous phase (usually inside the vesicle) provided that irreversible electron donors were present in the systems (see [9,14,15]). However we shall not discuss here such systems because they seem to be hardly applicable for light energy storage via accomplishing the complete cycle of water cleavage, which is the final goal of the studies discussed here.

Nowadays a new type of promising SOMA photocatalytic systems is under intensive study as well. They are based on using sophisticated "submolecular" structures as electron carriers. Note that in natural photosynthesis the electron phototransfer across chloroplast membranes is also provided by specially organized "submolecular" structures, i.e. a pile of "specially arranged" molecules of the electron-transfer chain. Recently small particles of semiconductors, e.g. CdS, immersed into the lipid bilayer have been tested as a first example of artificial "submolecular" photoelectron transferring structures connecting the internal and external surfaces of vesicles [17]. The data about the photocatalytic properties of such systems are still scarce. However, one can expect the appearence of numerous studies in this area in the nearest future.

Conjugation of photochemical charge separation with the steps of water reduction and oxidation. Development of united photocatalytic systems for closed-cycle water cleavage based on lipid vesicles. The scheme below gives a comparison of redox characteristics of intermediate species that participate in electron phototransfer in systems I, II and III. The scheme shows also the redox potentials for hydrogen and oxygen evolution from water at pH7 (redox potential E_e is respectively -0.44 and +0.8 V against N.H.E.).

Note that the scheme of electron phototransfer in systems I and II is quite similar to the "Z-scheme" of photosynthesis in plants (see Fig. 1a), while this scheme in system III resembles the scheme of charge separation and electron transfer in n-p-junctions of semiconductor solar cells.

It can be seen also that in all the systems the electron phototransfer produces particles which are capable (as far as their thermodynamic properties are concerned) of both reduction of water to dihydrogen (radical cations of viologens) and of oxidation of water to dioxygen (radical cation $ZnTPP^{+}$ and complex $Ru^{III}(bipy)_3$). It makes in principle possible to accomplish the closed cycle of water cleavage in systems I-III by introducing therein suitable catalysts for processes (2) and (3).

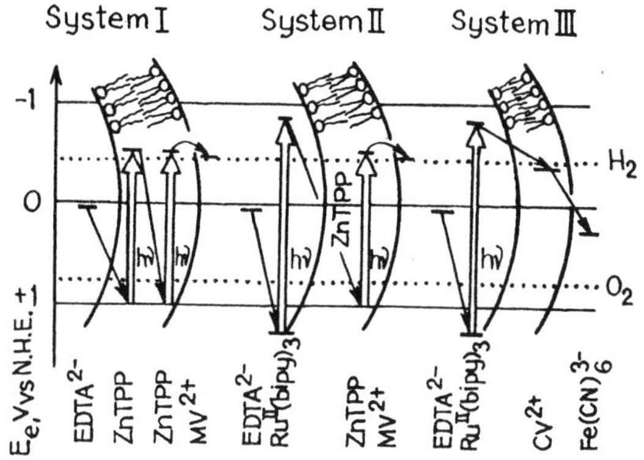

It should be stressed that all catalysts for dioxygen and dihydrogen evolution from water described above, have been initially developed and studied only in the situation, where particles A^- and D^+ together with the catalysts were introduced into the aqueous solution containing no vesicles. One should expect that, perhaps, some peculiarities concerning both activity and selectivity of the catalysts action can arise when using these catalysts in the presence of vesicles. For example, organic compounds that form microheterogeneous systems can be involved in side reactions with D^+ and A^- that can suppress evolution of O_2 and H_2. This latter fact imposes more strict requirements on the selectivity of the catalysts used in microheterogeneous systems.

In order to choose catalysts that are appropriate to accomplish reactions (2) and (3) in microheterogeneous systems (or, in other words, to create the catalytically active centers adjusted to these systems), numerous studies of these reactions in the so-called "sacrificial" systems, either for dihydrogen evolution from water (provided that a special irreversibly consumed donor-"sacrifice", but not water, is simultaneously oxidized by particles D^+) or for oxygen evolution from water (provided that an acceptor-"sacrifice" instead of water undergoes a simultaneous irreversible reduction, e.g., by particles A^-) are extensively studied nowadays.

Dihydrogen evolution in the outer volume of vesicles can be achieved rather easily in the case of systems I and II. In fact, the water-soluble MV^+ cation, generated in the outer volume of these systems, is known to be capable to evolve dihydrogen in the presence of enzymes such as hydrogenases as well as of some artificial catalysts. E.g. by using system I, the possibility was demonstrated of the photocatalytic dihydrogen evolution in the solution of vesicles as a result of electron phototransfer across the lipid membrane (from the irreversibly oxidized $EDTA^{2-}$ in the inner volume of vesicles). As the catalysts for H_2 evolution the hydrophylic hydrogenase of Thiocapsa roseopersicina or the heterogeneous rhodium-polymeric catalyst described above were used (see Fig. 2).

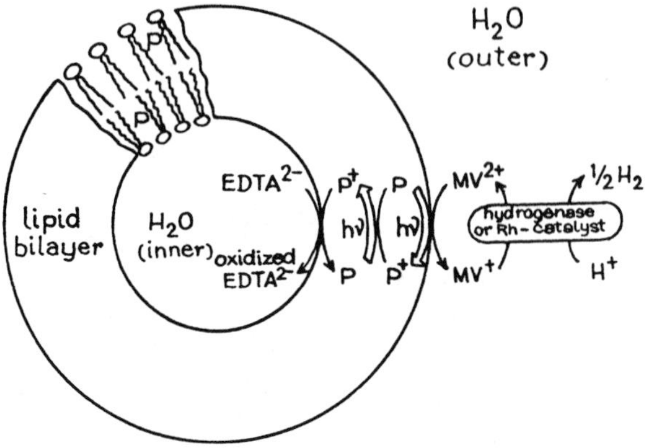

FIG. 2. Scheme of the photocatalytic dihydrogen generation conjugated with the vectorial electron phototransfer across the bilayer membrane of the lipid vesicles containing ZnTPP. Bi-tailed symbols denote molecules of the lipid and P stands for ZnTPP.

As far as dihydrogen evolution in system III is concerned, the situation is more complicated, since the radical cation CV^+ does not escape into the outer solution of the vesicles but is located within the lipid membrane near its surface. As found, this circumstance makes it difficult to use CV^+ (and similar lipophilic viologens [16]) as a specific electron donor for dihydrogen evolution catalyzed by the hydrophylic hydrogenase. This is due primarily to the fact that strong hydrophobic interactions of CV^+ with the membrane suppress the electron-donor activity of CV^+ in comparison with MV^+. A search is underway for the methods of conjugating such electron donors immersed in the lipid bilayer with catalysts of dihydrogen evolution from water.

Several photocatalytic systems based on lipid vesicles have been developed which allow dihydrogen evolution to occur in the outer volumes of the vesicles (without electron phototransfer across the membrane) [14,15].

Evolution of dioxygen inside lipid vesicles seems to be the most difficult problem to be solved. In principle the catalysts for this process when using oxidant of the $Ru^{III}(bipy)_3$ type are known (see above). Therefore one may hope to accomplish the photocatalytic dioxygen evolution at least in systems II and III upon substitution of the irreversible donor $EDTA^{2-}$ by the appropriate catalyst. A certain difficulty in accomplishing this process is associated with small (ca. 200 Å) sizes of inner volumes of the vesicles; homogeneous catalysts of dioxygen evolution seem to be more suitable for such systems than heterogeneous catalysts. In fact, recently homogeneous cobalt catalysts have been inserted into the inner volume of the vesicles and the possibility has been demonstrated of their conjugation with the chain of electron transfer in systems, that are similar to systems II and III [14,18,19]. Unfortunately the selectivity of the catalysts used in these works has been as yet insufficient to provide oxidation of H_2O in the presence of more easily oxidized organic compounds of which the vesicle is built. As a result, CO_2 -- the product of complete photocatalytic oxidation of organic compounds, as well as carbon monoxide and nitrogen oxides, but not O_2, was produced as the main gaseous products of photocatalytic oxidation process in such systems.

Recently the procedure was found for supporting quite efficient catalysts for dioxygen evolution onto the external surface of the lipid vesicles. This was accomplished via anchoring cobalt or manganese hydroxides to the external surface of the lipid bilayer [20-22]. The catalysts prepared via this procedure were found to provide dioxygen evolution from water upon photogeneration of the strong oxidant, $Ru^{III}(bipy)_3$, in the outer volume of the vesicles. However, attempts to conjugate these catalysts with the chain of electron phototransfer across the membrane have not so far been successfull.

On the ground of the above data, one can make the following conclusions about the present state of art and perspectives of developing united photocatalytic systems for closed cycle water cleavage in SOMA systems. Reasonably efficient catalysts as well as intermediate donors D and acceptors A, that can provide separately all the three necessary steps (1)-(3) of photocatalytic water cleavage reaction, have already been developed. The spectral and thermodynamic characteristics of these catalysts, donors and acceptors are such that they can provide a closed-loop water cleavage under the action of visible light. However, scientists have not yet succeeded in so adjusting the catalytic systems for reactions (1)-(3) that they could operate efficiently in the presence of each other.

The main difficulty to be overcome seems to increase the selectivity of the process of O_2 evolution. This can be done by the search of both better, i.e. more selective, catalysts for O_2 evolution, and more resistant (towards oxidation) substances for building organized molecular assemblies. Once such catalysts and/or substances are found, the laboratory scale molecular photocatalytic systems for the closed cycle water cleavage could be created. But, of course, a lot of further efforts will be needed to transform this hypothetic laboratory-scale system into the practically important one.

One of the problems to be solved here, e.g., is how to maintain a high quantum yield in a steady state operation of the closed cycle of water cleavage. In fact, the lipids are known to have rather high permeability towards H_2 and O_2. The corresponding migration of H_2 and O_2 through the lipid membranes will destroy the isolations of the inner and the outer water phases of vesicular suspensions, that may poison the catalytic systems for process (2) and (3) even if they were efficient enough when taken separately [22].

Production of dihydrogen by photocatalytic processing of non-traditional raw-materials. The development of photocatalytic systems mimicing the bacterial photosynthesis which use as substrates substances that can be oxidized more easily than water, was found to be a much simpler problem than photocleavage of water. As will be shown below, these systems can function without membranes which provide the spatial separation of particles A^- and D^+. By the moment a large number of both homogeneous and heterogeneous photocatalytic systems are already known, which mimic bacterial photosynthesis. These are the so-called "sacrificial" photocatalytic systems [3].

They provide a rather efficient evolution of dihydrogen owing to irreversible oxidation of a donor-"sacrifice" under the action of visible light.

The summary of the data obtained so far in this area can be found in review articles [3,5,11,23,24]. Most often various organic compounds or such inorganic compounds as H_2S and sulphite ion were used as donors-"sacrifices". Most of the works with organic donors were aimed mainly at fundamental goals -- modeling the principles of bacterial photosynthesis in systems that are chemically much simpler than photosynthetizing bacteria rather than at development of photocatalytic systems that can be of real potential interest for energy production. In contrast to this, many works with H_2S and sulphite ion as the donor-"sacrifice" were directly aimed at estimating their potentialities in energy production. So, we shall dwell here in detail only on the photocatalytic systems, utilizing sulfide substrate, i.e. systems based upon reaction

$$H_2S \longrightarrow H_2 + S . \qquad (6)$$

Potential practical interest in this process is caused by the following reasons. First, it allows one to produce a valuable fuel -- dihydrogen together with a valuable chemical -- elementary sulphur from non-traditional raw-materials -- waste water of some industrial (chemical, petrochemical and metallurgy) plants, geothermal water sources and sea-water which in some places (e.g. in the Black Sea) contains H_2S in a rather high concentration. Second, these processes can, perhaps, occur useful for purifying the waste-water of the above mentioned plants.

Reaction (6) was found to occur in the presence of a variety of photocatalysts, and seems to be one of the first artificial processes used to mimic bacterial photosynthesis (see reviews in refs. [3,11,23,24]). From a practical standpoint, the most interesting for this reaction appear to be photocatalysts based on colloids and suspensions of metal sulfide semiconductors, which are stable in sulfide-containing aqueous solution and have a large surface area [25-32]. The mechanism of photocatalytic decomposition of hydrogen sulfide is essentially similar to that of water cleavage (reactions (1)-(3)) with the only difference that, instead of a rather complicated 4-electron reaction (3) of water oxidation, oxidant D^+ provides a much simpler 2-electron reaction of oxidation of sulfide-anion (see Fig. 3). In case of semiconductor photocatalysts those are positively and negatively charged carriers of electric current -- hole h^+ and electron e^-, that are used as the oxidant D^+ and reductant A^-, respectively. These centers are produced in the charge separation process that occurs upon absorption of light quanta with the energy exceeding or equal to the width E_g of the bandgap of the semiconductor photocatalyst. The electron is further used to provide the two-electron reaction of dihydrogen evolution

$$2e^- + 2H^+ \rightarrow H_2 , \qquad (7)$$

while the hole provides the two-electron reaction of sulfide ion oxidation to sulphur:

$$2h^+ + S^{2-} \rightarrow S . \qquad (8)$$

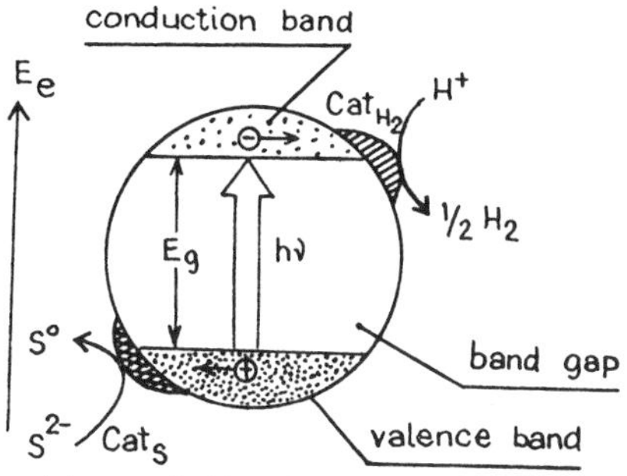

FIG. 3. Photogeneration of the holes and electrons at the excitation of a semiconductor particle by photons with $h\nu \geqslant E_g$ and the subsequent redox processes leading to hydrogen sulfide cleavage. $Cat._{H_2}$ and $cat._S$ are the catalysts for redox processes of H_2 and sulphur evolution, respectively; in the particular case of CdS as the photocatalyst, no additional catalyst $cat._S$ was found to be necessary.

Detailed studies have indicates that at pH 8-10 only metal sulfides with a bandgap width E_g exceeding 2.2 eV are photocatalytically active in reaction (6); these are, e.g., zinc(II), cadmium(II), indium(III) and tin(IV) sulphides with E_g = 3.7, 2.4, 2.4 and 2.2 eV, respectively. One can suppose that, likewise for semiconductor oxides [3-5], for metal sulfides there also exists a certain correlation between the flatband potential and width of the bandgap. The existence of this correlation imposes serious restrictions upon the application of the most efficient, from a standpoint of solar energy storage, narrow-band semiconductors as photocatalyst for dihydrogen production. This happens because with such a correlation, starting from a certain bandgap width, the electrochemical flatband potential of semiconductor in the substrate medium becomes more positive than the potential of the hydrogen electrode and, consequently, the reaction of water reduction to H_2 becomes thermodynamically forbidden. Nonetheless, even for semiconductors with the threshold energy $h\nu_o = E_g \approx$ 2.4 eV ($\lambda_o \approx 520$ nm) the maximum possible "commercial" efficiency η^* achieves ca. 10% (even if only combustion of H_2 is taken into account as the process, releasing the stored energy), which seems sufficient to justify the development of photocatalytic systems based even on such semiconductors.

Most efforts were associated with the development of photocatalysts based on finely dispersed (colloidal or suspended) CdS [25-32]. The rate-determining step that controls the overall efficiency of the photocatalytic process (6) in the presence of CdS seems to be water reduction to dihydrogen via reaction (7) on the surface of semiconductor particles. This process can be promoted, e.g., by supporting finely dispersed metals (Pt, Pd, Rh, Ni, Co, etc.) or ruthenium dioxide over the surface of CdS particles. Sulfide-anion is oxidized via reaction (8) without assistance of any additional catalyst. This is the adsorbed form of S^{2-} that participates in this reaction. Note that at pH12 this adsorbed form provides an almost monolayer coverage of CdS particles, if the total concentration of the sulfide-anion in solution exceeds 10^{-4}M. This latter observation indicates the possibility to accomplish an efficient photocatalytic oxidation of sulfide-anion even in diluted solutions.

The quantum yield of the photogenerated charge separation, i.e. formation of h^+ and e^- under the action of light, depends strongly upon the size of CdS particles and has two maxima. One of them is observed for the particles, whose size exceeds 1000-2000 Å . For such particles charge separation occurs in a near-surface layer where the jump of the electric potential takes place (see [5]). The second maximum is observed at the particle size of less than 50 Å . For such particles there occurs no jump of the potential in the near-surface layer, and a photogenerated hole, h^+, immediately reaches the CdS surface avoiding recombination in the bulk of the semiconductor. The maximum possible quantum yield of photoseparated charges on CdS surface is estimated to be 40% in the former case and to approach unity in the latter one [32]. There are also some evidences that oxidation of sulfide-anions occurs via preliminary oxidation, by h^+, of a surface center that serves as an adsorption site for S^{2-} anion rather than via a direct interaction of adsorbed S^{2-} with h^+. For the overall photocatalytic reaction (6) the quantum yields of upto 30-40% were, indeed, observed when using suspensions of free, i.e. unsupported particles of CdS promoted with metals or RuO_2 serving as catalysts for reaction (7). Unfortunately such suspensions to be efficient in the photocatalytic processes have to be stirred which results in mechanical destroying and coagulation of CdS particles which prevents their practical use.

This instability can be overcome via anchoring CdS species to polymers, e.g., films of cation-exchange forms of sulphited perfluoroethylenes of the "Nafion" type [25,26,31]. Due to slowed down diffusion of the reacting species from solution towards CdS particles and back inside the polymer matrix, the quantum yield observed for process (6) over such anchored CdS is smaller than for free CdS. But at λ = 436 nm it still reached 5 and 2% for CdS, promoted by finely dispersed RuO_2 and Pt, respectively. Note that here and below we present the figures for the quantum yield related to one electron transferred. Since reactions (7) and (8) are two-electron processes, the quantum yield for production of H_2 molecules and S atoms is twice smaller.

A substantial improvement of the overall efficiency of the photocatalytic process (6) was achieved upon switching from finely dispersed individual semiconductors as photocatalyst, to photocatalysts composed of two semiconductor phases forming microheterojunctions [31,33]. Such microheterojunctions are formed, e.g., upon contacting cadmium sulfide or zinc sulfide (or a mixed sulfide of these metals) with copper or silver sulfides. Particles with microheterojunctions can easily be prepared via consecutive precipitation of various semiconductor phases.

The presence of an appropriately chosen heterojunction is quite useful because of the following reasons. First, it is equivalent to the creation of an organized chain of vectorial electron transfer (cf. Figs. 1a and 4) and, thus, significantly improves conditions for photoseparation of electrons and holes. Second, it provides a substantial shift of the photocatalyst threshold energy $h\nu_0$ towards smaller values due to the fact that light quanta are now absorbed not only by a wide-band semiconductor, but also by a narrow-band one (see Fig. 4).

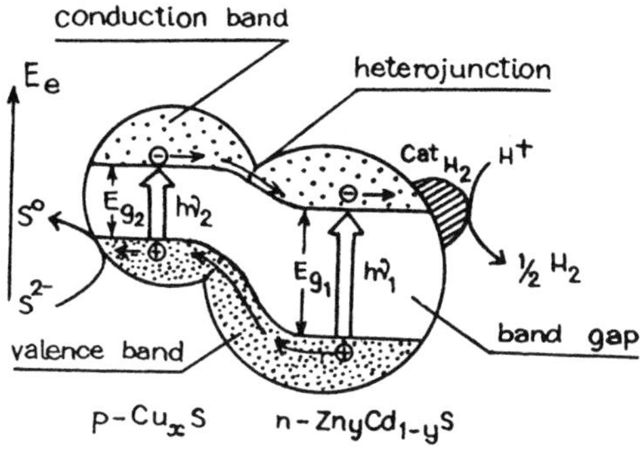

FIG. 4. Idealized energy diagram of electron transfer and the scheme of the photocatalytic dihydrogen formation for the photocatalyst particle with microheterojunction n-$Zn_yCd_{1-y}S$/p-Cu_xS.

A typical spectrum of photocatalytic action of the Pt-$Zn_{0.17}Cd_{0.83}S/Cu_xS$ catalyst that has a microheterojunction, is shown in Fig. 5.

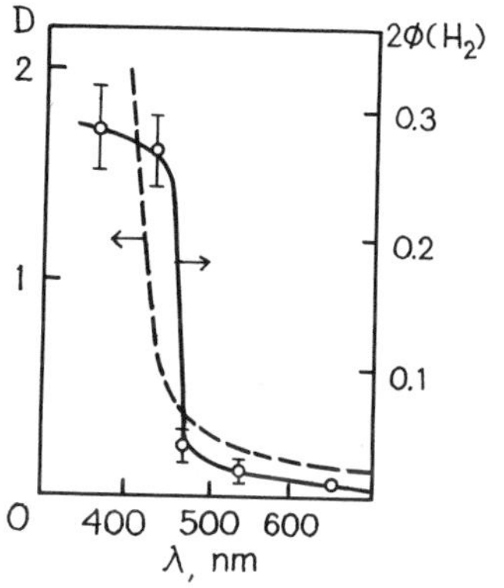

FIG. 5. Absorption spectrum (dashed line) for the Pt-$Zn_{0.17}Cd_{0.83}S/Cu_xS$ suspension and a typical spectrum of the photocatalytic action of the same suspension. D is optical density, $\phi(H_2)$ is the quantum yield of H_2 molecules (according to [33]).

It can be seen that the latter photocatalyst provides a noticeable quantum yield for reaction (6) even in the spectral region with $\lambda > 500$ nm, i.e. beyond the fundamental absorption band of $Zn_{0.17}Cd_{0.83}S$ semiconductor. Note that the Cu_xS semiconductor with $E_g < 1$ eV ($\lambda_o > 1000$ nm), when taken alone does not show any photocatalytic activity,as was mentioned above. In the optimal situation, upon illumination in the absorption band of the wide-band phase, for the above mentioned photocatalysts with microheterojunctions, having particle size of ca. 2-4 μm, the effective quantum yield of almost 50% has been obtained. This value exceeds considerably the quantum yield achieved so far for single-phase semiconductors, the high quantum yields being observed even for very low (ca. 10^{-4} M) concentrations of the sulfide anion in solutions. For illumination at $\lambda > 500$ nm a rather sufficient (up to 10%) quantum yield was registered as well.

In the course of reaction (6) the photocatalyst may be poisoned by elementary sulphur. Such poisoning can readily be eliminated in the presence of the sulphite-ion in solution. This ion cleans the photocatalyst surface via the "dark" reaction

$$S + SO_3^{2-} \longrightarrow S_2O_3^{2-} .$$

As found, in the presence of sulphite anion (in natural or waste waters it is typically present simultaneously with H_2S), practically complete decomposition of H_2S can be achieved over the above discussed photocatalysts.

Quite promising also seems to be preliminary data on photocatalytic decomposition of H_2S over a surface-modified crystalline silicon. This photocatalyst allows one to utilize a significant fraction of the whole spectrum of sun light. Interestingly, such photocatalysts can be prepared from low-quality silicon which is a waste of industry producing semiconductor devices.

A set of various pilot demostrational photocatalytic reactors for providing process (6) under the action of natural solar irradiation was designed and tested in the Institute of Catalysis (Novosibirsk). The surface area of photocatalytically active zone of these reactors exceeded several dm^2, that allowed one to produce several dm^3 of H_2 per day when using non-concentrated or several-fold concentrated solar light. The ways towards optimizing the design of such photocatalytic reactors taking into account the illumination conditions, heat release, conditions for the reagents input and the products output, etc., are under study. Upon certain improvements, such reactors could, perhaps, be used to create photocatalytic devices of practical interest. Thus, at present process (6) seems to be a candidate number one to become the first chemical reaction used in practice for direct photocatalytic conversion of solar energy into chemical energy of fuels.

THE STATE OF ART IN DEVELOPING THERMOCATALYTIC SYSTEMS FOR SOLAR ENERGY CONVERSION

Research aimed in constructing thermocatalytic systems for solar-to-chemical energy conversion has started later than that in the area of photocatalytic systems. However, progress in constructing actually working thermocatalytic systems occurred to be much more rapid. At present, efficiency (in its traditional meaning) of ca. 40% in solar-to-chemical energy conversion has been already achieved with thermocatalytic devices, and obvious ways are seen how to increase the efficiency to make it closer to the theoretical maximum of ca. 70% characteristic of thermocatalytic method [1].

Thermochemical conversion of solar energy is based on the possibility to shift thermodynamic equilibria in reacting systems toward formation of energy-rich compounds (fuels)due to heating of the systems by the sun light. Long storage of the chemical energy so obtained can be achieved only if the produced reaction mixture is reliably "hardened", i.e. it can preserve its composition after cooling. For this reason, thermochemical processes of energy conversion based on catalytic reactions seem to be most attractive for solar-to-chemical energy conversion, since in this case the reaction mixture can be "hardened" quite easily by just removing it from the catalytic reactor.

Thermocatalytic processes that are considered to be most suitable for solar-to-chemical energy conversion are listed in Table II.

TABLE II. Some reactions which can be used for thermochemical catalytic conversion of solar energy

Reactions[a)]	ΔG^o_{298}, kcal/mol	ΔH^o_{298}, kcal/mol	T^{*}[b)], K
$CH_4 + H_2O \longrightarrow 3H_2 + CO$	36.0	59.8	960
$CH_4 + CO_2 \longrightarrow 2H_2 + 2CO$	40.8	59.1	960
$C_6H_{12} \longrightarrow C_6H_6 + 3H_2$	23.4	49.3	570
$NH_3 \longrightarrow 1/2N_2 + 3/2H_2$	4.0	11.0	470
$SO_3 \longrightarrow SO_2 + 1/2O_2$	16.7	23.5	1030

a) Thermodynamic data are given for liquid H_2O, C_6H_{12}, C_6H_6; the other compounds are considered to be gases

b) T^{*} is the temperature at which $\Delta G^o = 0$.

All these processes are well known also as promising candidates to store (in the form of chemical energy) the heat that is produced at nuclear energy plants (see, e.g. [34-36]).

Experimental research in the area of thermocatalytic conversion of solar energy has started after it became clear from theoretical calculations, that the maximum possible efficiency of the non-quantum thermochemical method of conversion of the solar energy into the Gibbs' energy of chemical fuels is by no means inferior to that of the quantum photochemical methods; moreover, in some cases it can be even higher.

In fact, a thermodynamically allowed energetic efficiency, η, of thermochemical conversion of the solar energy into the Gibbs' energy of the products of an endothermic chemical reaction can be expressed as follows [1,37,38]:

$$\eta \approx \left(1 - \frac{T_o}{T}\right)\left(1 - \alpha \frac{T}{T_S}\right) . \qquad (9)$$

The first multiplier in the right-hand part of eq. (9) corresponds to the efficiency of conversion of heat with temperature T into the Gibbs' energy of energy-rich products of the reactive mixture, which was initially in the state of thermodynamic equilibrium at temperature T_o. The second multiplier characterizes the maximum possible efficiency of the concentrated sun light conversion into heat with temperature T. $T_S \approx 5800$ K corresponds to the radiative temperature of the sun; $\alpha \gtrsim 1$ is a dimensionless parameter whose value depends upon the shape of the light energy receiver.

From eq. (9) it follows that the maximum value of η is achieved at $T_{opt} = \sqrt{T_o T_S/\alpha}$ and equals ca. 70% at $T_o \approx 300$ K and $\alpha = 1$. This value exceeds notably the maximum possible efficiency (ca. 32%, see above) of any single-step quantum scheme of solar energy conversion into electricity or into the Gibbs' energy of chemical compounds. Note also that T_{opt} lies within the temperature region 900-1300 K, which is well assimilated by chemical industry. In this most optimal for solar energy conversion temperature region proceed, e.g., large scale industrial endothermic catalytic processes of steam and carbon dioxide reforming of methane into syn. gas. For these processes a significant shift of the thermodynamic equilibrium towards energy-rich mixture of dihydrogen and carbon monoxide occurs at temperatures above T* = 960 K (see Table II). Being reversible, these methane reforming reactions can serve as a basis of closed-loop thermochemical cycles of the "ADAM-EVA" type. For these cycles the chemical energy stored in syn. gas can then be released as heat at any moment by providing the back exothermic reaction of syn. gas methanation with a simultaneous regeneration of the initial composition of the reaction mixture (see Fig. 6 and [34-36,38,39]).

The releasing heat can be of high potential (with temperature up to ca. 900 K). Numerical calculations indicate that the overall energetic efficiency η_t of these cycles, measured as a ratio of the energy transferred as heat for the consumer to the energy of the concentrated light fallen on the solar energy receiver, can also be fairly high, i.e. exceeding 50% [38,39].

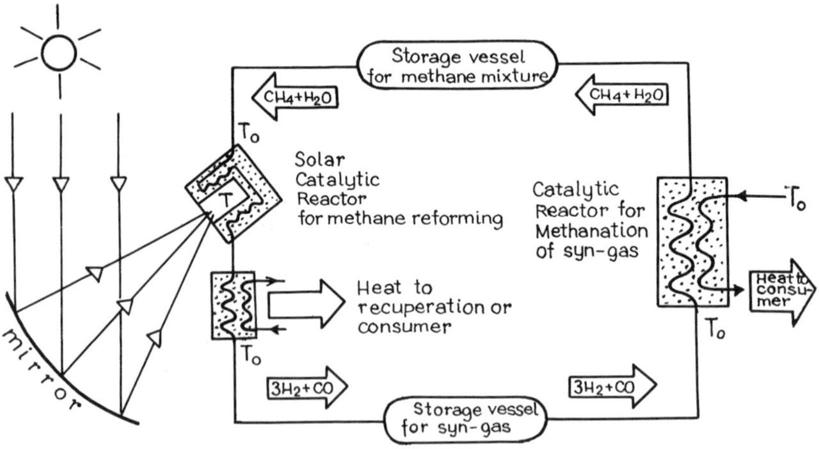

FIG. 6. A simplified principal scheme of a solar thermocatalytic power plant based on ADAM-EVA cycle

An essential advantage of thermocatalytic processes of energy conversion is the simplicity of their accomplishing which allows one to use both well elaborated concentrators of sun light and the available arsenal of methods of chemical engineering for heating the catalyst bed and the reagents. The main problems now are mostly technical ones and are associated primarily with the necessity of developing particular designs of solar catalytic reactors which would provide both the needed temperature of the catalyst bed during the course of the energy-storing process and the transfer of large heat energy fluxes across the reactor wall to the catalyst bed. Mathematical modeling and experimentsl studies have shown that in order to provide highly efficient solar energy conversion with the aid of the methane reforming reactions mentioned above, a solar catalytic reactor (SCR) should have the shape of a cavity inside which a 700-1000-fold concentrated sun light is fed [38,39].

Tests in natural conditions (1984) of a pilot steel solar catalytic reactor SCR-3, developed in the Institute of Catalysis (Novosibirsk) have proved the processes of thermocatalytic reforming of methane and other light hydrocarbons to be, indeed, highly potent for conversion of solar energy into chemical energy of syn. gas [38-40]. Some data for the best runs with this reactor are compiled in Table III. For example, in the steam reforming of methane a more than 40% efficiency of conversion of the concentrated solar energy into chemical energy (enthalpy) of the reaction products was achieved. The useful power of the SCR-3 exceeded 2 kWt, and the rate of dihydrogen production (as one of the syn. gas components) exceeded 2 m^3/h.

TABLE III. Summarized data on the best experimental runs of the SCR-3 solar catalytic reactor[a)]

Process tested	T_{cat}, K	W_{chem}/W, %	W_{chem}, kWt
$CH_4 + H_2O$	833	43	2.0
$CH_4 + CO_2$	933	34	1.43
$(0.17C_3H_8+0.83C_4H_{10})+H_2O$	853	10	0.38

a) T_{cat}, W_{chem} and W stand for, respectively, the averaged temperatures of the catalyst's bed during steady-state operation, power (enthalpy of chemical bonds per unit time), accumulated by the products of the methane reforming, and the total power of the concentrated solar flux entering the cavity of SCR-3.

In subsequent years successful natural conditions tests of pilot catalytic solar power plants based on the closed-loop thermochemical cycles of methane reforming into syn. gas (energy storage stage) and subsequent methanation of syn. gas (release of the stored energy) were also carried out. These power plants contained, along with a modernized steel solar catalytic reactor SCR-5, a pilot tube catalytic reactor of methanation, in which the stored energy was released as heat with temperature exceeding 920 K [39,41]. The overall efficiency, η_T, achieved by the moment with such pilot power plants exceeds 20% at ca. 1 kWt heat power transferred to the consumer, provided the energy device operates under steady-state conditions.

Thus, the natural conditions tests of pilot solar power plants based on closed-loop thermochemical cycles proved experimentally high efficiency of direct thermocatalytic methods of solar energy conversion and storage. Note that thermocatalytic solar energy conversion may, perhaps, occur useful not only for local consumers of moderate amounts of energy, but also for large scale solar energetics. In this latter case the solar catalytic reactors should be combined with sun light receivers of powerful (up to several MWt) solar powerplants of tower type [42,43]. Another promising application of solar thermocatalytic reactors could be processing of biogas (methane plus carbon dioxide mixture, i.e. an ideal rawmaterial for SCR) via syn. gas to more valuable products, e.g. synthetic liquid fuel. It seems that at present state of art in developing thermocatalytic devices for solar-to-chemical energy conversion, the date of the beginning of commercialization of such systems will be determined more by economic than by scientific or technical considerations.

Among important unsolved scientific and technological problems related with the development of thermocatalytic systems of solar energy conversion, the following problems could be mentioned. First, this is the development of special-purpose high active as well as thermally and mechanically stable catalysts, which will be resistant to sharp frequent changes of heating and cooling conditions typical for solar energy systems. Second, this is further improvement of the design of solar thermocatalytic reactors with the aim to increase their energetic efficiency and a specific power coefficient (i.e. solar light power converted into chemical energy per unit volume of the catalyst bed). For example, for steel reactors such as SCR-3 and SCR-5 the specific power coefficient of 2-3 kWt/dm^3 cat. was achieved. For these reactors in conditions of large light energy fluxes falling on their walls, the specific power coefficient was found to be limited by the rate of heat transfer from the wall of the reactor into the catalyst bed. The rate of heat transfer, and consequently the specific power coefficient of thermocatalytic solar reactors can be increased up to value 40 kWt/dm^3 or even higher, if one makes their wall transparent to solar light [38,44,45]. For such reactors the energy of solar flux can be directly transferred into the catalyst bed. Recently, in the Institute of Catalysis (Novosibirsk) first experiments with transparent quartz reactors have been carried out. In these experiments the value of ca. 10 kWt/dm^3 has been achieved for the specific power coefficient for light energy conversion in steam reforming of methane. Note that in experiments with direct heating by concentrated light, the heterogeneous catalyst is affected by the flux of visible and infrared light of extremely high intensity , up to several kWt/cm^2, which equals or even exceeds the fluxes from powerful lasers. It cannot be excluded that under these conditions photochemical and photocatalytic processes, along with common "dark" thermocatalytic reactions, take place on illuminated surface of the heated catalyst. Unfortunately, information about high temperature heterogeneous photochemical processes is as yet very scarce.

CONCLUSIONS

The above discussion shows that, in principle, both photocatalysis and thermocatalysis can provide a fairly efficient direct coversion and storage of solar energy in the form of the energy of chemical fuels.

The photocatalytic systems for light-to-chemical energy

conversion, which at present are being designed, actually mimic the function and, quite often, even the structural elements of photosynthetizing organisms, both green plants and bacteria.

A more difficult problem is the development of photocatalytic systems that mimic photosynthesis in green plants, i.e. ensure storage of solar energy in the form of chemical energy, e.g. dihydrogen, with a simultaneous evolution of dioxygen from water. For the water cleavage process some semiconductor photocatalytic systems (e.g., platinated suspensions of titanium dioxide) are already available. However, the efficiency of such systems is still rather small, since they provide a high quantum yield of the process only when illuminated in the UV-region rather than in the visible region. Serious efforts are being made to develop also molecular systems for accomplishing the same process. By the moment catalysts for all the three steps of water cleavage (i.e. photoseparations of charges under the action of visible light, dihydrogen evolution and dioxygen evolution) have already been found. Now the main problems that still have to be solved here appear to be, first, how to conjugate these three steps so that they could provide a closed cycle of water cleavage and, second, how to further improve selectivity and stability of catalysts for water oxidation to dioxygen as well as activity and stability of structurally organized molecular assemblies that provide photocatalytic charge separation.

From the view-point of potential practical application, much more progress has been achieved in developing the photocatalytic systems that mimic the photosynthesis in bacteria, i.e. provide production of fuels (usually also dihydrogen) coupled with an oxidation of some organic or inorganic compounds, that are much stronger electron donors than water. It seems that most interesting advances here have been achieved for the reaction of photocatalytic decomposition of hydrogen sulfide into dihydrogen and sulphur. Photocatalytic systems that are both active and stable have already been developed for this process. These photocatalytic systems seem to allow one to proceed from laboratory experiments to demonstration tests of a much larger scale.

Impressive achievements have been demonstrated in developing thermochemical catalytic solar-to-chemical energy converters. In accordance with the theoretical predictions, the thermochemical conversion of solar energy proceeds most efficiently within the temperature range from 900 to 1300 K. Quite promising for accomplishing solar-to-chemical energy conversion seems to be catalytic reforming of methane with water or carbon dioxide to syn. gas. These days the main efforts in developing thermochemical catalytic converters are already in the area of designing the appropriate reactors and their appropriate conjugation with the concentrators of solar flux in order to provide a higher efficiency. For the steam reforming of methane ca. 40% efficiency of conversion of concentrated solar energy to chemical energy (enthalpy) of reaction products has already been demonstrated. The useful power of the reactor (relative to stored energy) exceeded 2 kWt, and the rate of dihydrogen formation (as one of the syn. gas components) exceeded 2 nm^3/h. Solar energy devices which accomplish the closed-loop cycles of energy storage, transfer and release on the basis of reversible catalytic reactions of methane reforming and syn. gas methanation, have also been tested. The overall efficiency $\eta_T \gtrsim 20\%$ has been demonstrated in these experiments and ways are seen how to further improve it. Work is underway to create similar energy systems on a pilot-plant scale with a power of up to $\gtrsim$ 100 kWt.

Thus, the advances made seen to be quite encouraging. They allow one to expect that chemical methods of solar energy conversion (both photocatalytic and thermochemical catalytic ones) indeed may become important for the energetics of the future. Whether this actually happens or not, will depend primarily upon two factors. The first factor is of economic character. At present all the existing methods of energy production via convertion of solar energy are far more expensive than the traditional ones, such as combustion of fossil fuels at electric plants, nuclear fission, etc. In this situation the shift towards a wide use of solar energy is expected to occur gradually, following the trends of solar energy becoming cheaper with every new scientific and technological breakthrough, and traditional energy sources becoming more expensive due to their gradual exhaustion.

The second factor is of ecological character. Much will depend on how dangerous will actually be the global consequencies of the Earth pollution with manmade extra heat, chemicals, etc., associated with traditional types of energetics. Note that nuclear fusion energetics which sooner or later is anticipated to be developed, also is expected to pollute Earth with extra heat. If such pollution occurs intolerable, the development of solar energetics, which produces no extra heating of the Earth and in other respects also seems to be ecologically more pure than any other kind of energetics, may obtain high priorities.

One should note that a higher theoretical energetic efficiency of solar energy conversion in the case of thermocatalytic systems and a more advanced situation with their development does not mean that a photocatalytic way of solar energy conversion is a failure for practical application. In fact, for thermocatalytic conversion one can utilize only the direct, but not scattered solar radiation, since only the direct radiation can be concentrated to the needed extent. Note that the direct solar radiation in conditions of the turbid Earth' atmosphere may sometimes comprise only a small fraction of the total solar energy flux coming to an energy-converting device. Quite the opposite, photocatalytic energy-converting devices are able to utilize successfully even weak fluxes of scattered solar light. Moreover, such devices can operate at ambient temperatures. For this reason one can expect that both thermocatalytic and photocatalytic methods of solar energy conversion will be able to find its own "path" of practical applications. This has to be a stimulus for equally serious attention to further development of both these direct methods of solar energy conversion.

REFERENCES

1. V.N. Parmon and K.I. Zamaraev in: Photocatalysis, E. Pelizzetti and N. Serpone, eds. (John Wiley & Sons, New York 1988).
2. J.W. Warner and R. Berry, J. Phys. Chem., 91, 2216-2223 (1987).
3. M. Grätzel, ed., Energy Resources through Photochemistry and Catalysis (Academic Press, New York - London 1983).

4. K.I. Zamaraev and V.N. Parmon, eds., Fotokataliticheskoye Preobrazovaniye Solnechnoi Energii (Photocatalytic Conversion of Solar Energy), Parts 1 and 2 (Nauka, Novosibirsk 1985).
5. E. Pelizzetti and N. Serpone, eds., Photocatalysis (John Wiley & Sons, New York 1988).
6. J.O'M. Bockris and R.C. Kainthla in: Hydrogen Energy Progress VI, vol. 2, T.N. Veziroglu, N. Getoff and P. Weinzierl, eds. (Pergamon Press, New York - Oxford 1988) pp. 449-459.
7. I. Okura, Coord. Chem. Rev. 68, 53-99 (1985).
8. E.N. Savinov, E.R. Savinova, and V.N. Parmon in: Fotokataliticheskoye Preobrazovaniye Solnechnoi Energii, Part 2, K.I. Zamaraev and V.N. Parmon, eds. (Nauka, Novosibirsk 1985) pp. 107-151.
9. V.N. Parmon, S.V. Lymar, and K.I. Zamaraev in: Hydrogen Energy Progress VI, vol 1, T.N. Veziroglu, N. Getoff, and P. Weinzierl, eds. (Pergamon Press, New York - Oxford 1986) pp. 351-364.
10. G.L. Elizarova and V.N. Parmon in: Fotokataliticheskoye Preobrazovaniye Solnechnoi Energii, Part 2, K.I. Zamaraev and V.N. Parmon, eds. (Nauka, Novosibirsk 1985) pp. 152-185.
11. T.S. Dzhabiev and A.E. Shilov, ibid., Part 1, pp. 169-192.
12. G.L. Elizarova, L.G. Matvienko, and V.N. Parmon, J. Molec. Catal., 43, 171-181 (1987).
13. G.L. Elizarova, L.G. Matvienko, and V.N. Parmon, React. Kinet. Catal. Lett. (1988).
14. S.V. Lymar and V.N. Parmon in: Fotokataliticheskoye Preobrazovaniye Solnechnoi Energii, Part 2, K.I. Zamaraev and V.N. Parmon, eds. (Nauka, Novosibirsk 1985) pp. 186-244.
15. V.Ya. Shafirovich and A.E. Shilov, Zhurnal Vsesoyuznogo Khimicheskogo Obtschestva imeni D.I. Mendeleeva, No 6 pp. 542-550 (1986).
16. K.I. Zamaraev, S.V. Lymar, M.I. Khramov, and V.N. Parmon, Pure Appl. Chem.(1988).
17. Y.M. Tricot, A. Emeren, and J.H. Fendler, J. Phys. Chem. 89, 4721 (1985); 90, 3369 (1986).
18. I.M. Tsvetkov, S.V. Lymar, V.N. Parmon, and K.I. Zamaraev, Kinet. Katal., 27, 109-116 (1986).
19. I.M. Tsvetkov, O.V. Gerasimov, S.V. Lymar, and V.N. Parmon in: Homogeneous and Heterogeneous Catalysis, Novosibirsk, 1986, Yu. Yermakov and V. Likholobov, eds. (VNU Science Press, Utrecht, The Netherlands 1986) pp. 751-763.
20. E.I. Knerelman, N.P. Luneva, V.Ya. Shafirovich, and A.E. Shilov, Dokl. AN SSSR, 291, 632 (1986); J. C. S. Chem. Commun., 1504-1505 (1987).
21. O.V. Gerasimov, S.V. Lymar, I.M. Tsvetkov, and V.N. Parmon, React. Kinet. Catal. Lett., 36, 145-149 (1988).
22. N.P. Luneva, V.E. Majer, V.Ya. Shafirovich, and A.E. Shilov, Kinet. Katal., 28, 505 (1987).
23. K.I. Zamaraev and V.N. Parmon in: Fotokataliticheskoye Preobrazovaniye Solnechnoi Energii, Part 1, K.I. Zamaraev and V.N. Parmon, eds. (Nauka, Novosibirsk 1985) pp. 7-41.
24. V.N. Parmon, ibid., Part 2, pp. 6-106.
25. A.W.-H. Mau, Ch.-B. Huang, N. Kakuta, A.J. Bard, A. Campion, M.A. Fox, J.M. White, and S.E. Webber, J. Amer. Chem. Soc., 106, 6537-6542 (1984).

26. N. Kakuta, K.-H. Park, M.F. Finlayson, A.J. Bard, A. Campion, M.A. Fox, S.E. Webber, and J.M. White, J. Phys. Chem., 89, 5028-5031 (1985).
27. D. Meissner, R. Memming, and B. Kastening, Chem. Phys. Lett., 96, 34-37 (1983).
28. D.H.M.W. Thewissen, K. Timmer, E.A. van der Zouwen-Assink, and A.H.A. Tinnemans, J. Chem. Soc., Chem. Commun., 1485-1487 (1985).
29. M. Barbeni, E. Pelizzetti. E. Borgarello, N. Serpone, M. Grätzel, L. Balducci, and M. Visca, Int. J. Hydrogen Energy, 10, 249-253 (1985).
30. E. Borgarello, N. Serpone, M. Grätzel, and E. Pelizzetti, Int. J. Hydrogen Energy, 10, 737-741 (1985).
31. Yu.A. Gruzdkov, E.N. Savinov, and V.N. Parmon, Int. J. Hydrogen Energy, 12, 393-401 (1987).
32. Yu.A. Gruzdkov, E.N. Savinov, and V.N. Parmon, Khimicheskaya Fizika, 7, 44-50 (1988).
33. Yu.A. Gruzdkov, E.N. Savinov, and V.N. Parmon, Int. J. Hydrogen Energy, 13 (1988).
34. T.N. Veziroglu, N. Getoff, and P.Weinzierl, eds., Hydrogen Energy Progress VI (Pergamon Press, New-York - Oxford 1986).
35. E.K. Nazarov and A.Ya. Stolyarevskii in: Atomic Hydrogen Energetics and Engineering, Part 3, V.A. Legasov, ed. (Atomizdat, Moscow 1980) pp. 5-57.
36. E.E. Shpilrain, S.P. Malyshenko, and G.G. Kuleshov, Introduction in Hydrogen Energetics (Energoatomizdat, Moscow 1984).
37. V.N. Parmon, Zhurnal Fizicheskoi Khimii, 58, 993-995 (1983).
38. Yu.I. Aristov, V.I. Anikeev, V.A. Kirillov, V.N. Parmon, and K.I. Zamaraev in: Budutschiye Gorizonty v Katalize (Future Horizons in Catalysis), V.D. Sokolovskii, ed. (Institute of Catalysis, Novosibirsk 1986) pp. 16-66.
39. V.N. Parmon, V.I. Anikeev, V.A. Kirillov, and K.I. Zamaraev in : Hydrogen Energy Progress VI, T.N. Veziroglu, N. Getoff and P. Weinzierl, eds. (Pergamon Press, New York - Oxford 1986) pp. 582-599.
40. V.I. Anikeev, V.N. Parmon, Yu.I. Aristov, V.I. Zheivot, V.A. Kirillov, and K.I. Zamaraev, Dokl. AN SSSR, 289, 158-162 (1986).
41. V.I. Anikeev, V.A. Kuz'min, V.A. Kirillov, I.I. Bobrova, V.V. Pasichnyi, V.N. Parmon, and K.I. Zamaraev, Dokl. AN SSSR,293, 1427-1432 (1987).
42. R.B. Akhmedov and M.A. Berchenko in: Fotokataliticheskoye Preobrazovaniye Solnechnoi Energii, Part 1, K.I. Zamaraev and V.N. Parmon, eds. (Nauka, Novosibirsk 1985) pp. 58-68.
43. ISES Solar World Congress 1987, Hamburg, Book of Abstracts Vol. 2 (Hamburg, 1987) pp. 7.1.01-09.
44. Yu.I. Aristov, I.I. Bobrova, N.N. Bobrov, and V.N. Parmon, Limiting Values of Efficiency and Specific Power of Thermochemical Solar Energy Conversion Using Methane Steam Reforming (Institute of Catalysis, Novosibirsk 1987) 18 p.
45. Yu.I. Aristov, A.I. Fiksel, and V.N. Parmon, Geliotekhnika (1988).

STRATEGIES FOR THE PRODUCTION OF METHANE FROM INORGANIC MATERIALS

K. KRIST
Gas Research Institute, 8600 W. Bryn Mawr Ave.
Chicago, IL 60631

INTRODUCTION

GRI's inorganic SNG program addresses the need to secure for the future ample, reliable supplies of synthetic gaseous fuels in environmentally sound processes. The program's research also bears significantly on nearer-term objectives for the natural gas industry. This article describes the program's approach to the long-term inorganic fuel goal. It also outlines new areas that are being derived from the inorganic SNG program.

INORGANIC-BASED GASEOUS FUELS

Since CH_4 is more storable, transportable, and compatible with gas equipment than H_2, the objective of the inorganic SNG program is CH_4 production. However, H_2 could serve as an intermediate in CH_4 production, or could be added to pipelines in amounts up to 10-15% without unduly compromising the Btu value or other desirable properties of natural gas (1).

Inorganic SNG systems could produce gaseous fuels independently or be coupled to other processes. By adding sufficiently low-cost, inorganic-based H_2, or converting exhaust CO_2 to CH_4, an inorganic system could improve coal and biomass conversion. Inorganic SNG systems could also be coupled to sources of CO_2 in other industrial exhausts and natural reservoirs.

A number of reaction cycles could use sunlight to produce methane from inorganic materials. Some examples are given in table I. GRI's research is focused on the first two photochemical approaches, although new photochemical and thermochemical pathways continue to be evaluated.

Dark Fuel Formation Catalysis

Insufficiently rapid fuel and co-product formation is a key technical barrier to achieving efficient, durable, inorganic SNG systems. Two factors contribute to this issue. Light absorption creates energetic, excited states which can initiate thermodynamically favorable, auxiliary reactions that compete with the desired fuel reaction. Therefore, unless special kinetic properties at the high driving force of some photochemical systems can selectively inhibit undesired reactions, fuel formation rates will need to be high. On the other hand, the small molecule, multielectron, redox reactions generating H_2, CH_4, CO, and O_2 from inorganic materials tend to be slow, and catalysts for these reactions are either less than optimal or too expensive. Thus, both the rates and cost of surface and molecular catalysts used to generate gaseous fuels need to be greatly improved.

GRI is investigating a concept in its inorganic SNG program for the catalysis of gaseous fuel production under simplified, dark conditions. The approaches are primarily electrochemical to allow thermodynamic control of the reactions under conditions which separate the fuel and the

Photochemical Energy Conversion
James R. Norris, Jr. and Dan Meisel, Editors

TABLE I: INORGANIC ROUTES TO METHANE

	ΔG_{298} (kcal/mole)	ΔS_{298} (kcal/mole)	E° (V)
1. Photochemical			
$CO_2+2H_2O(1) \rightarrow CH_4+2O_2$	195.5	.0588	-1.06
2. Photochemical/Thermal			
$4H_2O(1) \rightarrow 4H_2+2O_2$	56.7(4)	.0393(4)	-1.23
$CO_2+4H_2 \rightarrow CH_4+2H_2O(1)$	-31.3	-.0984	.170
3. Photochemical Cycle			
$2H_2O(1) \rightarrow 2H_2+O_2$	56.7(2)	.0393(2)	-1.23
$2H_2+CO_2 \rightarrow CH_4+O_2$	82.1	-.0200	-.890
4. Photochemical Cycle			
$CO_2 \rightarrow CO+1/2O_2$	61.5	.0213	-1.33
$CO+2H_2O(1) \rightarrow CH_4+3/2\ O_2$	134.0	.0379	-.969
5. Thermochemical/Electrochemical			
$3CH_4+2H_2O+CO_2 \rightarrow 4CO+8H_2$	113.1	.221	-.307
$4CO+8H_2 \rightarrow 4CH_4+2O_2$	82.48	-.1624	-.224

oxygen coproduct. The list of dark catalytic systems being studied includes: electrogenerated and electrode-supported metal complexes; electrode-supported polymers and polymer-containing metals and metal complexes; metal and semiconductor electrocatalytic surfaces; solid polymer electrolyte/metal systems, intermediate temperature, proton and oxygen anion solid-electrolytes, and heterogeneous CO_2 hydrogenation catalysts.

Although extensive advances are still needed, this catalysis research has made substantial progress. For example, the research has:

- Demonstrated the first electrocatalytic conversion of CO_2/H_2O to CH_4, using ruthenium surfaces and determined the role of many factors influencing the reaction. The rates at low overpotentials were about a thousand times slower than needed for a practical process.

- Improved the rate of CH_4 formation by two hundred fold, using a cheap copper electrode/alkaline electrolyte cell. By depositing the Cu electrode in situ, 73% CH_4 and 25% ethylene could be generated for hourly periods with current densities of 8.3mA/cm^2 (but only at high overpotentials). The proposed model of this reaction involves electrochemical dissociation of adsorbed CO. The role of such factors as pretreatment, crystal orientation, electrode surface roughness, and temperature were determined.

- Extended the reaction to the gas phase, using a copper/solid polymer electrolyte - reduced the voltage by 60%, sustained the reaction for days, and devised a cheap method to regenerate the catalyst surface. The hydrocarbon yield is, however, 1/5 that in liquid electrolyte.

- Converted H^+/CO_2 to CO, using cobalt phthalocyanine electrodes, with the lowest voltages yet obtained at current densities of about 1 mA/cm^2. Supporting the catalyst improves its stability by a factor of a thousand relative to that in homogeneous solution.

- Determined the relative effectiveness of complexes having one and two metal centers, respectively, in the reduction of protons to H_2

- Developed a state-of-the-art, synthetic molecular catalyst for O_2 evolution from water that is 100X faster, and has much better durability than previous synthetic catalysts.

- Achieved room temperature and atmospheric pressure methanation of CO_2 by controlling metal catalyst/titania support properties.

Current research directions in this area include: improved mechanistic understanding and catalyst optimization; design of new synthetic molecular catalysts (including the use of computer-aided molecular design (CAMD) techniques); extension of fuel reactions to new metal and molecular catalytic systems; extension of the reactions to the gas phase, using solid electrolytes operating at temperatures between 150° and 600° C; and, the development of strategies to protect systems from poisons and carbon deposition.

Photochemical Systems

The research on dark catalysis supports three photochemical approaches. The approaches, some advances, and continuing directions for each approach are listed below.

- Photoelectrosynthetic devices based on semiconductors as electrodes, films, or dispersed particles.

  (a) Developed a durable, visible-light-absorbing, series connected, bipolar electrode array, made of relatively inexpensive materials, that splits water with efficiencies of 1% without electrical assistance.

  (b) Developed and characterized supported semiconductor/catalyst films that generate H_2 from sacrificial reagents.

  (c) Used visible light with SiC/Cu dispersions and InP/*in situ*-deposited Mn interfaces to produce CH_4 from H_2O/CO_2 in low yields.

  (d) Developed oxygen scavengers that improve the water splitting efficiency of dispersed semiconductor particles from 0.1 to 0.7%, and could enable fuel/oxygen separation in daily light/dark cycles. Improved dye sensitization of relatively stable but transparent semiconductors.

Research is continuing on: enlarging the size, improving the efficiency, and adapting to methane production, of the bipolar fuel synthesis arrays; and, characterizing, improving, and adapting to methane production, the dispersed semiconductor particle systems.

- Artifical photosynthetic membranes that make gaseous fuel directly, using molecular components.

  (a) Genetically-engineered the production and modification of structurally well-defined proteins to study the effect of protein environment on photo-induced electron transfer.

  (b) Applied Stark spectroscopic techniques to the determination of charge separation in the exited states of chlorophylls and other photosensitizers.

  (c) Deposited platinum catalysts exclusively within the pores of a zeolite, and used the system as a template for components that photochemically separate charge from the excited singlet state of an attached photosensitizer.

This area is continuing to investigate charge separation and catalysis by the techniques identified above.

- Living organisms (algae) that split water under special conditions.

  (a) Decomposed water with an efficiency of 10-12% in a green algal system operating at low light intensities and under special flow conditions.

  (b) Developed a laser assay of H_2 production from single algal colonies to speed up genetic selection of more effective algae.

Continuing investigations are attempting to increase the light intensity and remove the special flow conditions under which algae efficiently split water.

Distributed Photoreactors

Another key technical barrier for inorganic SNG systems is the design of a low-cost physical system that delivers sunlight photons within an array of fuel synthesis modules and transports reactants to, and products from, the modules.

The fuel synthesis plant will differ depending upon whether the system involves:

- Discrete intermediate electric power generation, as in photovoltaic (PV) electrolysis, or direct fuel synthesis at the light absorbing unit

- Thermochemical (high temperature) or photochemical (mild temperature) processes

- Light concentration or ambient sunlight

- Solar tracking or stationary light absorbing units

- A land or pond base

- Harvesting or non-harvesting of the fuel synthesis units

GRI is emphasizing systems that produce gaseous fuels without a discrete electric production step since such systems would not be used for electric power load management, and might compete effectively with PV approaches.

However, PV-electrolysis systems could be relevant if gaseous fuel transport resulted, or potential advantages (no inversion costs, higher reliability, less wiring, minimal fault detection) made these systems competitive with PV.

Figure 1 shows the effect of efficiency on the area requirements for different SNG systems. If inorganic systems achieve high efficiencies, the area requirements could be similar to those for coal gasification.

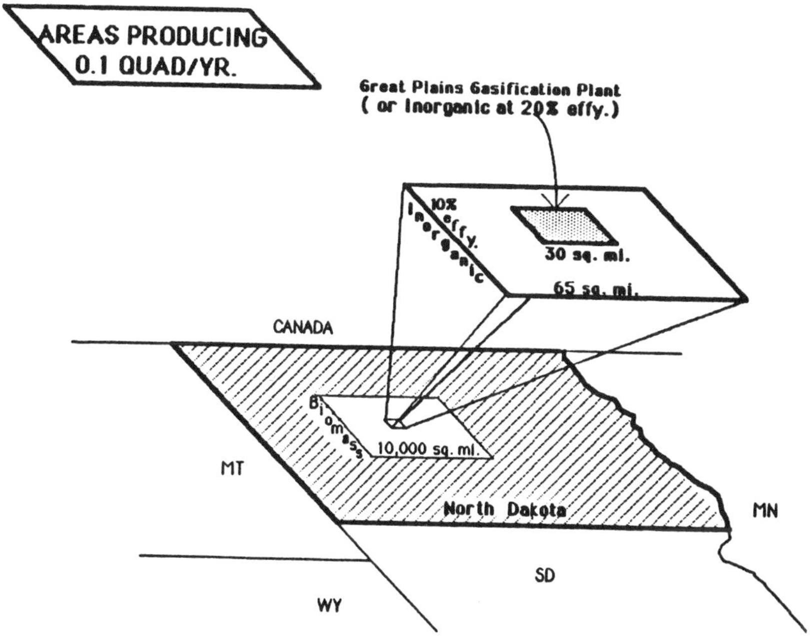

Figure 1 Area Requirements for The Annual Production of 0.1 Quad (100 BCF of SNG)

System Cost

Figure 2 (2) presents the allowed costs of inorganic SNG systems for different gas prices, solar intensities, and overall efficiencies, for a fixed charge plus operating and maintenance rate of 20%. For gas prices between \$5/MMBtu and \$30/MMBtu the allowed costs range from \$1/sq. ft. to \$20/sq. ft. Although the amounts and price could be higher depending on many possible developments, the current GRI baseline predicts that 0.2 quads of SNG will be sold in 2010 for a price of \$7.30/MMBtu (3).

Even with minimal fuel synthesis chemicals cost, the total system cost may exceed the allowed costs based upon the fuel value. For comparison, Figure 3 (4) presents the costs of a range of construction materials that might used in a photochemical array. The possibility of reaching these costs targets is uncertain.

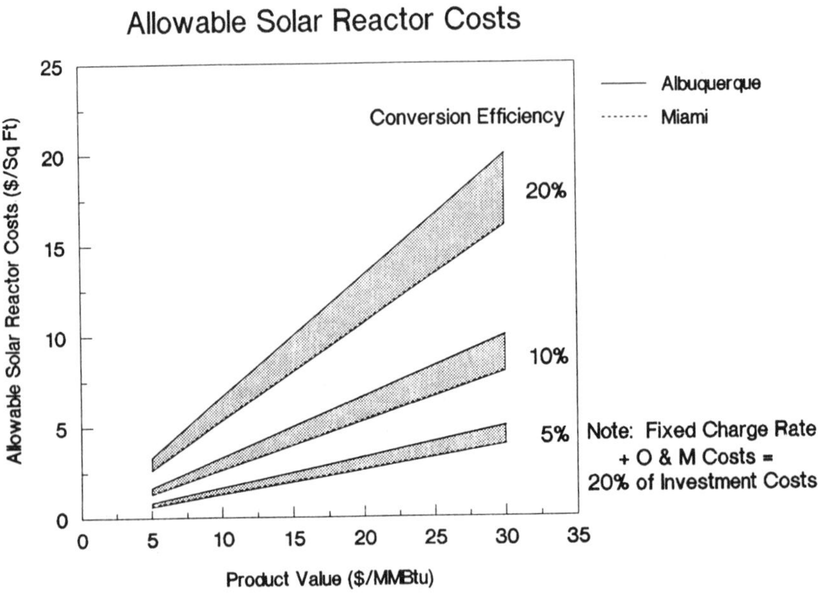

Figure 2

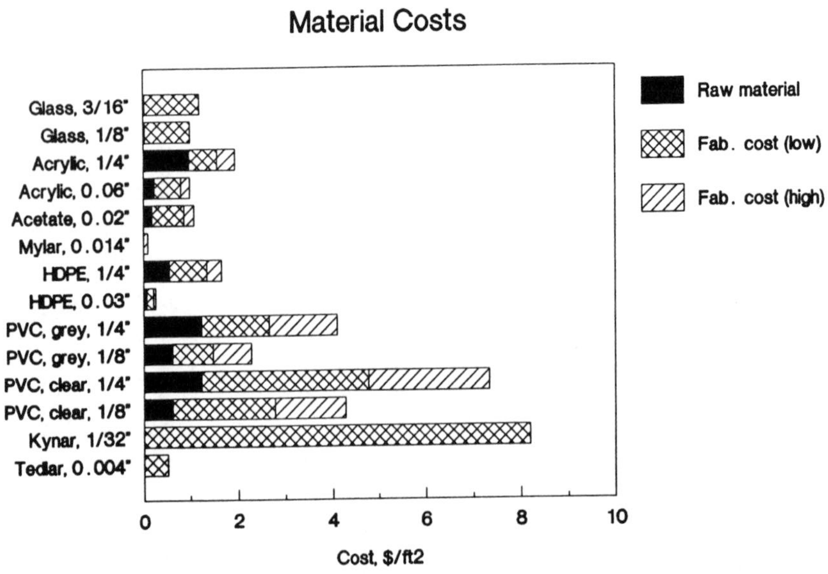

Figure 3

Figure 4 (5) shows the current and projected fuel costs of certain renewable systems as determined by their efficiencies and annual area costs. The costs of agriculture are only about $0.01/sq. ft/yr. The low costs may be due in part to a lower rate of return for agriculture than for utility systems. Nevertheless, they raise the possiblity that agriculture could have implications for achieving low inorganic SNG costs. An example that has been cited (6) is the use of harvestable, light weight, photochemical particles that would reduce the need for support and transport structures.

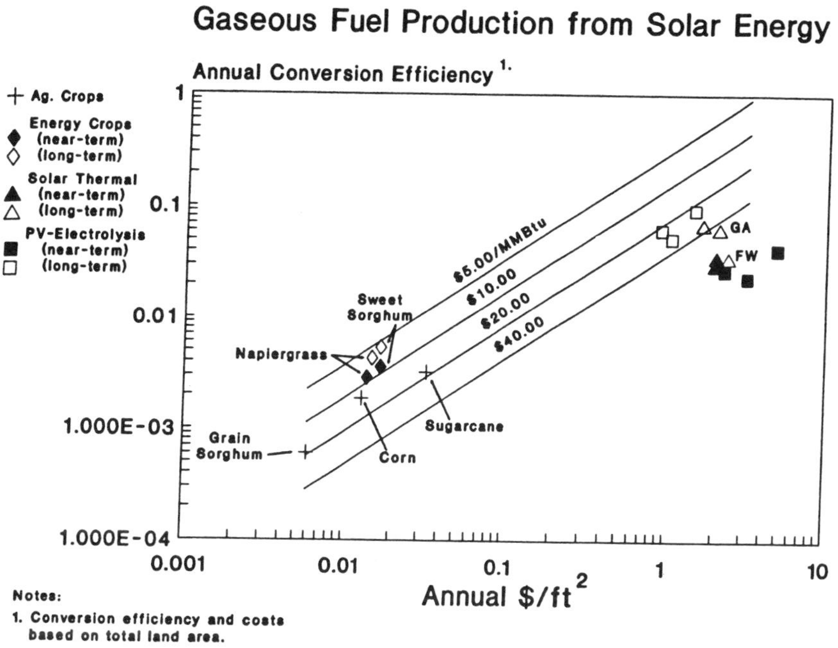

Figure 4 Annual Costs and Efficiencies for Gaseous Fuel Production Technologies

GRI is currently pursuing an effort to design and evaluate low-cost, photochemical arrays, including flat plate, pond, and concentrator systems. The project is examining the effects of scale, novel design, and improved fabrication on system costs.

CROSS-CUTTING GOALS

The basic capabilities in GRI's inorganic SNG research are also relevant to other gas technologies. Areas of inorganic SNG research capabilities include:

- The redox and binding processes of small molecules.
- Novel catalysis strategies

- Photophysical, photochemical and electrochemical approaches that can be used as analytic tools, and to assist difficult reactions and processes.

- Interfacial properties and processes.

These capabilities could be extended to improve gas conversion (gas conditioning, fuel cells, methane conversion, radiant processes), gas separation, and gas detection technologies. GRI is actively pursuing nearer-term opportunities that derive from inorganic SNG research. Below we discuss some examples of the cross-cutting significance of inorganic SNG research and the new areas that are deriving from it.

Fuel Cell and Methane Conversion Technologies

The complete oxidation of CH_4 to cogenerate heat and electricity, and the controlled oxidation of CH_4 to produce higher valued products are established but challenging research areas that might benefit from the novel approaches and insights of inorganic SNG research. GRI's objectives in these areas include:

- Direct CH_4 oxidation fuel cells that minimize fuel cell costs by avoiding a reforming step.

- Reactions of CH_4 that improve selectivity at acceptable conversions and space velocities.

Inorganic SNG research is related to these areas as follows:

- The electrocatalytic formation of CH_4 from H_2O/CO_2 is the reverse of direct methane oxidation, so that research on this reaction may be informative for fuel cell and fuel use chemistry.

- Electrochemical control of CH4 conversion could lead to more selective product formation. Some conversions of CH_4 to higher valued products, such as the formation of ethylene or acetylene and H_2, require relatively small energy inputs. The reactions typically need inconviently high temperatures or are not helped by increasing temperature. Minimal amounts of electrical energy might allow the reactions to occur under milder conditions with lowered overall costs. CH_4 conversion reactions could also be assisted with radiation, if greater than unit quantum yields can be achieved or if the radiation cost could be reduced sufficiently.

- The development of new catalytic strategies, involving supported/unsupported molecular and metal catalysts, lower-temperature, proton and oxygen anion conducting solid electrolytes, molecular control of catalytic environments, and computer-aided design of catalysts, may benefit fuel cell, and CH_4 conversion systems.

- Specific activation of weak inorganic oxidants may be critical to the control of CH_4 reactions. The inorganic SNG program is concerned with activation of "oxidants" such as water and CO_2. Such alternate oxidants may play a role in CH_4 conversion. CO_2, in particular, might be a source of carbon atoms in the conversion of CH_4 to higher hydrocarbons or oxygenates.

- GRI's inorganic SNG research has shown that catalytic metal surfaces can have unexpected properties for CO_2 reduction if the electrical bias, surface pretreatment, poisons, promoters, and other properties of the system are carefully controlled. The evolution of CH_4 on copper is an example of this since Cu is normally associated with methanol formation (Cu/ZnO catalysts). Such advances in the catalytic properties of metals have general significance for C1 chemistry.

- The inorganic SNG program's research on solid-polymer electrolyte fuel synthesis is relevant to similar fuel cell systems.

- Research in the inorganic SNG program recently demonstrated the cogeneration of electricity and small amounts of ethylene with a solid oxide electrolyte at 650-750° C.

We plan to start new research shortly on two projects that derive from the inorganic SNG program: "Electrochemical Methane Activation" and "Alternate Oxidants".

Hydrogen Sulfide Chemistry

New natural gas discoveries are expected to contain significant quantities of hydrogen sulfide. The Claus removal process oxidizes H_2S under severe conditions to steam and sulfur and is not economical on a small scale. Liquid processes, based upon redox cycles, for operating under mild conditions are limited in reliability and scale of operation. New processes might yield more valuable products, expand operation scale, improve operating conditions, avoid waste heat generation, and lower costs.

The inorganic SNG program has recently completed an evaluation novel H_2S chemistry. Fundamental research is planned on such concepts as:

- Low severity, simple redox cycles for decomposing H_2S

- Direct decomposition of H_2S using catalytic ceramic membrane reactors

- Heterogeneous catalytic oxidation of H_2S directly to S and H_2O, bypassing the Claus step (SO_2 formation)

- Improving the oxidation kinetics and other performance characteristics of metal centers used in liquid redox processes

- Improving solid sulfur formation kinetics

Solids Radiation

Improvements in the mechanical properties, radiant efficiency, and radiant selectivity of incandescent solids could enhance the use of gas as a source of radiant energy. Such systems would use tailored radiation, directly, or the auxiliary power produced when tailored radiation is coupled to a photovoltaic cell (a thermophotovoltaic (TPV) device). Potential applications include: improved visible lighting; better curing, drying or chemical processes using selective near IR radiation; auxiliary power for independent, gas-fired systems; and, cathodic protection of pipelines.

GRI recently completed an exploratory project that involved the use of rare earth oxide emitters to produce selective radiation. The research has shown that it is possible to:

Produce more durable, thermally cyclable, gas-fired, ceramic emitters of visible and near IR radiation

Convert gas energy to selective near-IR radiation with an efficiency in the range of 10-30%

GRI is planning to continue fundamental and applied research on solids radiation.

Electrochemical Oxygen Separation

Combustion of natural gas with pure O_2 or O_2 enriched air raises the combustion temperature. O_2 enriched combustion could make gas high temperature processes such as metal casting and heat treating, and glass manufacturing more efficient. Emerging technologies for high temperature, "tailored" radiation could also use low cost oxygen. The high temperature flue gases in these radiators represent a large loss that requires the use of heat recirculation. O_2-enriched combustion would improve efficiency by reducing the mass of flue gas produced. Burning natural gas in pure O_2 would eliminate the formation of NOX and other emissions. Using low-cost O_2 could be cheaper than other, post combustion treatments for controlling NOX.

Energy is a major cost of O_2 separation. Two commercial O_2 concentration processes, pressure swing adsorption (PSA) and cryogenic separation, typically require more than 0.2 kW-hr/lb of oxygen in large scale plants and many times this in small scale plants. Since the theoretical energy requirement is 0.03 kW-hr/lb, an opportunity exists to reduce the cost. Electrochemical processes can use energy selectively in removing O_2 and, therefore, may be able to lower energy consumption.

Approaches to electrochemical O_2 separation include redox carriers, solid electrolytes, and molten salts. Recent work has shown that certain redox carrier molecules will selectively bind O_2 in one oxidation state and release the bound O_2 in another, giving rise to a possible O_2 concentrating cycle. Solid electrolyte and molten salt systems that oxidize and reduce O_2 directly might also minimize energy consumption.

Redox carrier molecules must be stable, readily bind O_2 in one electronic state and release it in another. The process must operate at suitably high current densities with minimal polarization losses. The electrochemical carrier could transport several molecules of O_2 per pass through the cell to avoid fixed losses. Systems must be developed that avoid solvent loss. Solid electrolyte and molten salt systems also need to improve their stabilities and current densities.

GRI recently issued an exploratory RFP on electrochemical O_2 separation and is planning to continue fundamental research in this area.

Surface Treatment of Metals

The vents and secondary heat exchangers of high efficiency furnaces, and the boilers of absorption heat pump systems, are subject to corrosion. High chromium, 29-4C alloy steel used to prevent chloride ion corrosion in these systems is expensive. Organic coatings, added to less expensive substrates, operate over a limited temperature range, and can be mechanically compromised. Diffusional coatings formed by boronizing, aluminizing, or chromizing, or other inorganic coatings may improve the cost and durability of furnace and heat pump components.

The inorganic SNG program recently submitted a proposal for exploratory research on diffusional coatings and is planning to continue fundamental research in this area.

SUMMARY

GRI's inorganic SNG program is addressing the long-term solar, and possibly nuclear, option for the gas industry while also actively pursuing opportunities for nearer-term, cross-cutting research as they derive naturally from the program. In the near term, the program is focusing on two key technical barriers, the dark evolution of gaseous fuel and oxygen co-product, and the design of low-cost, wide area, physical plants for fuel synthesis. The program is making significant fundamental advances in fuel synthesis catalysis and photochemistry that also have implications for other gas technologies. Exploratory research in areas that derive from the inorganic SNG program has led to advances of practical significance to the gas industry, such as the development of durable mantles for visible gas lighting, and thermophotovoltaic devices that are being evaluated as sources of auxiliary power for gas lights and appliances.

REFERENCES

1. (a) J. D'Acierno and M. Beller, "Analysis of the Use of Hydrogen as a Supplement to Natural Gas", Office of Advanced Energy Systems Research, Division of Energy Storage, DOE, Publication Nos. DE82006844 and BNL-51442, September, (1981).

   (b) Public Service Electric and Gas Co., "Blending of Hydrogen in Natural Gas Distribution Systems", NTIS Report Nos. CONS-2925-1, CONS-2915-2, CONS-2915-3, June 1, 1976 - April 30, 1978.

2. M. Karpuk, "February, 1988 Progress Report" to Dr. Kevin Krist, Gas Research Institute, for GRI Contract No. 5087-267-1543, May 9, 1988.

3. T. J. Woods, "The Long-Term Trends in U.S. Gas Supply and Prices: The 1987 GRI Baseline Projection of U.S. Energy Supply and Demand to 2010", Gas Research Insights December, 1987.

4. M. Karpuk, "April, 1988 Progress Report" to Dr. Kevin Krist, Gas Research Institute, for GRI Contract No. 5087-267-1543, May 9, 1988.

5. M. Karpuk, "May, 1988 Progress Report" to Dr. Kevin Krist, Gas Research Institute, for GRI Contract No. 5087-267-1543, June 9, 1988.

6. J. E. Guillet, "Studies in Solar Photochemistry", Proceeding of the Tenth DOE Solar Photochemistry Research Conference, Division of Chemical Sciences, BES, DOE, June 8-12, 1986.

ACKNOWLEDGEMENT

GRI acknowledges M. Karpuk and R. Copeland of Technology Development Associates, Inc. for their contribution to the analysis of this paper.

PANEL DISCUSSIONS: PHOTOELECTROCHEMISTRY

Communicated by: M. Tomkiewicz[1] (Panel Chairman)

Panel Members: M. Archer[2], F. Decker[3], A. Fujishima[4], B. Parkinson[5], Ke Xiao[6]

Modern photoelectrochemistry was born out of the happy marriage of solid state physics with electrochemistry. To some degree, the International Conferences on Photochemical Conversion and Storage of Solar Energy (IPS) may be viewed as matchmakers for the union of photoelectrochemistry with other areas of photochemistry to jointly contribute to the commonly sought goal of solar energy utilization. One product of this union is the expanded area of particulate photochemistry, titled in this volume as "Heterogeneous Photocatalysis". It is, therefore, of interest to note the great skepticism that Edmond Becquerel's original observation had to face among its contemporaries. A brief account of this historical perspective is provided below by M. Archer[2].

The Controversial Origin of Photoelectrochemistry (M. Archer)

All photoelectrochemists with a sense of history duly acknowledge Edmond Becquerel[7] as the father of their field, but it is perhaps less well known that the announcement of his discovery was followed by a lively controversy.

---

[1]Department of Physics, Brooklyn College of CUNY, Brooklyn, NY 11210
[2]Department of Physical Chemistry, University of Cambridge, Lensfield Road, Cambridge CB2 1EP, United Kingdom
[3]Instituto de Fisica, UNICAMP, 13081 Campinas, SP, Brazil
[4]Department of Synthetic Chemistry, Faculty of Engineering, University of Tokyo, Hongo, Bunkyo-ku, Tokyo 113, Japan
[5]E. I. DuPont de Nemours & Co., Central Research and Development Department, Experimental Station E328/216B, Wilmington, DW 19898
[6]National Laboratory for Interfaces Science; P. O. Box 425, Tianjin-9, Peoples' Republic of China
[7]E. Becquerel, *Compte Rendu* **9**, 1-567 (1839)

Published 1989 by Elsevier Science Publishing Co., Inc.
Photochemical Energy Conversion
James R. Norris, Jr. and Dan Meisel, Editors

On 30 July 1839, Edmond's father, Antoine César Becquerel, professor of physics at the *Musée d'Histoire Naturelle* in Paris, communicated to the *Académie* a brief note by his son[8] on a novel electrochemical means of estimating the rate of a photochemical reaction at the illuminated interface of two liquid phases such as ether and aqueous ferric perchlorate. Edmond had observed that a small current flowed between two platinum electrodes immersed one in each layer when the vessel was irradiated. He experimented with various color filters and, in reporting his observations, assumed a proportionality between the rate of the interfacial photochemical reaction, the observe photocurrent and the '*nombre des rayons chimiques*'.

This drew a quick response from a senior member of the Academy, Jean Baptiste Biot, professor of physics in the *College de France* and a long-established authority on optical dispersion and polarization. Correctly noting that the observation of a physical effect does not necessarily provide a measure of its cause, Biot commented that Becquerel had failed to establish that '*l'effect chimique des radiations sera measuré par l'intensité du courant électrique produit dans l'action de la lumiére*'[9] .

Edmond was silent for a few months and then in November he communication, again through his father, the longer paper[7] containing the first observations of a photoelectrochemical effect on coated electrodes. He had noted that a small effect was produced by the use of two clean platinum electrodes of different temperatures, but that a much more substantial effect was obtained if one electrode had previously been heated to redness. Reasoning that a layer '*des corpuscules d'une tenuité extrême*' on the metal surface might be responsible, he coated silver electrodes with thin layers of bromine or iodine, and platinum electrodes with silver halides, and observed substantial photocurrents. Again he asserted that '*des effects électriques...peuvent servir a déterminer le nombre des rayons chimiques actifs,*' causing Biot to dissent the following week[10]

---

[8]E. Becquerel, *Compte Rendu* **9**, 145-149 (1839)
[9]J. B. Biot, *Compte Rendu* **9**, 169-173 (1839)
[10]J. B. Biot, *Compte Rendu* **9**, 579-581 (1839)

A fortnight later, Edmond was back in print defending his hypothesis with vigor[11], drawing a quick and heavily sarcastic response from Biot[12], '*Il ne peut y avoir rien de tel, de ma part, envers un jeune homme dont j'apprécie le zéle et l'esprit inventif, autant j'aime sa personne...*'. The controversy rumbled on, incapable of settlement until the quantization of light and electricity were established some decades later ...*plus ca change, plus c'est la même chose!*

Current Areas of Activity

With this perspective we may now return to contemporary photoelectrochemistry. On the occasion of the 7th IPS, it is appropriate to review the areas of activity and the outstanding problems for future directions as reflected at this meeting. Among the more than 50 papers contributed to the conference, six areas of activity were evident. These include:

1. Kinetics, Mechanisms and Surface Modification.
2. Applications of Photoelectrochemical Cells (PEC's).
3. High Efficiency Devices
4. New Electrode Materials
5. New Systems and New Preparation Methods
6. Techniques

No revolutionary developments were identified in the field of photoelectrochemistry. On the other hand, some skepticism was expressed by the panel members regarding high-efficiency devices when open circuit voltages and short circuit currents are measured in systems where battery electrode is connected to a PEC.

---

[11]E. Becquerel, *Compte Rendu* **9**, 711-712 (1839)
[12]J. B. Biot, *Compte Rendu* **9**, 719-726 (1839)

Outstanding Problems and Future Directions

Our terminology and much of our fundamental understanding are based on solution of one dimensional transport and field equations on the semiconductor side and on the electrochemistry of mercury electrode from the electrolyte side. Yet all the plenary lectures on photoelectrochemistry summarized in this section deal with complex structures; structures for which the simplistic one dimensional theoretical machinery cannot possibly be adequate. These papers represent a general trend in photoelectrochemistry where the systems are application oriented with optimized functionality as the main driving force in the construction of the system. The reason for this general trend is our present inability to design general "model" systems with controlled chemistry, spatial distribution and energetics on the semiconductor side, in contact with the electrolyte where we can control the absorption properties. There is a definite progress in this direction, particularly with Si and InP. However, without such "model systems" it will be difficult to develop a realistic theoretical understanding of currently loosely exploited terms such as "surface states" and "effective" parameter; terms that could be identified, verified and characterized experimentally. Alternatively, a parallel approach might be to use the currently evolving statistical tools to describe composite media, for a better understanding of the application driven structures.

Future applications (A. Fujishima): In addition to the more conventional potential applications to solar energy conversion, new possibilities are emerging in the following fields:

a. Corrosion: Corrosion products of metals such as Fe and Cu are oxides, such as Fe_2O_3, Fe_3O_4, Cu_2O and CuO, with semiconducting properties. Photoelectrochemical techniques such as light-induced current-voltage characteristics are being used in the study of the mechanism of the corrosion processes, the effects of inhibitors and the topography of the corrosion processes.

b. Pigments: Most pigments, including the white ones (n-TiO_2 anatase or rutile) are semiconductors. Their photoelectrochemical reactions with their polymeric binders in the presence of oxygen determine their stability to sunlight through oxidation of the binder by photogenerated holes and formation of peroxides by reducing oxygen with photogenerated electrons. Judicious introduction of charge recombination centers reduces these effects and enhances the stability of the pigments.

c. Area Selective Photoreactions: Our ability to selectively induce photoreactions in micro-targeted areas may find applications in the following fields: (1) image formation; (2) pattern formation; (3) electrochemical photoetching; (4) photoetching and scanning tunneling microscopy.

d. Sensors: The sensitivity of electric field distribution and/or transport properties of semiconductors to the ambient plays a key role in their application in sensor technology. One example in which a mechanism was demonstrated is the application of electrochromic materials such as WO_3 and MoO_3 as alcohol sensors by utilizing the current doubling properties of these interfaces.

e. Biological Treatments: Two applications of photoelectrochemical techniques to biological applications have been recently demonstrated: (1) water purification (photoelectrochemical disinfection); (2) photoelectrochemical therapy (possibilities for cancer treatments were demonstrated).

Future Prospects of PEC's in Energy Conversion (B. Parkinson):

a. Surface modified photoelectrodes cannot compete with hybrid of solid state photovoltaic (a-Si:H, $CuInSe_2$, Si GaAs concentrators) coupled to an electrolysis cell.

b. PEC configuration is still useful for screening new materials for solid state applications due to easy access to junctions for modification.

c. A major challenge is to identify reactivity which is unique to semiconductor/electrolyte interfaces (multielectron catalysis). Intermediates with energies in the band gap might be stabilized.

Future Trends (F. Decker):

a. New Systems and New Techniques: (1) superlattices and quantum well structures; (2) particulate systems: colloids, aggregates and films with quantum size effects; (3) Films with dynamic optical properties (electrochromic, thermochromic, "passive solar").

b. New Characterization Techniques: (1) development of new techniques (e.g., scanning tunneling microscopy); (2) new applications of existing techniques, e.g., UHV *in situ* and/or *ex-situ*, X-ray diffraction (Synchrotron radiation, etc.), electroreflectance and electroluminescence, Raman and Surface Enhanced Raman spectroscopies.

Index

H

I

K

L

M

W

X

Z